Sustainable Supply Chains: Strategies, Issues, and Models

Usha Ramanathan · Ramakrishnan Ramanathan
Editors

Sustainable Supply Chains: Strategies, Issues, and Models

Springer

Editors
Usha Ramanathan
Nottingham Business School
Nottingham Trent University
Nottingham, UK

Ramakrishnan Ramanathan
University of Bedfordshire Business School
Luton, UK

ISBN 978-3-030-48875-8 ISBN 978-3-030-48876-5 (eBook)
https://doi.org/10.1007/978-3-030-48876-5

This Springer imprint is published by the registered company Springer Nature Switzerland AG
The registered company address is: Gewerbestrasse 11, 6330 Cham, Switzerland

Preface

This book was originally designed as a sequel to our earlier book on a similar subject and title (available at https://www.springer.com/gp/book/9781447153511). Our aim was to prepare a similar title that includes chapters focussing on newer developments. We thank the contributors of this book for helping us achieve this aim with chapters involving very interesting, relevant, and innovative themes.

Over the last few years, a number of research themes that would help improve value in the management of sustainable global and local supply chains have been developed. They include measuring triple-bottom-line performance of supply chains, reducing waste especially in food supply chains, improving agility especially in fashion supply chains, and developing smart supply chains with the emergence of newer Industry 4.0 concepts of big data and the Internet of Things (Ghadimi et al., 2019; Koberg and Longoni, 2019; Manavalan and Jayakrishna, 2019; Yadav et al, 2020). Research in sustainable supply chains has further focussed on utilization of renewable resources, reduction of pollution, use of newer technologies for efficient utilization of available resources, and efficient management of resources in general. While most of the current literature seem to consider developments in the field of sustainable supply chains and in the field of Industry 4.0 as two distinct entities, we believe that there is significant synergy in bringing these two distinct fields together; the use of big data, artificial intelligence, and Internet of Things technologies in supply chains could help improve their sustainability. We are happy that several chapters in this volume aim to focus on this synergy in greater detail.

Our very Chap. 1 focuses on this synergy via a comprehensive literature review from both academic and industrial standpoints. This chapter proposes a case for Sustainable Supply Chain 4.0 (SSC 4.0) by combining SSC and Industry 4.0. In this chapter, key features from previous related literature are synthesized and knowledge gaps to benefit future researchers are highlighted. A dynamic framework is proposed consistent with the benefits, drawbacks, and boundaries of current research works.

The rest of the book is structured into two sections. The first section, comprising five chapters, focuses on management of sustainability/Industry 4.0 on supply chains as a whole, while the second section focuses on issues related to the application of SSC in specific application sectors.

Part I: Management of Sustainability/Industry 4.0 on Supply Chains

Chapter 2 focuses on the applications of Industry 4.0 in logistics and supply chains. Based on exploratory case studies from seven logistics firms, a new Value-Adding-Input-Output (VAIO) framework is presented to support understanding of the use of Industry 4.0 in logistics and supply chains. The framework is integral to SSC as well and provides a practical basis for supply chains to derive value from their Industry 4.0 investments.

Commercial Returns, End-of-Use Returns, and End-of-Life Returns for remanufacturing components are integral parts of sustainable supply chains. This reverse logistics process is the prime focus of Chap. 3. Using a variety of mathematical programming tools, authors attempt to model the returns and remanufacturing processes in order to generate interesting insights on the reverse logistics part of the SSC.

Chapter 4 focuses on sustainability metrics derived from CSR reports of companies. Using data on the triple-bottom-line information from CSR reports, a sustainability score is defined and computed in this chapter for Indian companies. The paper discusses the patterns in the sustainability scores of these companies over time.

Chapter 5 introduces the ideas of Six Sigma Marketing as a useful tool for achieving goals of SSC. Six Sigma is essentially a process improvement approach. We believe that practicing sustainability will provide competitive advantage to companies in the long run and increase market share. As the author highlights, Six Sigma Marketing is a fact-based data-driven approach to grow market share by providing targeted product/markets with superior value. An illustration is used to show the usefulness of Six Sigma Marketing for SSC.

Long-term aspects of SSC are the topics covered in the sixth chapter. Since a number of studies on SCC focus primarily on environmental aspects, a call is made in this chapter for integrating the other two, namely, social and economic aspects, more effectively. Evolutional game theory is used to build a strong theoretical framework to integrate, explain, and predict sustainability for supply chains using cross-disciplinary effort. An example application to explain and predict social and economic sustainability (in tandem) for a public health insurance supply chain using evolutionary game theory is illustrated in this chapter. By focussing on a specific sectoral application, this chapter links logically to the chapters in the next section that focusses on applications in various sectors.

Part II: Applications of SSC in Specific Sectors

Chapters in this section apply the principles of SSC in specific sectors including the restaurant industry in the UK (Chap. 7), SMEs in the manufacturing industry in India (Chap. 8), the furniture Industry (Chap. 9), and UK food manufacturing (Chap. 10).

The chapters in this volume use a variety of research methods including modeling, statistical analysis, and case-study-based qualitative analysis. They cover a range of application areas including multiple sectors (restaurant, manufacturing, logistics, furniture, food, and insurance), domains (supply chains, logistics, marketing, and reverse logistics), and multiple country contexts (UK and India). The potential links between sustainability and the recent technological innovations from Industry 4.0 have been highlighted in at least two chapters. Thus, we believe that this book will go a long way in instilling interests in sustainable supply chains in line with the recent calls (e.g., Koberg and Longoni, 2019; Ghadimi et al., 2019) for more research on this interesting, fast emerging, innovative topic.

Nottingham, UK Usha Ramanathan
Luton, UK Ramakrishnan Ramanathan

References

Ghadimi, P., Wang, C., & Lim, M. K. (2019). Sustainable supply chain modeling and analysis: Past debate, present problems and future challenges. *Resources, Conservation and Recycling, 140*, 72–84.

Koberg, E., & Longoni, A. (2019). A systematic review of sustainable supply chain management in global supply chains. *Journal of Cleaner Production, 207*, 1084–1098.

Manavalan, E., & Jayakrishna, K. (2019). A review of Internet of Things (IoT) embedded sustainable supply chain for industry 4.0 requirements. *Computers & Industrial Engineering, 127*, 925–953.

Yadav, G., Luthra, S., Jakhar, S., Mangla, S. K., & Rai, D. P. (2020). A framework to overcome sustainable supply chain challenges through solution measures of industry 4.0 and circular economy: An automotive case. *Journal of Cleaner Production*, 120112.

Contents

Chapter 1
Management of Sustainable Supply Chain and Industry 4.0: A Literature Review

Ali Roozbeh Nia, Anjali Awasthi, and Nadia Bhuiyan

Abstract This review aims to investigate the advanced collected works on sustainability and Industry 4.0 in the supply chain (SC) management from both academic and industrial standpoints. Hence, a review of the literature from 2010 to 2018 has been presented, knowledge gaps and all areas of application in the assumed investigation topic are highlighted, and the key features of the former study are associated. Furthermore, a dynamic framework for this topic is proposed consistent with the benefits, drawbacks, and boundaries of current research works, and the term "Sustainable Supply Chain 4.0" (SSC 4.0) is proposed. The suggested dynamic framework aims to distinguish the characteristics, elements and technology enablers, achievement aspects and challenges for evolving an SSC 4.0. Therefore, the current study and dynamic framework can provide awareness to academics and industrial specialists in their application of SSC 4.0.

Keywords Supply chain (SC) · Sustainable supply chain 4.0 (SSC 4.0) · Literature review · Industry 4.0 · Sustainability

1 Introduction

Traditional supply chains (SCs) comprise tangible facilities distributed with respect to geography support, create, and sustain shipping connections among them. SCs are explained in the functional chain of interrelated actions that include the direction, scheduling, and checking of services and goods among clients and providers (see Fig. 1). These managerial configurations are no more independent as a result of industrial progress (Büyüközkan and Göçer 2018). There are some factors which influence SC management such as performance, technology, environmental policy,

A. Roozbeh Nia (✉) · A. Awasthi
CIISE, Concordia University, Montreal, Canada
e-mail: ali.roozbehnia@mail.concordia.ca; ali.roozbehnia@gmail.com

N. Bhuiyan
Department of Mechanical, Industrial and Aerospace Engineering (MIAE), Concordia University, Montreal, Canada

U. Ramanathan and R. Ramanathan (eds.), *Sustainable Supply Chains: Strategies, Issues, and Models*, https://doi.org/10.1007/978-3-030-48876-5_1

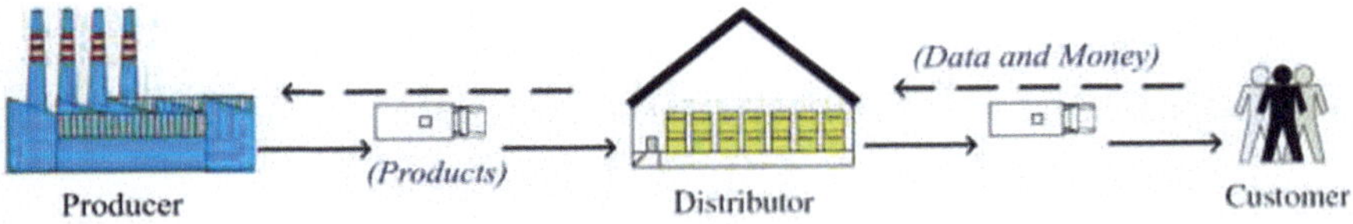

Fig. 1 An illustration of a traditional supply chain (Roozbeh Nia et al. 2014)

economics, SC collaboration, competition, strategy, customer engagement, real-time information, procurement, and zero errors (Manavalan and Jayakrishna 2019).

On one hand, several companies take on uncertainty because of an increasing marketplace demand for different goods and services at the end of the twentieth century. On the other hand, economically beneficial types of manufacture cause durable influences on the environment and civilization (Rajeev et al. 2017). Therefore, the combination of economic with social and environmental concerns, or sustainability, must be a shared interest among researchers and practitioners (Brandenburg et al. 2014; Seuring and Müller 2008). Progressively, environmental problems were used as a basis for strategic transformation (Aragón-Correa et al. 2008). Recently, environmental aspects attract many researchers in the literature (Roome, and Hinnels 1993; Noci and Verganti 1999; Schiederig et al. 2012). In addition, eco-innovation methods such as life cycle valuations, cleaner manufacture, and eco-design are employed in many companies (Huber 2008; Van Hemel and Cramer 2012).

The management of sustainable supply chains (SSC) is described by Seuring and Müller (2008) as "the administration of substance, information and assets streams in addition to teamwork among corporations alongside the SC whereas choosing objectives from entirely three elements of sustainable progress, namely, environmental, economic and social, which are come from client and shareholder necessities". For about 20 years, the issue of SSC has been given significant attention by researchers and specialists (Craig and Easton 2011; Beske et al. 2014; Brandenburg et al. 2014; Ghadimi et al. 2016, 2019; Seuring and Müller 2008; Seuring 2013). SSC management is fundamentally a part of green supply chain management (GSC), that is, the combination of ecological philosophy in the management of SC (Srivastava 2007) which encompasses environmental, economic, and social interests (Yan et al. 2016). An SSC is designated by a company's ability to decrease the consumption of energy, substances, or water and to discover solutions that are further eco-efficient by enhancing the administration of their SCs (Lopes de Sousa Jabbour et al. 2015).

Nowadays, most initiatives are undergoing digitization Industry 4.0. The emphasis of the digital revolution is placed mostly on manufacture; consequently, the names for instance "Smart Factory" or "Factory of the Future" are employed and compared with this idea (Kayikci 2018). In 2011, the German association "Industrie 4.0" invented the term Industry 4.0. The association is made of managers, academics, and legislators, who suggested a fourth industrial revolution is created on the digitization of organization procedures (Kagermann et al. 2011). In fact, the key impression motivating Industry 4.0 is to guide companies by implementing digital technologies that know how to assist in generating links among their process, provide systems, manufacturing capabilities, finished goods, and clients with the purpose of collection, and distribute real-time functioning and marketplace information with stakeholders (Ardito et al. 2019). The digitization in SCs is established based on six features: connectivity, cooperation, integration, adoption, cognitive improvement, and autonomous control (Kayikci 2018). Furthermore, for Industry 4.0, the empowering technologies include additive manufacturing, advanced manufacturing, augmented reality, cloud computing, simulation, industrial IOT, big data analytics, cybersecurity, and customer profiling (Ardito et al. 2019).

Industry 4.0 has been demonstrated to be successful in offering several business advantages containing operational optimization and value chain optimization (Strange and Zucchella 2017). Accordingly, Industry 4.0 is widely adopted by German companies, for instance, Volkswagen, Daimler, and BMW. In addition, the Government of China has likewise presented the "Made in China 2025" strategy which focuses on enhancing manufacturing through speeding up digitalization in China. Similar plans have also been started by the USA, French, UK, Japanese, and Singaporean governments (Bag et al. 2018). More precisely, the objective of Industry 4.0 is to improve the digitization and, therefore, the combination of business procedures mutually horizontally (that is through functional parts) and vertically (i.e., through the whole value chain, from goods procuring to production, delivery, and customer service). Along these lines, entire data-concerning processes, inbound/outbound logistics, marketplace requirements, and product–customer relations will be accessible in real time. Consequently, digital initiatives will operate jointly with clients and providers in an industrial digital ecosystem that permits them to superior handle the line among SC management and promotion purposes (Schrauf and Berttram 2016; Ranganathan et al. 2011).

There exist several explanations for taking into consideration the digitalization influences in SCs and the significance of SC in Industry 4.0. The main potentials of this idea allow real-time definite from providers to clients, small orders quantity, various goods changes, linked decentralized procedures, and autonomous administration. These advantages cannot be attained just by manufacture besides the whole of SC, though. Furthermore, SCs must achieve a bigger foresight to accomplish the necessities of Industry 4.0 as sustainable and as probable in expressions of using suitable technologies and improving horizontal and vertical combination with the SC associates (Kayikci 2018). SC with Industry 4.0 is transformed into a value-driven, smart, effective procedure to produce novel outlines of income and commercial value for administrations and to influence innovative methods with novel technological and systematic procedures as well. SC within the Industry 4.0 is not about if products and facilities are physical or digital, it is about the manner in what way SC procedures are administered by an extensive diversity of innovative technologies, such as "Big Data" (BD), "Augmented Reality" (AR), "Cloud Computing" (CC), "Sensor Technology" (ST), "Robotics" (R), "Omni Channel" (OC), "Internet of Things" (IOT),

"Unmanned Aerial Vehicle" (UAV), "Self-Driving Vehicles" (SDV), "Nanotechnology" (N), and "3D Printing" (3DP), but a few to mention (Büyüközkan and Göçer 2018).

This survey is prepared in this way: the next section reviews and classifies associated publications, in addition to clarifying the methodology assumed in this study. Reviewing the idea of SSC and Industry 4.0, its aspects and elements to shape a dynamic conceptual framework that is resulting from the current literature are represented in Sect. 3. In Sect. 4, the benefits and challenges of Industry 4.0 for SCs are described. A dynamic framework for SSC and Industry 4.0 is proposed in Sect. 5. Finally, the article's concluding remarks, the limitations, as well as possible directions for SSC and Industry 4.0, are presented in Sect. 6.

2 Review of Literature on SSC and Industry 4.0

This review of earlier research works is built on arrangement procedure which offers in what way the literature is considered to be a foundation for the abstract framework. Primarily, the arrangement used in this study is described and afterward, the procedure of the literature review is presented.

2.1 Method of Reviewing

Related research works are detected with the help of a comprehensive online exploration, besides the aim to gather, classify, and synthesize current SSC and Industry 4.0 knowledge. Recognized articles span some sorts of connected fields comprising management, marketing, operations management, industrial engineering, management science, and SC management. Owing to the deficiency of exact keywords describing the issue, we put a considerable attempt to sort papers by studying their titles, abstracts, and texts. Typically, this stage can be accomplished through aiming noticeable journals, books, and conferences. It is not true for SSC and Industry 4.0 because this new topic has appeared only a couple of years ago and associated publication networks are not dispersed yet. The literature is reviewed for the period 2010–2018 by exploring the main databases of scientific and common search engines such as Thomson Reuter's Web of Science, Taylor & Francis online, Elsevier's Scopus, IEEE Explore, Emerald Insight, ProQuest (ABI/INFORM), and Science Direct (Elsevier). We examine and organize the related research works to meet a vision of SSC and Industry 4.0. The overall review methodology for SSC and Industry 4.0 papers is as follows:

Phase1: Identifying the sources (online databases)
Phase2: Search keywords

Phase3: Taxonomy, and analysis based on journal papers, conference papers, books, theses, and so on.
Phase4: Implications and issues include SC, sustainability, Industry 4.0, features, component and technologies, challenges and successes factors.
Phase5: Survey outcomes: a framework for the development and identification of future work.

2.2 *Academic Literature on SSC and Industry 4.0*

Industry 4.0 has commenced obtaining considerable concern from companies throughout the world as it makes greater advantages to many businesses. Our review on Industry 4.0 and SSC literature signifies a gap between the theory and practice in SCs. At present, there exist a restricted number of surveys on Industry 4.0 and SSC. There is also SC focused articles which discuss Industry 4.0 and SSC in expressions of their functions. Based on literature review, 55 areas of application for SSC and Industry 4.0 and their Nomenclature are determined in Appendix Table 10. The existing research papers and conference papers related to Industry 4.0 and SSC along with their application areas, method, and objective are classified in details in Appendix Tables 11 and 12, respectively. In addition, with the consideration of the highest number of citations, top-ten research papers and top-three conference papers in SSC and Industry 4.0 are presented in Tables 1 and 2, respectively.

Recently, Lopes de Sousa Jabbour et al. (2018) recommended a master plan to improve the function of the circular economy (CE) notions in businesses by Industry 4.0 methodologies. They contributed to the literature through presentation on what way diverse Industry 4.0 tools could support CE approaches, and to organizations by directing those tools as a foundation for the policymaking of sustainable operations' management. The key outcomes of their research were as follows: (a) an argument on the equally advantageous connection between Industry 4.0 and the CE; (b) a detailed recognition of the possible influences of smart manufacture equipment to the ReSOLVE model of CE business models; (c) an investigation outline for research on the grouping of CE principles and Industry 4.0 based on the best-related administration principles.

Ginige et al. (2016) established a concept of environmental-precise feasible information which allows the customer to perform with the smallest quantity and more administration. User-centered agriculture ontology was established to change distributed quasi-static information to feasible information. They used "empowerment theory" to make empowerment-oriented farming ways to encourage agriculturalists to act on this information and collected the transaction data to create situational information. This method helps agriculturalists for producing various kinds of yields to meet sustainable agriculture production by means of harvest change.

Table 1 Top-ten research papers with the highest number of citations in SSC and Industry 4.0

Citations	Year	Authors	Objective	Area of application	Publisher
203	2012	Davis et al.	Introducing smart manufacturing, manufacturing intelligence, and demand-dynamic performance	Networked information-based technologies	Elsevier
170	2014	Gebler et al.	This study represents the first comprehensive assessment of 3DP from a global sustainability perspective	3D technologies	Elsevier
56	2014	Holmström and Partanen	The purpose of this paper is to explore the forms that combinations of digital manufacturing, logistics, and equipment use are likely to take and how these novel combinations may affect the relationship among logistics service providers (LSPs), users, and manufacturers of equipment	Manufacturing and logistics	Emerald
53	2013	Baumers et al.	This article investigated whether the adoption of additive manufacturing (AM) technology can be used to reach transparency in terms of energy and financial inputs to manufacturing operations	Manufacturing and logistics	Yale University
46	2015	Yue et al.	This paper described the development and character of ICPS. Then, it presented a service-oriented ICPS model	Information communication technology (ICT)	Elsevier
31	2016	Lom et al.	This paper proposed the conjunction of the Smart City Initiative and the concept of Industry 4.0	Smart City	IEEE

(continued)

Table 1 (continued)

Citations	Year	Authors	Objective	Area of application	Publisher
24	2015	Prause	The paper addressed the research question of how new and sustainable business models and structures for Industry 4.0 might look like and in which direction existing traditional business concepts have to be developed to deploy a strong business impact of Industry 4.0	E-Residency	Elsevier
24	2017	Prause and Atari	The paper investigated the relationship between networking, organizational development, structural frame conditions, and sustainability in the context of Industry 4.0	Manufacturing and logistics	VsI Entrepreneurship and Sustainability Center
24	2017	Brofman Epelbaum and Martinez	This paper presented a theoretical framework grounded on the resource-based view (RBV) of the firm to determine the strategic impacts of the technological evolution of food traceability systems	Food industry	Elsevier
19	2018	Lopes de Sousa Jabbour et al.	The paper extended the state-of-the-art literature by proposing a pioneering roadmap to enhance the application of CE principles in organizations by means of Industry 4.0 approaches	Circular economy (CE)	Springer

Table 2 Top-three conference papers with the highest number of citations in SSC and Industry 4.0

Citations	Year	Authors	Objective	Area of application	Publisher
8	2016	Ginige et al.	Developing a notion of context-specific actionable information which enables the user to act with the least amount of further processing	Agriculture sector	IEEE
4	2017	Tan et al.	Discussing how organizations can investigate and implement techniques for their modern enterprise with a focus on how advanced big data tools can be applied to quality analytics for monitoring and improving quality in the electronics industry	Electronic industry	IEEE
2	2010	Price et al.	Presenting a project structure which has been designed to address these issues using at its core, a digital framework for the creation and management of performance parameters related to the lifecycle performance of thermoplastic composite structures	Thermoplastic composite structures	Mark A Price

2.3 Published Books on Industry 4.0 and SSC

To the finest of our information, there exist six books that focused on Industry 4.0 and SSC (see Table 3). Recently, Abdi et al. (2018) developed manufacturing ideas and further functions than tangible manufacture for a broader industrial value chain integrating external shareholders that include providers of raw matters and pieces, clients, manufacturing service suppliers, cooperating manufacturing companies, and environmental organizations. They highlighted the two advanced concepts of reconfigurable manufacturing systems (RMS) and Industry 4.0 together with their joint progress. They presented disputes of mass-customization and active variations in the

Table 3 Literature review of SSC and Industry 4.0 (Books)

Year	Authors	Objective	Method	Area of application
2012	Goodship and Stevels	Drawing lessons for policy and practice from all over the world	Review	Waste electrical and electronic equipment (WEEE)
2014	Xu	Describing the setup of digital enterprises and how to manage them, focusing primarily on the important knowledge and essential understanding of digital enterprise management required by managers and decision makers in organizations	Modeling	Digital enterprises
2015	Fiorini and Lin	Providing an overview of current topics in intelligent and green transportation on the land, sea, and in flight, with contributions from an international team of leading experts	Review	Intelligent transport systems (ITS)
2016	Kiritsis	This book not only explains in detail what LEAP is and how to use it but also provides LEAP case studies from sectors such as auto manufacturing and offshore engineering	Case study	Manufacturing and logistics
2017	Handfield and Linton	Addressing the changes that have occurred and are still unfolding at various organizations that are involved in building real-time SCs	Review	Global economy
2018	Abdi et al.	Developing manufacturing concepts and applications beyond physical production and toward a wider manufacturing value chain incorporating external stakeholders that include suppliers of raw materials and parts, customers, collaborating manufacturing companies, manufacturing service providers, and environmental organizations	Modeling	Re-configurable manufacturing systems (RMS)

SC background by concentrating on advancing novel methods connected to integrity, scalability, and re-configurability at the system level and engineering readiness in names of the practical, and commercial feasibility of RMS. The authors applied decision support systems (DSS) for the collection of families' product and optimizing product-process configuration. Their suggested models were explained across real case studies in applicable manufacturing firms.

2.4 *Published Book Chapter on Industry 4.0 and SSC*

Authors recognized seven book chapters on the issue of Industry 4.0 and SSC which are presented in Table 4. Recently, Jirsak (2018) examined the influence of Industry 4.0 revolutions on SC management. The writer offered the results achieved in the investigation of recent essential variations and presented a contrast with a preceding conversion of the paradigm. This chapter suggested a revolution that the business SC system has to go over to re-establish its competitive situation in an era of Industry 4.0. In addition, the chapter offered case study of 3PL (demand planning, production planning, and supply planning) insight about Industry 4.0 founded on detailed meetings performed among the major global 3PLs operating in the Czech Republic.

2.5 *Analysis of Industry 4.0 and SSC Literature*

It is vital to emphasize that the recognition of 83 research works motivated the outcomes of this review, which are informed in Sects. 2.2, 2.3 and 2.4. (Also see the appendix for detailed information.) The keywords were not prearranged ahead of the search; however, they have progressively appeared through the wide-ranging reading procedure that happens during the preparation of this paper. The last list of keywords is as follows: Sustain, Sustainable, Sustainability, Green, Industry 4.0, Smart factory, Digital, Supply chain, and Logistic. In a few conditions, the research under review will still be used to explain the outcomes and meet an improved comprehension of the subject. Figure 2 demonstrates the important results by offering a complete sum-up in phrases of types of SSC and Industry 4.0 publications. The highest amount of publication was in the form of "research article" with about 45 papers, and the second highest level belongs to "conference paper" with 15 papers. "Book chapter" and "Books" were at the lowest level with 7 and 6 papers, respectively. In addition, the trend for a number of related publications per year from 2010 to 2018 is presented in Fig. 3 and the percentage of publications per year is demonstrated in a pie chart in Fig. 4. With regard to these figures, the number of publications has a small fluctuation between 2010 and 2016 (about 2–7 papers or 2–9%), while the trend increased dramatically in 2017 to hit the highest point in 2018 with 26 papers (or 31%) out of 83. Moreover, in total, about 17 papers out of 83 were open access and 66 papers have access as an abstract only (see Fig. 5). It has been mentioned that we find out about

Table 4 Literature review of sustainable SC and Industry 4.0 (Book chapters)

Year	Authors	Objective	Method	Area of application
2010	Ndou and Sadguy	Suggesting that digital marketplaces could provide a viable model for SME networking; however, the successful path toward networking requires harmonization of the digital marketplace business model with SC characteristics	Modeling	SME networking
2013	Montreuil et al.	Providing insights on the foundations of the physical internet that has been introduced as a solution to the global logistics sustainability grand challenge of improving by an order of magnitude the economic, environmental, and social efficiency and sustainability of the way physical objects are moved, stored, realized, supplied, and used across the world	Modeling	Physical Internet
2013	Kückelhaus et al.	Addressing how visibility solutions based on digital product memories (DPMs) developed in the SemProM project can be demonstrated in the logistics domain to guarantee the carbon offset of transport and integrity control within SCs	Modeling	SemProM project
2015	Kagermann	Discussing the impact, challenges, and opportunities of digitization and concludes with examples of recommended policy action	Modeling	Digitization

(continued)

55 areas of application in SSC and Industry 4.0 literature (see Appendix Table 10). Furthermore, we recognized top-five areas of application in the literature (see Table 5 and Fig. 6) that include manufacturing and logistics, food industry, circular economy, agriculture sector, and clothing industry with 8, 6, 5, 4, and 3 papers, respectively.

We agreed to create Tables 1, 2, 3 and 4 with the author's names and publication year. Listed in the rows, the objective explains the aim of the research works, the

Table 4 (continued)

Year	Authors	Objective	Method	Area of application
2015	Kamarulzaman and Eglese	Providing relevant e-procurement solutions with respect to the MPOI and will provide comprehensive purchasing activities of different types of products along the SC through e-procurement technologies	Case study	E-procurement technologies
2017	Meera et al.	Proposing a framework for rice extension strategies that integrate knowledge, technology, and markets which helped to provide better, faster, and cheaper solutions to reach out to rice farmers and integrate knowledge, technologies, and markets	Modeling	Agriculture sector
2017	Kasemsap	Introducing the roles of lean supply chain management (SCM) strategies and green SCM strategies in the global business environments	Modeling	Global economy
2018	Jirsak	Presenting an impact of Industry 4.0 transformation on logistics and SC management	Case study	Global 3PLs operating

method indicates the approach which is used in the research, the application identifies the area of papers concentrations; citations demonstrated the number of citations related to the paper while the publisher indicated the name of the institute which publishes the paper. These features have been designated based on authors' proficiency in the subject and the applicable investigation. When the relevant works on Industry 4.0 and SSC are combined and studied completely, they show reliable benefits to the readers. These employed benefits explain the master plan for creating the Industry 4.0 and SSC framework in the succeeding sections created on the summary of the content, scope, and outcomes of designated literature. Based on the research in this paper classification, next sections use and recognize the main restrictions and projections in Industry 4.0 and SSC, encapsulate the previous investigation to detect knowledge gaps through offering benefits, drawbacks, and boundaries of specific approaches and present a development dynamic framework as a master plan for upcoming study.

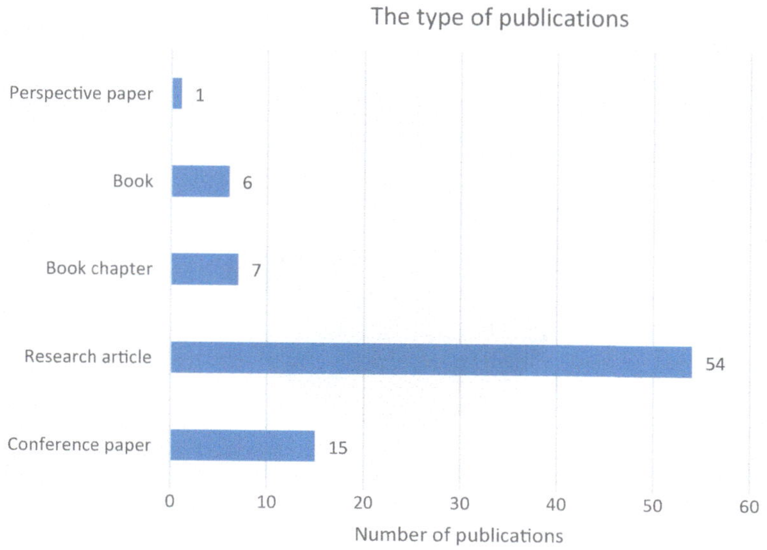

Fig. 2 Types of publications for SSC and Industry 4.0

Fig. 3 The number of publications for SSC and Industry 4.0 per year

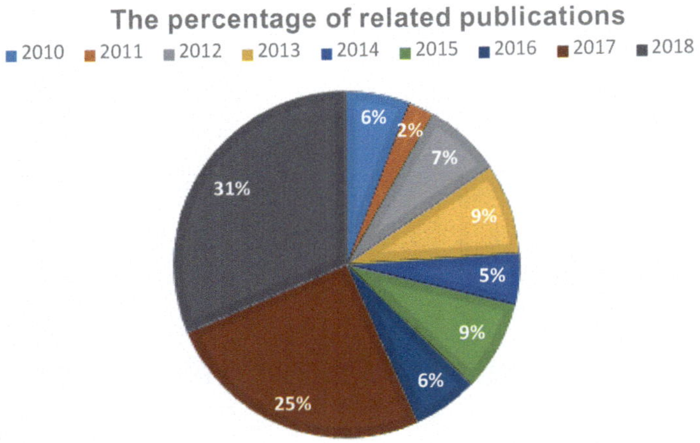

Fig. 4 The percentage of publications for SSC and Industry 4.0 per year

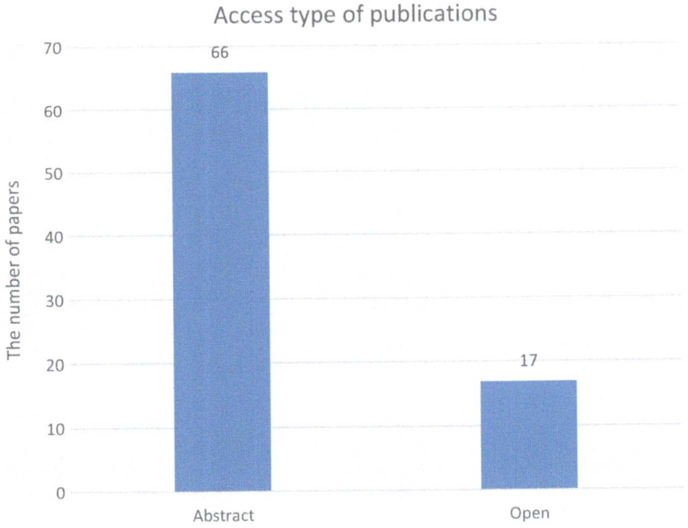

Fig. 5 The type of access for publications in SSC and Industry 4.0

3 Enabling Technologies and Key Elements of Industry 4.0 for SSC

Several characteristics are not presented in traditional SC while they are required in today's and tomorrow's commercial environment. The conventional SC has a chain of disconnected stages, mostly. Converting a conventional SC into Industry 4.0 and SSC breaks down these walls with the purpose of the chain converts into an integrated

Table 5 Top-five areas of application in the literature of SSC and Industry 4.0

Number of papers	Area of application	Source
8	Manufacturing and logistics	Baumers et al. (2013), Blümel (2013), Holmström and Partanen (2014), Kiritsis (2016), Prause and Atari (2017), De Carolis et al. (2017), Luthra and Mangla (2018), Forkel et al. (2018)
6	Food industry	Brofman Epelbaum and Martinez (2014), Clear et al. (2013), Pilinkienė et al. (2017), Zhong et al. (2017), Todorovic et al. (2018), Gružauskas et al. (2018)
5	Circular economy (CE)	Jensen and Remmen (2017), Lopes de Sousa Jabbour et al. (2018), Garcia-Muiña et al. (2018), Tseng et al. (2018), Bressanelli et al. (2018)
4	Agriculture sector	Ginige et al. (2016), Kalogianni et al. (2017), Meera et al. (2017), Bucci et al. (2018)
3	Clothing industry	Papahristou and Bilalis (2016), Pal and Sandberg (2017), Papahristou and Bilalis (2017)

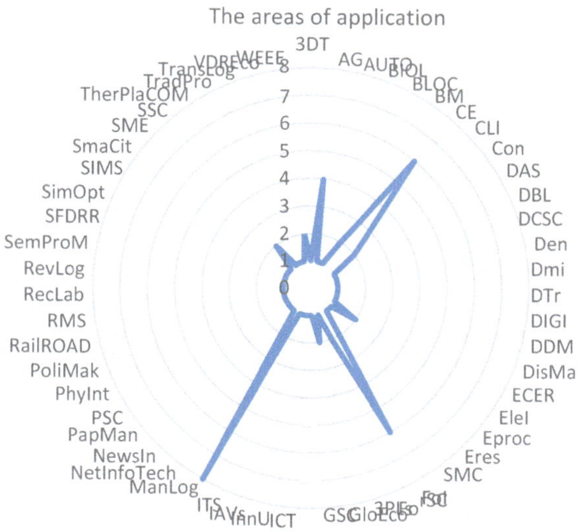

Fig. 6 The areas of application for SSC and Industry 4.0

system that operates perfectly. Therefore, Industry 4.0 and sustainability allow the succeeding invention of SCs evolve and present mutually productivity and flexibility (Ardito et al. 2019; Büyüközkan and Göçer 2018). Bag et al. (2018) demonstrated the Industry 4.0 enablers of SSC management, which are governmental support; support of research institutes and universities; law and policy about employment; improved IT security and standards; management commitment; focus on human capital; change management; horizontal integration; vertical integration; standardization and reference architecture; and corporate governance and third-party audits. Since Industry 4.0 and sustainability solutions are interrupting conventional SC, there exist several noticeable elements that are practically related to every Industry 4.0 and SSC. These distinct advantages are gathered into some key elements that Industry 4.0 and SSC would like to make. The key elements include speed, flexibility, global connectivity, real-time inventory, intelligent, transparency, cost-effective, scalability, innovative, proactive, and eco-friendly (Büyüközkan and Göçer 2018).

4 Challenges, Success Factors, and Research Gaps in Industry 4.0 and SSC

Without considering the companies' size, they need to investigate advancing several kinds of Industry 4.0 and sustainability association competencies since businesses will contest on SCs at worldwide level eventually.

4.1 Challenges and Concerns About Executing Industry 4.0 and SSC

Many problems can happen along the SC. Xu (2014) described the chief challenges of building Industry 4.0 and SSC on collecting totally needed data from various suppliers, certifying the correctness of that information, and building up a software design and policy that can use the information to administer and perform the SC. Because the dimension of chain includes inside and outside associates, it will be time-consuming and tend to mistake. Furthermore, the current great quantities of stock cannot be capable of fulfilling the demand, and SCs substructure can be inadequate and the characteristic of products can be difficult to check (Büyüközkan and Göçer 2018). In Table 6 we presented 24 recognized challenges for Industry 4.0 and SSC and described each of them briefly.

4.2 Success Aspects for Industry 4.0 and SSC

The execution measurement in Industry 4.0 and SSC are especially significant. This measure could be studied by the capability of satisfying requests tills due date, distribution schedule, provider consistency, the budget of chain or postponements, among several others. As stated by a current investigation, over 33% of 2000 respondents have launched employing Industry 4.0 in their SCs, and entirely 72% supposed to have completed so in five years (Schrauf and Berttram 2016). Some motives why Industry 4.0 and SSC implementation has been slowed being the absence of consciousness between the staff and shareholders about digital instruments, and the absence of essential abilities among workers and shareholders (Büyüközkan and Göçer 2018). Consequently, the widespread adoption of Industry 4.0 and SSC will depend on the recognizable aspects of these significant successes so listed in Table 7.

4.3 Research Gaps in Industry 4.0 and SSC

One of the key goals of our investigations is to present a review of the literature of Industry 4.0 and SSC. The literature reviews also presented that few papers were really directed on Industry 4.0 and SSC simultaneously (see Tables 1, 2, 3 and 4 and Appendix); however, most of them were directed on empowering of its concentration on SCs. Our study displayed that since there exists a steady growth of research issued on the subject as 2010 for example, from five articles printed in 2010 to 26 articles in 2018, the strong mainstream of papers is still "research papers" (see Fig. 2). Hence, more investigation on Industry 4.0 and SSC is required to be done by industries and organizations.

More extensively, research should emphasis on the development frameworks to convert, employ, and accept Industry 4.0 and sustainability in the context of SCs. In spite of the current attention in the Industry 4.0 and SSC subject because of its vast possibilities, the studies that report this topic advantages and contests are in their

Table 6 Challenges and issues of implementing Industry 4.0 and SSC

Challenges	Description
Lack of vision and strategy	Industry 4.0 describes an innovative approach to the digital transformation, which requires a clear digital operations vision and mission (Erol et al. 2016)
Lack of planning	Deficiency of proper demand plan and guidelines and tools (Xu 2014; Schrauf and Berttram 2016)
Financial constraints	In Industry 4.0, financial constraints are considered to be a very important challenge in terms of advanced equipment and machines, facilities, and sustainable process innovations (Dawson 2014; Theorin et al. 2017; Nicoletti 2018)
Lack of competency in adopting/applying new business models	As it is not necessary that all the new insights of Industry 4.0 will be workable and only some events are interesting out of million events, so revealing these insights is a challenge for data scientists to write suitable algorithms in adopting/applying new business models (Khan et al. 2017; Saucedo-Martínez et al. 2017)
Lack of collaboration and coordination	Deficient collaboration with external associates and deficient input from internal functions (Penthin and Dillman 2015; Xu 2014; EY 2016; Lee et al. 2014; Duarte and Cruz-Machado 2017; Pfohl et al. 2017)

(continued)

Table 6 (continued)

Challenges	Description
Poor existing data quality	Data quality is one of the foremost requirements in making decisions in successful Industry 4.0 adoption and so inaccurate over-optimistic forecasts for demand, inventory, production, and other data are key challenges (Xu 2014; Carter et al. 2009; Richey et al. 2016; Santos et al. 2017)
Security issues	Security is the prime requirement to transform a factory into a smarter factor and an SC into smarter value chains (Sommer 2015; Wang et al. 2016; Pereira et al. 2017)
Lack of global standards and data sharing protocols	The industries are deficient in standards and protocols, data transfers, adopting sustainability-oriented modern information interface technologies and in business networks (Branke et al. 2016)
Lack of information sharing	Companies' reluctance on information sharing (Xu 2014; Nowak et al. 2016)
Lack of infrastructure and internet-based networks	High infrastructure, information technology-based facilities, and technologies are crucial in the effective adoption of Industry 4.0 concepts (Leitão et al. 2016; Bedekar 2017; Pfohl et al. 2017)
Low management support and dedication	In order to develop an effective Industry 4.0 concept, management support and dedication to accept the changes are very crucial (Gökalp et al. 2017; Savtschenko et al. 2017; Shamim et al. 2017)
Silver bullet chase	The belief that everything will be fine (Xu 2014; Hines 2004)
Poor research and development (R&D) on Industry 4.0 adoption	Lack of focused research on addressing the various aspects of Industry 4.0 adoption (Schmidt et al. 2015a; Hermann et al. 2016)
Lack of knowledge	Deficiency of SC management training and skills (Xu 2014; Hines 2004)
Lack of digital culture	Industry 4.0, generally of interdisciplinary in nature, requires digitization to connect different elements of a network (Ras et al. 2017; Schuh et al. 2017)
Low understanding of Industry 4.0 implications	There is a very low understanding of Industry 4.0 implications among both the researchers and practitioners (Almada-Lobo 2016; Hofmann and Rüsch 2017)
Agility and flexibility	Lack of required flexible and agile SC management (Penthin and Dillman 2015; Xu 2014; Hines 2004; Richey et al. 2016; Nabben 2016)
High volatility	Lack of knowledge and skills in dealing with volatility in SC management (Xu 2014; EY 2016; Hines 2004)
Overconfidence in suppliers	Relying on certain suppliers in certain parts of the globe (Xu 2014; Hines 2004)
Profiling and complexity issues	The lack of roadmaps and guides supporting its implementation, as well as its high complexity makes "Industry 4.0" too uncertain for achieving sustainability in SCs (Erol et al. 2016; Ras et al. 2017)

(continued)

Table 6 (continued)

Challenges	Description
Lack of integration	Deficient view on the integration of digital and non-digital SC management. The integration of technology is very essential in effective communication and higher productivity (Zhou et al. 2015; Penthin and Dillman 2015; Xu 2014; EY 2016; Hines 2004)
Unclear economic benefit of digital investments	The lack of clearly defined return on investment could be seen as one of the major challenges to Industry 4.0 initiatives for accomplishing sustainability in the SC (Kiel et al. 2017; Marques et al. 2017)
Lack of governmental support and polices	Policy analysts and government bodies have not revealed the roadmap for transforming the traditional business functions into smarter and sustainable processes (BRICS Business Council 2017)
Legal issues	Data privacy and security issues need to be considered in developing data-driven sustainable business models of Industry 4.0 (Schröder 2018; Muller et al. 2017a)

initial phases. There exist some study gaps in the present sources about Industry 4.0 and SSC that can be condensed as follows:

- Lack of research on implementing Industry 4.0 and sustainability in different industries. As it is mentioned before, we recognized 55 areas of application, but the number of papers related to each of them is only one. It means there is a vast context for researchers to investigate Industry 4.0 and sustainability in different sectors.
- Deficiency of development frameworks that offer advice for Industry 4.0 and SSC implementation in a perspective with roadmaps and obvious plans. This issue may help in directing executives about what phases and which place in SCs can leaders use Industry 4.0 and SSC, assumed that SCs may be at various stages of the Industry 4.0 and SSC employment. Moreover, development frameworks may offer support in shifting the administration preparations in the SCs.

Table 7 Success factors for Industry 4.0 and SSC

Success factor	Description
Continuous collaboration	Capabilities are harmonized within and beyond physical boundaries to increase collaboration between involved actors of the SC (CapGemini et al. 2016; Hines 2004; Accenture 2014)
Real-time visibility	Dynamic, secure, and interactive visibility across the entire SC will improve the management of Industry 4.0 and sustainable SC (Cecere 2014; Guarraia et al. 2015; CapGemini et al. 2016; Hines 2004; Accenture 2014)
Integration	Building the integration of digital and non-digital SCs so that a unified and whole view of inventory across the firm can be achieved (Xu 2014; Raj and Sharma 2014; Schmidt et al. 2015a)

(continued)

Table 7 (continued)

Success factor	Description
Alignment of suppliers	Aligning the interest of all the firms in the SC with your own to create incentives for better performance and developing trust (alignment) (Xu 2014; Raj and Sharma 2014; Schmidt et al. 2015a; CapGemini et al. 2016)
Highly evolved operating models	Product and service functions can be altered easily to meet customers' changing demands (Raj and Sharma 2014; Hanifan et al. 2014; Accenture 2014).
Shared information	Industry 4.0 and sustainable SC allows easier information sharing on sales forecast and production data (Xu 2014; Raj and Sharma 2014; Cecere 2014; Schmidt et al. 2015a; CapGemini et al. 2016; Hines 2004)
Automated execution	Seamless human–machine interactions increase operational efficiency (Raab and Griffin-Cryan 2011; Raj and Sharma 2014; Schmidt et al. 2015a; Rakowski 2015; CapGemini et al. 2016; GTnexus 2016; Accenture 2014)
Adopting advanced analytics and analytics tools	Advanced data analysis improves decision making. Gaining better understanding and forecasting of the demand and solve previously unsolvable and even unknown problems along the SC (e.g., BD and Data Analytics, etc.) (Xu 2014; Raj and Sharma 2014; Hanifan et al. 2014; Accenture 2014)
Maximum efficiency	Seamless integration of people, processes, and technology (Raj and Sharma 2014; Rakowski 2015)
Enhanced and accelerated innovation	Digital SCs inspire and abet innovations in designs, operations, and customer relationships (Xu 2014; Raj and Sharma 2014; Cecere 2014; Schmidt et al. 2015a; Accenture 2014)
Personalized experiences, customer-centric	Channel-centric supply networks support customized products and services (Penthin and Dillman 2015; Xu 2014; Raj and Sharma 2014; Schmidt et al. 2015a; Hanifan et al. 2014; Accenture 2014)
Organizational flexibility	Digital plug-and-play capabilities make it easier to configure and re-configure (Raab and Griffin-Cryan 2011; Raj and Sharma 2014; Cecere 2014)
Proactive prevention	Decision support systems driven by predictive analytics can strengthen adaptability and reliability (Xu 2014; Raj and Sharma 2014; Hanifan et al. 2014; Accenture 2014)
Enhanced responsiveness	Better information and sophisticated analytics can help accelerate responses to competitors' moves, technology shifts, and changing demand and supply signals (Xu 2014; Raj and Sharma 2014;Cecere 2014; Schmidt et al. 2015a; CapGemini et al. 2016; Hanifan et al. 2014; Accenture 2014)
Last mile postponement	Swiftly repurposing organizational assets assists in ensuring that the supplies are aligned with evolving demands (Xu 2014; Raj and Sharma 2014; Hanifan et al. 2014; Accenture 2014)

- Deficiency of technologies and instruments that deal with SCs challenges in the context of Industry 4.0 and sustainability because this context is dissimilar from that of a usual SC. Decisions in an Industry 4.0 and SSC circumstances need novel technologies and tools. Industry 4.0 and SSC will change many processes such as maintenance, quality control, inventory control, planning for manufacture, and purchasing.
- There are many obstacles for the execution of Industry 4.0 and SSC from mutually managerial and technological outlooks. There's rather a substantial transformation occurring in the world. Businesses are at the edge of a contest to change their SCs to Industry 4.0 and sustainable context (CapGemini et al. 2016). Consequently, Industry 4.0 and SSC problems and concerns presented in Subdivision 4.1 require to be answered through the support of Industry 4.0 and SSC success elements resulting from existing literature. There exist not considerable research papers on in what way to deal successfully with these.

In the next section, we will employ all the understanding, and information collected from the investigated literature to set up an advanced different framework.

5 Developing Industry 4.0 and SSC: A Dynamic Framework

Effective SCs work with a well-defined foresight, benefiting frameworks and master plan that describe the path advancing. The present collected works are condensed into four main phases and their sub-objectives. Figure 7 displays the dynamic framework in a graphical arrangement created on analysis of the related research works. The dynamic of this framework is related to the Deming's PDCA's (plan-do-check-act) process which is a cyclic four-step model for continuous improvement (CI) in commercial procedure management. It means that the PDCA's cycle should be applied in all four main stages as well as their sub-goals and interaction among them. Therefore, this framework evolves and improves over time via performing corrective actions for eliminating causes of non-conformities. With regard to this dynamic framework, the term "Sustainable Supply Chain 4.0 (SSC 4.0)" is proposed to show the integration of four separated domains in a real-world environment. Actually, SSC 4.0 gradually permits SCs to turn into an integral section of decision-making and tactical developing. Organizations can influence and develop SSC 4.0 to harmonize various aspects of their policies and more successfully direct their definite requirements. The fact is that the vision on the SSC 4.0 literature and projected framework for the development of SSC 4.0 brings the problem of how it can be appropriately applied and proved in regular SC. It should be mentioned that every SC will have a slightly dissimilar set of SSC 4.0 development objectives with diverse main concern. Along with reconsidering and reforming whole SCs, the crucial required assessment

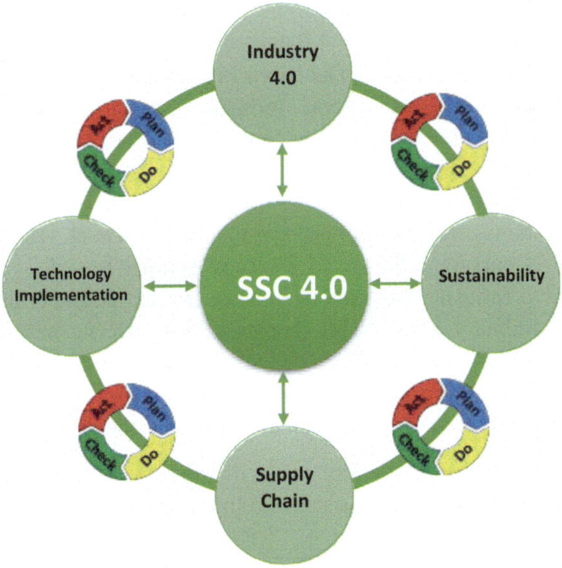

Fig. 7 A dynamic framework for the development of sustainable supply chain 4.0

objectives for SCs regularly map the fields of Industry 4.0, technology implementation, sustainability, and SC management, which are key phases for administrative arrangement.

By applying this dynamic framework, most of the SC executives will be acquainted with the basic SSC 4.0 methodologies, evaluating the SCs' existing Industry 4.0 and sustainability state, founding a foresight for technology implementation, and expanding a revolution plan for SC management in the novel atmosphere. Explanation of these subjects, their decomposition, and creation of their arrangement along with the PDCA cycle in all stages and interactions are the core of SSC 4.0 employments in usual SC. The disintegrated framework for progressing SSC 4.0 is presented in Tables 8 and 9.

6 Conclusions, Limitations, and Further Research Trends

This review investigates the transformation of SCs to a sustainable supply chain 4.0 (SSC 4.0), an issue of vast interest mutually for specialists and academics. It is absorbing and prepared in a reliable arrangement, so that to show the key suggestions, and created on a method for the progress of an SSC 4.0.

To the best of author's knowledge, there is only one paper that considered a literature review for Industry 4.0 and supply chain sustainability. In short, the highlights of the differences of this book chapter with the literature review by Bag et al. (2018) are as follows:

- A dynamic framework is proposed for sustainability and Industry 4.0 in the supply chain (SC) management
- The term "Sustainable Supply Chain 4.0" (SSC 4.0) is proposed.
- A summary of the documents such as research articles, conference papers, books, and book chapters from 2010 to 2018 is investigated.
- Use different major online databases to get a vast insight into the issues.
- All areas of application in the literature for sustainability and Industry 4.0 in the supply chain (SC) management is recognized.
- Challenges, success factors, and research gaps are determined.

The outcomes of this survey target to response the inquiries for example, what the recent position of SSC 4.0 is in the theoretical, and engineering investigations, along with what the SSC 4.0 future developments seem, and how the present importance of Industry 4.0 and sustainability can be integrated into SC or logistics, and so on. With the intention of illustrating the advancement of inside the SSC 4.0 issues, a review of the research works is offered, learning mismatches in the specified investigation issue

Table 8 Decomposed framework for SSC 4.0

Sustainability	Supply chain	Technology implementation	Industry 4.0
Economic (Pisching et al. 2015a, b)	Process (Counsil 2004; Turhan et al. 2011)	Technology enablers (Ibem and Laryea 2014) formation of technology infrastructure (Najmi et al. 2016; Klievink 2015)	Virtualization (MacDougall 2014) interoperability (Saldivar et al. 2015) decentralization (Gilchrist 2016)
Environmental (Badurdeen et al. 2009)	Integration (Sahin and Robinson 2002, 2005; Bagchi et al. 2005)	Human and technology relationship (Oyekan et al. 2017)	Real-time capability (Vogel-Heuser and Hess 2016)
Social (Wittstruck and Teuteberg 2011)	(Alfalla-Luque et al. 2013; Lee 2000) Responsive (Banchuen et al. 2017) automation (Barratt 2016; Viswanadham 2002) re-configuration (Buyukozkan and Gocer 2018) Transpiration and logistics (Speranza 2018) Analytics (Schmidt et al. 2015b; Sahay and Ranjan 2008) Information systems (Agus and Ahmad 2017) Collaboration (Cao and Zhang 2011)	Project management (Yee and Oh 2013)	Service orientation (Sanders et al. 2017) modularity (Peres et al. 2017)

Table 9 Decomposed sustainability for SSC 4.0

Sustainability dimensions	Sustainability criteria
Economic	Logistics cost; delivery time; transport delay; inventory reduction; loss/damage; frequency of service; forecast accuracy; reliability; flexibility; transport volumes; applications (Anderson 2007; Fahimnia et al. 2017; Genovese et al. 2017; Sauer and Seuring 2017; Zeng et al. 2017; Monnet and Le Net 2011; Dougados et al. 2013; Gebler et al. 2014; Schrauf and Berttram 2016)
Environmental	Resource efficiency; process energy; process emissions; waste; pollutions; land use impact (Nowak et al. 2016; Coyle et al. 2015; Dam and Petkova 2014; Zhu et al. 2011; Gebler et al. 2014; Monnet and Le Net 2011)
Social	Development benefits; impacts; health; safety; labor patterns; acceptance (Krause, et al. 2009; Mani, et al. 2016; Kogg and Mont 2012; Gebler et al. 2014; Monnet and Le Net 2011; Schrauf and Berttram 2016)

are recognized and the features of the previous study are established. Together with this broad analysis of upcoming developments on SSC 4.0, a dynamic framework for SSC 4.0 is settled consistently with the benefits, drawbacks, and restrictions of current SSC 4.0 literature. It is determined to meet the mismatches of former investigations concerning the creation of a comprehensive abstract or academic framework. The recommended dynamic framework goals are at recognizing the characteristics, factors, and technology enablers, success aspects, and disputes for advancing an SSC 4.0. Therefore, the current study and dynamic framework can make available visions to mutually academicians and practitioners in their function of SSC 4.0.

6.1 Limitations

With regard to the above-mentioned issues, this review has some restrictions. The following topics summarize these possible restrictions:

- Classified papers in this review of literature are grounded mostly on results from academic journals (consider Fig. 2). Adding more industrial reports in the forthcoming can improve this analysis's outcomes.
- The study results are constructed on the exploration of the point out databases by running the entered keywords. As exploration is vastly responsive to these keywords, reviews which take a little diverse enters may be neglected.
- The fact is that in this study a systematic literature review procedure has been employed in which every database is independently explored and the gathered papers are picked just prior to the examination phases. Another method can be used for organizing these papers obtained in the database.
- We considered a period of previous eight years (2010–2018). We believe these associated research works are friendly on approaches for SSC 4.0. Although the

results are not full, we think that they are wide-ranging as they embrace several extremely graded scholarly journals.
- The demonstrated SSC 4.0 dynamic framework aims at employing the integration of Industry 4.0, along with sustainability, technology implementation, and SC management which evolve over time. We have not involved in the extra disintegration of the SSC 4.0 model because it is away from our study.

6.2 Future Research Topics

With regards to the above-mentioned restrictions, the succeeding upcoming investigation fashions on SSC 4.0 are constructed on a detailed review of literature along with the previous operational proficiency of writers. Additional examination of these recommendations can produce new awareness and strong concepts in the subject. Hence, the succeeding topics are presented:

- This review proposes a supplementary investigation into manufacturing real-case purposes for the offered SSC 4.0 dynamic framework, demonstrated in Fig. 7.
- Businesses from various engineering circumstances are affected by their particular approaches for SSC 4.0, subject to their particular reason of utilizing new Industry 4.0 technologies. Consequently, significant fashions for upcoming SSC 4.0 require a clear plan for each to enhance the revolution of its SSC 4.0 tasks. The given category can, consequently, be supplementarily improved to enlighten mutually theoretically with experts knowledge by creation of sub-frameworks for every business.
- Industry 4.0 and sustainability will change the approach of SCs. For the purpose of appropriate application and confirmation of the development framework, the offered steps should be understood and evaluated in typical SC.
- Although the benefits and restrictions of SSC 4.0 have been examined at a theoretical level, more developments are needed yet in some parts of SSC 4.0 so that a strong, consistent, and flexible solution is obtained for useful execution of SSC 4.0 into engineering real-case functions.
- Additionally, the advantages and disputes of SSC 4.0 can be investigated for the superior recognition of the possibility and success of the recommended dynamic framework.
- To summarize, the equipment and elements of SSC 4.0 can be joined to further the current SC-associated investigations in equally scholar journals and engineering reports. SSC 4.0 is vastly far from implementing its greatest ability, and as stated in this study, there exist many fields (55 topics are recognized by this study) that need urgent consideration.

Appendix

See Tables 10, 11 and 12.

Table 10 The areas of application and nomenclature

The areas of application	Nomenclature
3D technologies	3DT
Agriculture sector	AG
Automotive	AUTO
Biologicalization	BIOL
Blockchain technology	BLOC
Business model	BM
Circular economy (CE)	CE
Clothing industry	CLI
Construction industry	Con
Digital assorting system (DAS)	DAS
Digital business landscape	DBL
Digital China Company's SC integration system	DCSC
Digital enterprises	Den
Digital mine	Dmi
Digital training method	DTr
Digitization	DIGI
Direct digital manufacturing (DDM)	DDM
Distributed manufacturing	DisMa
E-CRM strategy	ECER
Electronic industry	EleI
E-procurement technologies	Eproc
E-Residency	Eres
FoFdation Smart Machine Controller (SMC)	SMC
Food industry	FoI
Forest-based supply chains	ForSC
Global 3PLs operating	3PLs
Global economy	GloEco
Green supply chain	GSC
Information communication technology (ICT)	ICT
Innovation union	InnU
Intelligent autonomous vehicles (IAVs)	IAVs
Intelligent transport systems (ITS)	ITS
Manufacturing and logistics	ManLog

(continued)

Table 10 (continued)

The areas of application	Nomenclature
Networked information-based technologies	NetInfoTech
Newspaper industry	NewsIn
Paper manufacturing	PapMan
Pharmaceutical supply chain (PSC)	PSC
Physical internet	PhyInt
Policy-making	PoliMak
Rail-road intermodal transport network	RailROAD
Re-configurable manufacturing systems (RMS)	RMS
Record labels and retail outlets	RecLab
Reverse logistics	RevLog
SemProM project	SemProM
Sendai framework for disaster risk reduction (SFDRR)	SFDRR
Simulation optimization	SimOpt
Small-scale intelligent manufacturing system (SIMS)	SIMS
Smart city	SmaCit
SME networking	SME
Supply chain sustainability	SSC
Thermoplastic composite structures	TherPlaCOM
Trade promotion	TradPro
Transport and logistics	TransLog
Virtual digital retail ecosystem	VDREco
Waste electrical and electronic equipment (WEEE)	WEEE

Table 11 Literature review of sustainable SC and Industry 4.0 (Research papers)

Year	Authors	Objective	Method	Area of application
2010	Dzopalic et al.	Introducing E-CRM strategy	Case study	E-CRM strategy
2011	Norris	Introducing a new program called the Newspaper Industry Environmental Vision which is gathering a critical mass of newspaper publishers and printers calling for increased efforts in industry best practices and sustainability	Modeling	Newspaper industry

(continued)

Table 11 (continued)

Year	Authors	Objective	Method	Area of application
2011	Hajdul and Cudzilo	Presenting how the Common Framework supports interoperability between commercial actors and communication to authorities and transportation network responsible	Case study	Transport and logistics
2012	Agarwal et al.	Studying the consumer return behavior of end-of-life goods at different incentive levels and make an attempt to incorporate the latest research practices	Modeling	Waste electrical and electronic equipment (WEEE)
2012	Lin et al.	Creating digital mine and key technologies in China	Modeling	Digital mine
2012	Davis et al.	Introducing smart manufacturing, manufacturing intelligence, and demand-dynamic performance	Modeling	Networked information-based technologies
2013	Hentza et al.	Describing the on-going work with a specific focus on the definition and implementation of the FoFdation smart machine controller (SMC) in an adaptable architecture that satisfies both commercial and open source CNC controllers	Modeling	FoFdation smart machine controller (SMC)
2013	Zhu et al.	Investigation of BIM digital technology in the construction industry	Review	Construction industry
2013	Blümel	Studies means by which digital engineering and virtual and augmented reality technologies can support the creation of sustainable smart manufacturing and smart logistics processes as well as on-the-job training and qualification and knowledge transfer	Modeling	Manufacturing and logistics

(continued)

Table 11 (continued)

Year	Authors	Objective	Method	Area of application
2013	Clear et al.	Brings together participants from a diverse range of disciplines to develop an understanding of existing food consumption practices, and how this domain can profit from novel Ubicomp technology and interaction designs	Modeling	Food industry
2013	Baumers et al.	Studying whether the adoption of additive manufacturing (AM) technology can be used to reach transparency in terms of energy and financial inputs to manufacturing operations	Modeling	Manufacturing and logistics
2014	Gebler et al.	Representing the first comprehensive assessment of 3DP from a global sustainability perspective	Modeling	3D technologies
2014	Holmström and Partanen	Exploring the forms that combinations of digital manufacturing, logistics, and equipment use are likely to take and how these novel combinations may affect the relationship among logistics service providers (LSPs), users, and manufacturers of equipment	Modeling	Manufacturing and logistics
2014	Brofman Epelbaum and Martinez	Offering a theoretical framework grounded on the resource-based view (RBV) of the firm to determine the strategic impacts of the technological evolution of food traceability systems	Case study	Food industry
2015	Yue et al.	Defining the development and character of ICPS and offering a service-oriented ICPS model	Modeling	Information communication technology (ICT)
2015b	Chen, R.-Y.	Simulating complex system by connected physical and digital objects with relationships while enhancing decision-making performance efficiency for green inventory management	Modeling	Green supply chain

(continued)

Table 11 (continued)

Year	Authors	Objective	Method	Area of application
2015	Prause	Responding to the research question of how new and sustainable business models and structures for Industry 4.0 might appear and in which direction existing traditional business concepts have to be developed to deploy a strong business impact of Industry 4.0	Modeling	E-Residency
2016	Papahristou and Bilalis	Analyzing the challenges, the threats, and the opportunities across the SC partners emerging to reduce the environmental footprint	Modeling	Clothing industry
2016	Ranzo et al.	Studying new mobility and manufacturing concepts carried out in the framework of a research project funded by the Regional Government of Campania for innovative development of the automotive SC	Report	Automotive supply chain
2016	Lom et al.	Suggesting the conjunction of the smart city initiative and the concept of Industry 4.0	Modeling	Smart City
2017	Lin et al.	Using a descriptive analysis with descriptive statistics under the innovation policy framework proposed by Rothwell and Zegveld. Moreover, informing a comparative policy analysis across China and Taiwan	Case study	Policy-making
2017	De Carolis et al.	Illustrating a "tool" for building a maturity assessment method to measure the digital readiness of manufacturing firms	Modeling	Manufacturing and logistics
2017	Man and Strandhagen	Considering potential sustainable business scenarios, and proposes an agenda for research into how Industry 4.0 can be used to create sustainable business models	Modeling	Business model
2017	Palm	Studying recent trends of vinyl traffic and critique a prominent feature of contemporary vinyl culture: Record store day	Review	Record labels and retail outlets

(continued)

Table 11 (continued)

Year	Authors	Objective	Method	Area of application
2017	Rauch et al.	Considering the actual state-of-the-art in distributed manufacturing	Review	Distributed manufacturing
2017	Jensen and Remmen	Investigating how different "product stewardship" and "end-of-life" strategies can support the circular economy and what the challenges and benefits are from an original equipment manufacturer perspective	Modeling	Circular economy (CE)
2017	Lee et al.	Explaining to what extent the business sectors involved in and how to safeguard the cross-border trade and investments with safer and smarter regional strategies in the digital age with large-scale disasters	Modeling	Sendai framework for disaster risk reduction (SFDRR)
2017	Zhong et al.	Studying food SC management (FSCM) in terms of systems and implementations	Modeling	Food industry
2017	Paul and Zhou	Studying an empirical case of a leading paper manufacturing company in central Java, Indonesia, in their way of building their maintainable innovation capability in their SC by applying a combination of various existing models	Case study	Paper manufacturing
2017	Strandhagen et al.	Studying the challenges of Industry 4.0, current trends, and offering a model to understand and relate the different elements of business operations	Review	Business model
2017	Prause and Atari	Exploring the relationship between networking, organizational development, structural frame conditions, and sustainability in the context of Industry 4.0	Case study	Manufacturing and logistics
2017	Papahristou and Bilalis	Considering the relationship between corporate social responsibility (CSR) and collective actions on sustainability and the environmental impact of the new model of fast and accelerating fashion	Modeling	Clothing industry

(continued)

Table 11 (continued)

Year	Authors	Objective	Method	Area of application
2018	Tombido et al.	Reviewing the literature on the entry and use of third parties in reverse logistics with the objective of providing researchers with future research directions for this fast-emerging topic	Review	Reverse logistics
2018	Byrne et al.	Studying the meaning and implications of "Biologicalization" from the perspective of the design, function and operation of products, manufacturing processes, manufacturing systems, SCs, and organizations	Modeling	Biologicalization
2018	Saberi et al.	Studying blockchain technology and smart contracts with potential application to SC management	Modeling	Blockchain technology
2018	Scholz et al.	Considering digital technologies in forest-based SCs and summarizing the state-of-the-art digital technologies for the real-time data collection on forests, product flows, and forest operations, along with planning systems and other decision support systems in use by SC actors	Review	Forest-based supply chains
2018	Banks et al.	Explaining enhancing high-rise residential construction through design for manufacture and assembly	Case study	Construction industry
2018	Luthra and Mangla	Recognizing key challenges to Industry 4.0 initiatives and key challenges for SC sustainability in emerging economies by taking Indian manufacturing industry perspective	Review	Manufacturing and logistics
2018	Nascimento et al.	Studying how rising technologies from Industry 4.0 can be integrated with a circular economy (CE) practices to establish a business model that reuses and recycles wasted material such as scrap metal or e-waste	Modeling	Circular economy (CE)

(continued)

Table 11 (continued)

Year	Authors	Objective	Method	Area of application
2018	Bag et al.	Recognizing the Industry 4.0 enablers of SC sustainability and further attempt to propose a research framework to bridge the theoretical gaps	Review	Supply chain sustainability
2018	Lopes de Sousa Jabbour et al.	Offering a pioneering roadmap to improve the application of CE principles in organizations by means of Industry 4.0 approaches	Case study	Circular economy (CE)
2018	Bechtsis et al.	Providing a framework that obtains the main software architecture elements for developing highly customized simulation tools that support the effective integration of intelligent autonomous vehicles (IAVs) in sustainable supply networks, as an emerging field in the operations management agenda	Modeling	Intelligent autonomous vehicles (IAVs)
2018	Sendlhofer and Lernborg	Studying how workers are trained on their labor rights with a digital training method	Case study	Digital training method
2018	Gružauskas et al.	Examining the limited possibilities to reach cost-effective performance and sustainability	Review	Food industry
2018	Sun et al.	Offering an agent-based simulation that models the micro-level protocols of mobile recourse units and their interaction with the physical infrastructure in a rail–road intermodal transport network	Modeling	Rail–road intermodal transport network
2018	Ding	Recognizing the potential sustainability barriers of PSC and examining how Industry 4.0 can be applied in the sustainable PSC paradigms	Review	Pharmaceutical supply chain (PSC)
2018	Bucci et al.	Presenting an overview of worldwide development and status of precision agriculture, starting from 2000 until to date	Review	Agriculture sector

(continued)

Table 11 (continued)

Year	Authors	Objective	Method	Area of application
2018	Forkel et al.	Investigating smart interoperable logistics and additive manufacturing—Modern technologies for digital transformation and Industry 4.0	Modeling	Manufacturing and logistics
2018	Todorovic et al.	Studying how the SFSC could be designed from the aspects of innovative logistics modes and contemporary information and communication technologies, with the final aim to outline and evaluate different food distribution scenarios toward greater sustainability	Modeling	Food industry
2018	Holmström et al.	Investigating how current and future direct digital manufacturing (DDM)-based operational practices can be used to advance products and processes	Modeling	Direct digital manufacturing (DDM)
2018	Wu et al.	Studying how to provide trade promotions in a sustainable manner when consumer demand is disrupted	Modeling	Trade promotion
2018	Garcia-Muiña et al.	Exploring the phases of the transition from a linear to a circular economy and suggesting a procedure for the principles of sustainability (environmental, economic, and social) in a manufacturing environment, through the design of a new circular business model (CBM)	Modeling	Circular economy (CE)
2018	Bressanelli et al.	Recognizing the main challenges that companies have to face when they want to redesign their SC according to CE principles, i.e., to implement a circular SC	Review	Circular economy (CE)

(continued)

Table 11 (continued)

Year	Authors	Objective	Method	Area of application
2018	Delina et al.	Suggesting a framework for innovation-driven SC ecosystem based on interoperability between commercial and public innovation procurement organization and research environment. Moreover, developing single digital infrastructure for supporting critical issues in requirements analysis, sourcing, negotiation, contract execution, and post-contractual phase to build sustainable, motivational, and trusted innovation-driven environment	Modeling	Innovation Union
2018	Dallasega and Sarkis	Examining the nexus of Industry 4.0 and greening SCs with using proximity analysis	Modeling	Supply chain sustainability

Table 12 Literature review of sustainable SC and Industry 4.0 (Conference papers)

Year	Authors	Objective	Method	Area of application
2010	Pöltner and Grechenig	Offering a concept for the establishment of a future virtual digital retail ecosystem	Modeling	Virtual digital retail ecosystem
2010	Price et al.	Proposing a digital framework for the creation and management of performance parameters related to the lifecycle performance of thermoplastic composite structures	Modeling	Thermoplastic composite structures
2010	Kang and Diao	Analyzes the route choice of information technology (IT) which enterprises can obtain long-term competitive advantages	Modeling	Digital China Company's supply chain integration system
2012	Ji and Niu	Studying variety and high-frequency necessities of modern cold chain logistics, and application of digital assorting system (DAS) in cold chain logistics warehousing system to meet the JIT	Modeling	Digital assorting system (DAS)

(continued)

Table 12 (continued)

Year	Authors	Objective	Method	Area of application
2013	Bjorn et al.	Explaining how a general operating model of re-use of electrical and electronic equipment (EEE), and specifically for PCs in developing countries, deal with the challenges and opportunities of increasing e-waste awareness	Case study	Electronic industry
2015	Tzoulis et al.	Combining data on timber trade in Greece and also studying how the economic crisis has affected the forest, its products, and how it has affected trade (imports and exports)	Case study	Forest-based supply chains
2016	Ginige et al.	Studying a notion of context-specific actionable information which allows the user to act with the least amount of further processing	Case study	Agriculture sector
2017	Kalogianni et al.	Studying an efficient monitoring and control software tool (MCT) for assessing the operation data of an olive oil production facility	Modeling	Agriculture sector
2017	Yu and Solvang	Presenting a new concept: small-scale intelligent manufacturing system (SIMS), and the comparison with previous concepts and the benefits of SIMS are discussed in this paper	Modeling	Small-scale intelligent manufacturing system (SIMS)
2017	Pilinkienė et al.	Investigating a case study of the European Union food industry by modeling different logistic network scenarios, and implemented a competitiveness strategy based on the Industry 4.0 concept and lean philosophy	Case study	Food industry
2017	Tan et al.	Considering how organizations can investigate and implement techniques for their modern enterprise with a focus on how advanced big data tools can be applied to quality analytics for monitoring and improving quality in the electronic industry	Modeling	Electronic industry

(continued)

Table 12 (continued)

Year	Authors	Objective	Method	Area of application
2017	Crowley et al.	Literature reviewing and discussing core learnings in relation to impacts on sourcing and supplier management in a digital business landscape	Review	Digital business landscape
2017	Pal and Sandberg	Studying the inter-organizational value creation, in apparel SC context, through circularity and digitalization for sustainability, by gathering evidence from vivid research experiences	Modeling	Clothing industry
2018	Alrabghi	Investigating the key elements of simulation optimization frameworks that will facilitate the transformation to Industry 4.0	Review	Simulation optimization

References

Abdi, M. R., Labib, A., Edalat, F. D., & Abdi, A. (2018). *Integrated reconfigurable manufacturing systems and smart value chain: Sustainable infrastructure for the factory of the future*. Cham, Online: Springer. ISBN 978-3-319-76846-5.

Accenture. (2014). Big data analytics in supply chain: hype or here to stay? In *Accenture global operations megatrends study* (pp. 1–20).

Agarwal, G., Barari, S., & Tiwari, M. K. (2012). A PSO-based optimum consumer incentive policy for WEEE incorporating reliability of components. *International Journal of Production Research, 50*(16), 4372–4380.

Agus, A., & Ahmad, S. (2017). The significant impact of customer relations practices (CRP), information technology (IT) and information sharing between supply chain partners (IS) on product sales. *Gading Journal for the Social Sciences, 12*(01), 65–85.

Alfalla-Luque, R., Medina-Lopez, C., & Dey, P. K. (2013). Supply chain integration framework using literature review. *Production Planning & Control, 24*, 800–817. https://doi.org/10.1080/09537287.2012.666870.

Almada-Lobo, F. (2016). The Industry 4.0 revolution and the future of manufacturing execution systems (MES). *Journal of Innovation Management, 3*(4), 16–21.

Alrabghi, A. (2018) Simulation based optimization frameworks as key enablers for the transformation to Industry 4.0. In *Proceedings of International Conference on Computers and Industrial Engineering, CIE*.

Anderson, M. S. (2007). An introductory note on the environmental economics of the circular economy. *Sustainability Science, 2*, 133–140.

Aragón-Correa, J. A., Hurtado-Torres, N., Sharma, S., & García-Morales, V. J. (2008). Environmental strategy and performance in small firms: A resource-based perspective. *Journal of Environmental Management, 86*(1), 88–103.

Ardito, L., Petruzzelli, A. M., Panniello, U., & Garavelli, A. C. (2019). Towards Industry 4.0: Mapping digital technologies for supply chain management-marketing integration. *Business Process Management Journal, 25*(2), 323–346. https://doi.org/10.1108/BPMJ-04-2017-0088.

Badurdeen, F., Iyengar, D., Goldsby, T. J., Metta, H., Gupta, S., & Jawahir, I. S. (2009). Extending total life-cycle thinking to sustainable supply chain design. *International Journal of Product Lifecycle Management, 4*(1–3), 49–67.

Bag, S., Telukdarie, A., Pretorius, J. H. C., & Gupta, S. (2018). Industry 4.0 and supply chain sustainability: framework and future research directions. *Benchmarking: An International Journal*. https://doi.org/10.1108/BIJ-03-2018-0056.

Bagchi, P. K., Chun, H. B., Skjoett-Larsen, T., & Boege, Soerensen L. (2005). Supply chain integration: A European survey. *The International Journal of Logistics Management, 16*(2), 275–294. https://doi.org/10.1108/09574090510634557.

Banchuen, P., Sadler, I., & Shee, H. (2017). Supply chain collaboration aligns order winning strategy with business outcomes. *IIMB Management Review, 29*(2), 109–121.

Banks, C., Kotecha, R., Curtis, J., Dee, C., Pitt, N., & Papworth, R. (2018). Enhancing high-rise residential construction through design for manufacture and assembly—A UK case study. *Proceedings of Institution of Civil Engineers: Management, Procurement and Law, 171*(4), 164–174.

Barratt, R. (2016). How automation is changing the supply chain. Supply Demand Chain Exec. Retrieved October 15, 2016, from https://www.sdcexec.com/warehousing/article/12267524/how-automation-is-changing-the-supply-chain.

Baumers, M., Tuck, C., Wildman, R., Ashcroft, I., Rosamond, E., & Hague, R. (2013). Transparency built-in: Energy consumption and cost estimation for additive manufacturing. *Journal of Industrial Ecology, 17*(3), 418–431.

Bedekar, A. (2017). Opportunities & Challenges for IoT in India. https://technology.siliconindiamagazine.com/viewpoint/cxoinsights/opportunities-challenges-for-iot-in-india-nwid-1060.html.

Bechtold, J., Kern, A., Lauenstein, C., & Bernhofer, L. (2014). Industry 4.0—The Capgemini consulting view: Sharpening the picture beyond the hype.

Bechtsis, D., Tsolakis, N., Vlachos, D., & Srai, J. S. (2018). Intelligent autonomous vehicles in digital supply chains: A framework for integrating innovations towards sustainable value networks. *Journal of Cleaner Production, 181*, 60–71.

Beier, J. (2016). *Simulation approach towards energy flexible manufacturing systems*. Cham: Springer International Publishing.

Beske, P., Land, A., & Seuring, S. (2014). Sustainable supply chain management practices and dynamic capabilities in the food industry: A critical analysis of the literature. *International Journal of Production Economics, 152*, 131–143.

Bjorn, C., Olivier, V. E., & Stijn, V. (2013) Impact of e-waste on the operating model of a close the digital divide organization. In *ECIS 2013—Proceedings of the 21st European Conference on Information Systems*.

Blümel, E. (2013). Global challenges and innovative technologies geared toward new markets: Prospects for virtual and augmented reality. *Procedia Computer Science, 25*, 4–13.

Brandenburg, M., Govindan, K., Sarkis, J., & Seuring, S. (2014). Quantitative models for sustainable supply chain management: Developments and directions. *European Journal of Operational Research, 233*(2), 299–312.

Branke, J., Farid, S. S., & Shah, N. (2016). Industry 4.0: A vision for personalized medicine supply chains? *Cell Gene Therapy Insights, 2*(2), 263–270.

Bressanelli, G., Perona, M., & Saccani, N. (2018) Towards the circular supply chain: A literature review of challenges. *Proceedings of the Summer School Francesco Turco*, 171–178.

BRICS Business Council. (2017). Skill development for Industry 4.0. In *A White Paper by BRICS Skill Development Working Group*. BRICS Business Council, India Group. Retrieved October 22, 2017, from http://www.globalskillsummit.com/Whitepaper-Summary.pdf.

Brofman Epelbaum, F. M., & Martinez, M. G. (2014). The technological evolution of food traceability systems and their impact on firm sustainable performance: A RBV approach. *International Journal of Production Economics, 150*, 215–224.

Bucci, G., Bentivoglio, D., & Finco, A. (2018). Precision agriculture as a driver for sustainable farming systems: State of art in litterature and research. *Quality—Access to Success, 19*(S1), 114–121.

Büyüközkan, G., & Göçer, F. (2018). Digital supply Chain: Literature review and a proposed framework for future research. *Computers in Industry, 97,* 157–177.

Byrne, G., Dimitrov, D., Monostori, L., Teti, R., van Houten, F., & Wertheim, R. (2018). Biologicalisation: Biological transformation in manufacturing. *CIRP Journal of Manufacturing Science and Technology, 21,* 1–32.

Cao, M., & Zhang, Q. (2011). Supply chain collaboration: Impact on collaborative advantage and firm performance. *Journal of Operations Management, 29*(3), 163–180.

CapGemini, Infor, & GTNexus. (2016). The current and future state of digital supply chain transformation. http://www.gtnexus.com/resources/papers-and-reports/current-and-future-state-digital-supply-chain-transformation.

Carter, P. L., Monczka, R. M., Ragatz, G. L. (2009) Supply chain integration: Challenges and good practices.

Cecere, L. (2014). Digital supply chain insights on driving the digital supply chain transformation.

Chen, F., Deng, P., Wan, J., Zhang, D., Vasilakos, A. V., & Rong, X. (2015). Data mining for the internet of things: Literature review and challenges. *International Journal of Distributed Sensor Networks, 11*(8), 1–14.

Chen, R.-Y. (2015). Intelligent IoT-enabled system in green supply chain using integrated FCM method. *International Journal of Business Analytics, 2*(3), 47–66.

Clear, A. K., Ganglbauer, E., Comber, R., Hazas, M., Friday, A., & Rogers, Y. (2013). Green food technology: Ubicomp opportunities for reducing the environmental impacts of food. In *UbiComp 2013 Adjunct—Adjunct Publication of the 2013 ACM Conference on Ubiquitous Computing* (pp. 553–557).

Craig, R. C., & Easton, P. L. (2011). Sustainable supply chain management: Evolution and future directions. *International Journal of Physical Distribution & Logistics Management, 41*(1), 46–62.

Crowley, C., Carcary, M., Doherty, E., & Conway, G. (2017). Rethinking IT sourcing and supplier management for the digital age. In *Proceedings of the 11th European Conference on Information Systems Management, ECISM 2017* (pp. 64–72).

Counsil, S. C. (2004). SCOR model, supply chain operations reference model. http://www.supply-chain.org.

Coyle, J. J., Thomchick, E. A., & Ruamsook, K. (2015). Environmentally sustainable supply chain management: An evolutionary framework. In *Marketing dynamism & sustainability: Things change, things stay the same* (pp. 365–374). Cham: Springer.

Cukier, K. (2018). Big data and the future of business. In Turner (Ed.), *The book reinventing the company in the digital age, BBVA* (pp. 37–51).

Dallasega, P., & Sarkis, J. (2018). Understanding greening supply chains: Proximity analysis can help. *Resources, Conservation and Recycling, 139,* 76–77.

Dam, L., & Petkova, B. N. (2014). The impact of environmental supply chain sustainability programs on shareholder wealth. *International Journal of Operations & Production Management, 34*(5), 586–609.

Dawson, T. (2014). Industry 4.0 opportunities and challenges for smart manufacturing. Retrieved April 28, 2019, from https://ihsmarkit.com/research-analysis/q13-industry-40-opportunities-and-challenges-for-smart-manufacturing.html.

Davis, J., Edgar, T., Porter, J., Bernaden, J., & Sarli, M. (2012). Smart manufacturing, manufacturing intelligence and demand-dynamic performance. *Computers & Chemical Engineering, 47,* 145–156.

De Carolis, A., Macchi, M., Negri, E., & Terzi, S. (2017). A maturity model for assessing the digital readiness of manufacturing companies. *IFIP Advances in Information and Communication Technology, 513,* 13–20.

Del Giudice, M. (2016). Discovering the internet of things (IoT) within the business process management: A literature review on technological revitalization. *Business Process Management Journal, 22*(2), 263–270.

Delina, R., Húska, P., & Klézl, V. (2018). Transparency for innovation driven supply chain ecosystem—National approach, IDIMT 2018: Strategic Modeling in Management. In *Economy and Society—26th Interdisciplinary Information Management Talks* (pp. 311–318).

Ding, B. (2018). Pharma Industry 4.0: Literature review and research opportunities in sustainable pharmaceutical supply chains. *Process Safety and Environmental Protection, 119,* 115–130.

Dougados, M., Ghioldi, S., Van Doesburg, R., & Subrahmanyam, K. V. J. (2013). *The missing link supply chain and digital maturity.* White Paper: Capgemini Consulting.

Duarte, S., & Cruz-Machado, V. (2017). Exploring linkages between lean and green supply chain and the Industry 4.0. In *International Conference on Management Science and Engineering Management* (pp. 1242–1252). Cham: Springer.

Dzopalic, M., Zubovic, J., & Bradic-Martinovic, A. (2010). Effective implementation of E-CRM strategy. *Polish Journal of Management Studies, 1,* 52–63.

Erol, S., Schumacher, A., & Sihn, W. (2016). Strategic guidance towards Industry4.0-A three-stage process model. In *International Conference on Competitive Manufacturing, COMA 2016; Proceedings Stellenbosch* (pp. 495–500).

EY. (2016). Digital supply chain: It's all about that data.

Fahimnia, B., Sarkis, J., Gunasekaran, A., & Farahani, R. (2017). Decision models for sustainable supply chain design and management. *Annals of Operations Research, 250*(2), 277–278.

Fiorini, M., & Lin, J.-C. (2015). Clean mobility and intelligent transport systems. *The Institution of Engineering and Technology.* ISBN: 978-184919896-7, 978-184919895-0.

Forkel, E., Baum, J., Schumann, C.-A., & Mueller, E. (2018) Smart interoperable logistics and additive manufacturing—Modern technologies for digital transformation and Industry 4.0. In *SAE Technical Papers.*

Garcia-Muiña, F. E., González-Sánchez, R., Ferrari, A. M., & Settembre-Blundo, D. (2018). The paradigms of Industry 4.0 and circular economy as enabling drivers for the competitiveness of businesses and territories: The case of an Italian ceramic tiles manufacturing company. *Social Sciences, 7*(12), 255.

Gebler, M., Schoot Uiterkamp, A. J. M., & Visser, C. (2014). A global sustainability perspective on 3D printing technologies. *Energy Policy, 74*(C), 158–167.

Genovese, A., Acquaye, A. A., Figueroa, A., & Koh, S. L. (2017). Sustainable supply chain management and the transition towards a circular economy: Evidence and some applications. *Omega, 66,* 344–357.

Gilchrist, A. (2016). *Industry 4.0: The industrial internet of things.* Apress.

Ginige, A., Walisadeera, A. I., Ginige, T., De Silva, L., Di Giovanni, P., Mathai, M., et al. (2016). Digital knowledge ecosystem for achieving sustainable agriculture production: A case study from Sri Lanka. In *Proceedings—3rd IEEE International Conference on Data Science and Advanced Analytics, DSAA 2016* (Vol. 7796947, pp. 602–611).

Ghadimi, P., Azadnia, A. H., Heavey, C., Dolgui, A., & Can, B. (2016). A review on the buyer—supplier dyad relationships in sustainable procurement context: Past, present and future. *International Journal of Production Research, 54*(5), 1443–1462.

Ghadimi P., Wang C., & Lim M. K (2019). Sustainable supply chain modeling and analysis: Past debate, present problems and future challenges. *Resources, Conservation & Recycling, 140,* 72–88.

Gibson, I., Rosen, D., & Stucker, B. (2014). *Additive manufacturing technologies: 3D printing, rapid prototyping, and direct digital manufacturing.* New York, NY: Springer.

Gružauskas, V., Baskutis, S., & Navickas, V. (2018). Minimizing the trade-off between sustainability and cost effective performance by using autonomous vehicles. *Journal of Cleaner Production, 184,* 709–717.

GTnexus. (2016). The internet of things supply chain. GT NEXUS (pp. 1–6).

Goodship, V., & Stevels, A. L. N. (2012). *Waste electrical and electronic equipment (WEEE) handbook Woodhead Publishing*. eBook ISBN: 9780857096333.
Gökalp, E., Şener, U., Eren, P. E. (2017). Development of an assessment model for industry 4.0: Industry 4.0-MM. In *International Conference on Software Process Improvement and Capability Determination* (pp. 128–142). Cham: Springer.
Guarraia, P., Gerstenhaber, G., Athanassiou, M., & Boutot P.-H. (2015). The intangible benefits of a digital supply chain.
Hajdul, M., Cudzilo, M. (2011). One common framework for information and communication systems in transport and logistics—Case study. *Environmental Science and Engineering (Subseries: Environmental Science)*, 501–513.
Handfield, R., & Linton, T. (2017). *The LIVING supply chain: The evolving imperative of operating in real time*. Wiley. ISBN: 978-111930802-7, 978-111930625-2.
Hanifan, G., Sharma, A., & Newberry, C. (2014). The digital supply network: A new paradigm for supply chain management.
Hentza, J. B., Nguyen, V. K., Maederb, W., Panaresec, D., Gunninkd, J. W., Gontarze, A., et al. (2013). An enabling digital foundation towards smart machining. *Procedia CIRP, 12*, 240–245.
Hermann, M., Pentek, T., & Otto, B. (2016). Design principles for industrie 4.0 scenarios. In: *2016 49th Hawaii International Conference on System Sciences (HICSS)* (pp. 3928–3937), IEEE.
Hines, T. (2004). *Supply chain strategies: Customer-driven and customer-focused*.
Hofmann, E., & Rüsch, M. (2017). Industry 4.0 and the current status as well as future prospects on logistics. *Computers in Industry, 89*, 23–34.
ström, J., & Partanen, J. (2014). Digital manufacturing-driven transformations of service supply chains for complex products. *Supply Chain Management, 19*(4), 421–430.
Holmström, J., Liotta, G., & Chaudhuri, A. (2018). Sustainability outcomes through direct digital manufacturing-based operational practices: A design theory approach. *Journal of Cleaner Production, 167*, 951–961.
Huber, J. (2008). Technological environmental innovations (TEIs) in a chain-analytical and life-cycle-analytical perspective. *Journal of Cleaner Production, 16*(18), 1980–1986.
Ibem, E. O., & Laryea, S. (2014). Survey of digital technologies in procurement of construction projects. *Automation in Construction, 46*, 11–21.
Jensen, J. P., & Remmen, A. (2017). Enabling circular economy through product stewardship. *Procedia Manufacturing, 8*, 377–384.
Ji, S., Niu, X. (2012). Application of DAS in cold chain logistics warehousing system based on WMS. *Lecture Notes in Electrical Engineering, LNEE, 113*, 1603–1610.
Jirsak, P. (2017). The roles of lean and green supply chain management strategies in the global business environments. In *Operations and service management: Concepts, methodologies, tools, and applications* (pp. 250–271).
Jirsak, P. (2018). Analyzing an impact of Industry 4.0 on logistics and supply chain. In *Analyzing the impacts of Industry 4.0 in modern business environments* (pp. 121–146).
Kagermann, H., Lukas, W., & Wahlster, W. (2011). Industrie 4.0: Mit dem internet der Dinge auf dem Weg zur 4. industriellen Revolution, VDI nachrichten, No. 13.
Kagermann, H. (2015). Change through digitization—Value creation in the age of Industry 4.0. *Management of Permanent Change*, 23–45.
Kalogianni, E. P., Salvaras, P., Bechtsis, D., & Stergiopoulos, F. (2017). A monitoring and control software tool to assess process and production data in olive oil production units. *CEUR Workshop Proceedings, 2030*, 284–292.
Kamarulzaman, N. H., & Eglese, R. W. (2015). Managing purchasing with different e-procurement solutions. In *Optimization of supply chain management in contemporary organizations* (pp. 246–279).
Kang, S., & Diao, X. (2010). Study on the sustainable competitive advantages of the managing skills which incorporate the IT resources and SCM—Analysis and annotation on the case of information integrate system of supply chain of DCC. In *Proceedings—3rd International Conference on*

Information Management, Innovation Management and Industrial Engineering, ICIII 2010 (Vol. 3, 5694680, 57–61).

Kasemsap, K. (2016). The roles of lean and green supply chain management strategies in the global business environments. In S. Joshi, & R. Joshi (Eds.), *Designing and implementing global supply chain management* (pp. 152–173).

Kückelhaus, M., Magerkurth, C., & Baus, J. (2013). A SemProM use case: Tracking & tracing for green logistics and integrity control. *Cognitive Technologies, 9783642373763,* 311–327.

Kayikci, Y. (2018). Sustainability impact of digitization in logistics. *Procedia Manufacturing, 21,* 782–789.

Khan, M., Wu, X., Xu, X., & Dou, W. (2017). Big data challenges and opportunities in the hype of Industry 4.0. In *2017 IEEE International Conference on Communications(ICC)* (pp. 1–6), IEEE.

Kiel, D., Müller, J., Arnold, C., & Voigt, K. I. (2017). Sustainable industrial value creation: Benefits and challenges of Industry 4.0. In *ISPIM Innovation Symposium, The International Society for Professional Innovation Management (ISPIM).*

Kipper, G., & Rampolla, J. (2012). *Augmented reality: An emerging technologies guide to AR.* Waltham, MA: Elsevier Science.

Kiritsis, D. (2016). *Taking the LEAP: The methods and tools of the linked engineering and manufacturing platform (LEAP)* (1st edn), Academic Press. eBook ISBN: 9780444637543.

Klievink, B. (2015). Understanding public-private collaboration configurations for international information infrastructures. In E. Tambouris et al. (eds.), *Electronic government. EGOV 2015. Lecture Notes in Computer Science* (Vol. 9248). Cham: Springer.

Kogg, B., & Mont, O. (2012). Environmental and social responsibility in supply chains: The practice of choice and inter-organizational management. *Ecological Economics, 83,* 154–163.

Krause, D. R., Vachon, S., & Klassen, R. D. (2009). Special topic forum on sustainable supply chain management: Introduction and reflections on the role of purchasing management. *Journal of Supply Chain Management, 45*(4), 18–25.

Lee, J., Kao, H. A., & Yang, S. (2014). Service innovation and smart analytics for Industry 4.0 and big data environment. *Procedia CIRP, 16,* 3–8.

Lee, H. L. (2000). Creating value through supply chain integration. *Supply Chain Management Review, 4*(4), 30–36.

Lee, Y., Watanabe, K., & Li, W.-S. (2017). Enhancing regional digital preparedness on natural hazards to safeguard business resilience in the Asia-Pacific. *IFIP Advances in Information and Communication Technology, 501,* 170–182.

Leitão, P., Colombo, A. W., & Karnouskos, S. (2016). Industrial automation based on cyber-physical systems technologies: Prototype implementations and challenges. *Computers in Industry, 81,* 11–25.

Li, S., Xu, L. D., & Zhao, S. (2015). The internet of things: A survey. *Information Systems Frontiers, 17*(2), 243–259.

Lin, H., Liu, P., Li, W., Zhang, L., & Ji, Y. (2012). Construction of digital mine and key technologies. *Advanced Materials Research, 524–527,* 413–420.

Lin, K. C., Shyu, J. Z., & Ding, K. (2017). A cross-strait comparison of innovation policy under Industry 4.0 and sustainability development transition. *Sustainability* (Switzerland), *9*(5), 786.

Lopes de Sousa Jabbour, A. B., Cortegoso de Oliveira Frascarelli, F., & Chiappetta Jabbour, C. J. (2015). Green supply chain management and firms' performance: Understanding potential relationship and the role of green sourcing and some other green practices. *Resources, Conservation and Recycling, 104,* Part B, 366–374.

Lopes de Sousa Jabbour, A. B., Jabbour, C. J. C., Godinho Filho, M., & Roubaud, D. (2018) Industry 4.0 and the circular economy: A proposed research agenda and original roadmap for sustainable operations. *Annals of Operations Research, 270*(1–2), 273–286.

Lom, M., Pribyl, O., & Svitek, M. (2016). Industry 4.0 as a part of smart cities. In *2016 Smart Cities Symposium Prague, SCSP 2016,* 7501015.

Luthra, S., & Mangla, S. K. (2018). Evaluating challenges to Industry 4.0 initiatives for supply chain sustainability in emerging economies. *Process Safety and Environmental Protection, 117,* 168–179.

MacDougall, W. (2014). Industrie 4.0: Smart manufacturing for the future. Germany Trade & Invest.

Man, J. C. D., & Strandhagen, J. O. (2017). An Industry 4.0 research agenda for sustainable business models. *Procedia CIRP, 63,* 721–726.

Manavalan, E., & Jayakrishna, K. (2019). A review of internet of things (IoT) embedded sustainable supply chain for Industry 4.0 requirements. *Computers & Industrial Engineering, 127,* 925–953.

Mani, V., Gunasekaran, A., Papadopoulos, T., Hazen, B., & Dubey, R. (2016). Supply chain social sustainability for developing nations: Evidence from India. *Resources, Conservation and Recycling, 111,* 42–52.

Marques, M., Agostinho, C., Zacharewicz, G., & Jardim-Gonçalves, R. (2017). Decentralized decision support for intelligent manufacturing in Industry 4.0. *Journal of Ambient Intelligence and Smart Environments, 9*(3), 299–313.

Meera, S.N., Gandhi, R., & Padaria, R. (2017) Accelerating impact through rice innovation systems: Integrating knowledge. In *Technology, and markets, the future rice strategy for India* (pp. 435–460).

Monnet, J.-M., & Le Net, E. (2011). Assessment of logistics concept to sustainability: Development of a common approach to transport issues. Deliverable D3.3.3, EFORWOOD Project, European Forest Institute. http://www.efi.int/files/attachments/publications/eforwood/efi_tr_75.pdf.

Montreuil, B., Meller, R. D., & Ballot, E. (2013). Physical internet foundations. *Studies in Computational, Intelligence, 472,* 151–166.

Müller, J., Dotzauer, V., & Voigt, K. I. (2017a). Industry 4.0 and its impact on reshoring decisions of German manufacturing enterprises. In *Supply Management Research* (pp. 165–179). Wiesbaden: Springer Gabler.

Nabben, D. H. (2016). 12 Trends that are shaping the future of logistics. *Supply Chain Digital.* Retrieved April 28, 2019, from https://www.supplychaindigital.com/logistics/12-trends-are-shaping-future-logistics.

Najmi, A., Shakouri, H., & Nazari, G. S. (2016). An integrated supply chain: A large scale complementarity model for the biofuel markets. *Biomass and Bioenergy, 86,* 88–104. https://doi.org/10.1016/j.biombioe.2016.01.010.

Nascimento, D. L. M., Alencastro, V., Quelhas, O. L. G., Caiado, R. G. G., Garza-Reyes, J. A., Lona, L. R., et al. (2019). Exploring Industry 4.0 technologies to enable circular economy practices in a manufacturing context: A business model proposal. *Journal of Manufacturing Technology Management, 30*(3), 607–627.

Ndou, V., & Sadguy, N. (2010). Digital marketplaces as a viable model for SME networking. In *E-business issues, challenges and opportunities for SMEs: Driving competitiveness* (pp. 118–130).

Nicoletti, B. (2018). The future: Procurement 4.0. In *Agile procurement* (pp. 189–230). Cham: Palgrave Macmillan.

Noci, G., & Verganti, R. (1999). Managing 'green' product innovation in small firms. *R&D Management, 29*(1), 3–15.

Norris, K. (2011). A new program in North America is attracting a lot of attention among newspaper publishers and printers a new vision for newspapers. *PPI Pulp and Paper International, 53*(11), 22–24.

Nowak, G., Maluck, J., Sturmer, C., & Pasemann, J. (2016). The Era of digitized trucking: Transforming the logistics value chain, PWC, White Paper.

Oyekan, J., Prabhu, V., Tiwari, A., Baskaran, V., Burgess, M., & Mcnally, R. (2017). Remote real-time collaboration through synchronous exchange of digitised human-work piece interactions. *Future Generation Computer Systems, 67,* 83–93.

Pal, R., & Sandberg, E. (2017). Sustainable value creation through new industrial supply chains in apparel and fashion. *IOP Conference Series: Materials Science and Engineering, 254*(20), 202007.

Palm, M. (2017). Analog backlog: Pressing records during the vinyl revival. *Journal of Popular Music Studies, 29*(4), e12247.
Papahristou, E., & Bilalis, N. (2016). A new sustainable product development model in apparel based on 3D technologies for virtual proper fit. *Smart Innovation, Systems and Technologies, 52,* 85–95.
Papahristou, E., & Bilalis, N. (2017). Should the fashion industry confront the sustainability challenge with 3D prototyping technology. *International Journal of Sustainable Engineering, 10*(4–5), 207–214.
Paul, J., & Zhou, Y. (2017). How to build sustainable innovation capability in supply chain management. *International Journal of Business and Globalisation, 19*(4), 456–476.
Penthin, S., & Dillman, R. (2015). *Digital SCM*. Germany. www.bearingpoint.com.
Pereira, T., Barreto, L., & Amaral, A. (2017). Network and information security challenges within Industry 4.0 paradigm. *Procedia Manufacturing, 13,* 1253–1260.
Peres, R. S., Rocha, A. D., Coelho, A., & Oliveira, J. B. (2017). A highly flexible, distributed data analysis framework for Industry 4.0 manufacturing systems. In *Service orientation in holonic and multi-agent manufacturing* (pp. 373–381). Cham: Springer.
Pfohl, H. C., Yahsi, B., & Kurnaz, T. (2017). Concept and diffusion-factors of Industry 4.0 in the supply chain. In *Dynamics in Logistics* (pp. 381–390), Springer International Publishing.
Pilinkienė, V., Gružauskas, V., & Navickas, V. (2017). Lean thinking and Industry 4.0 competitiveness strategy: Sustainable food supply chain in the European Union. In *Trends and Issues in Interdisciplinary Behavior and Social Science—Proceedings of the 5th International Congress on Interdisciplinary Behavior and Social Science, ICIBSOS 2016*, (No. 210489, pp. 15–20).
Pisching, M. A., Junqueira, F., Santos Filho, D. J., & Miyagi, P. E. (2015b). An architecture for organizing and locating services to the Industry 4.0. In *Proceedings of 23rd ABCM International Congress of Mechanical Engineering* (Vol. 8), r462TX4E.
Pisching, M. A., Junqueira, F., Santos Filho, D. J., & Miyagi, P. E. (2015a). Service composition in the cloud-based manufacturing focused on the Industry 4.0. *DoCEIS* (pp. 65–72).
Pöltner, P., & Grechenig, T. (2010). A joint infrastructure of "digital corporate organisms" as facilitator for a virtual digital retail ecosystem. In *4th IEEE International Conference on Digital Ecosystems and Technologies—Conference Proceedings of IEEE-DEST 2010, DEST 2010* (No. 05610626, pp. 327–332).
Prause, G. (2015). Sustainable business models and structures for Industry 4.0. *Journal of Security and Sustainability Issues, 5*(2), 159–169.
Prause, G., & Atari, S. (2017). On sustainable production networks for Industry 4.0, *Entrepreneurship and Sustainability Issues, 4*(4), 421–431.
Price, M.A., Murphy, A., Butterfield, J., McCool, R., & Fleck, R. (2010) Integrating digital manufacturing, processing and design of composite structures. In *10th AIAA Aviation Technology, Integration and Operations Conference 2010, ATIO 2010* (p. 1).
Raab M., & Griffin-Cryan B. (2011). Digital transformation of supply chains: Creating value—When digital meets physical.
Raj S., & Sharma A. (2014). Supply chain management in the cloud. https://www.accenture.com/tr-en/insight-supply-chain-management-cloud.
Rajeev, A., Pati, R. K., Padhi, S. S., & Govindan, K. (2017). Evolution of sustainability in supply chain management: A literature review. *Journal of Cleaner Production, 162,* 299–314.
Rakowski, E. (2015). TechTrends 2016, Part II: The supply chain goes digital.
Ranganathan, C., Teo, T. S. H., & Dhaliwal, J. (2011). Web-enabled supply chain management: Key antecedents and performance impacts. *International Journal of Information Management, 31*(6), 533–545.
Ranzo, P., Scarpitti, C., & Veneziano, R. (2016). Evolutive scenarios for a new concept of sustainable mobility. *Smart Innovation, Systems and Technologies, 52,* 39–47.
Ras, E., Wild, F., Stahl, C., & Baudet, A. (2017). Bridging the skills gap of workers in Industry 4.0 by human performance augmentation tools: Challenges and roadmap. In *Proceedings of*

the 10th International Conference on Pervasive Technologies Related to Assistive Environments (pp. 428–432), ACM.
Rauch, E., Dallasega, P., & Matt, D. T. (2017). Distributed manufacturing network models of smart and agile mini-factories. *International Journal of Agile Systems and Management, 10*(3–4), 185–205.
Richey, R. G., Morgan, T. R., Lindsey-Hall, K., & Adams, F. G. (2016). A global exploration of big data in the supply chain. *International Journal of Physical Distribution & Logistics Management, 46*(8), 710–739.
Roome, N., & Hinnels, M. (1993). Environmental factors in the management of new product development: Theoretical framework and some empirical evidence from the white goods industry. *Business. Strategy and Environment, 2*(2), 12–27.
Roozbeh, Nia A., Hemmati, Far M., & Niaki, S. T. A. (2014). A fuzzy vendor managed inventory of multi-item economic order quantity model under shortage: An ant colony optimization algorithm. *International Journal of Production Economics, 155,* 259–271.
Saberi, S., Kouhizadeh, M., Sarkis, J., & Shen, L. (2019). Blockchain technology and its relationships to sustainable supply chain management. *International Journal of Production Research, 57*(7), 2117–2135.
Sahay, B. S., & Ranjan, J. (2008). Real time business intelligence in supply chain analytics. *Information Management & Computer Security, 16*(1), 28–48.
Sahin, F., & Robinson, E. P. (2002). Flow coordination and information sharing in supply chains: Review, implications, and directions for future research. *Decision Science, 33*(4), 505–536.
Sahin, F., & Robinson, E. P. J. (2005). Information sharing and coordination in make-to-order supply chains. *Journal of Operations Management, 23*(6), 579–598.
Saldivar, A. A. F., Li, Y., Chen, W. N., Zhan, Z. H., Zhang, J., & Chen, L. Y. (2015). Industry 4.0 with cyber-physical integration: A design and manufacture perspective. In *2015 21st international conference on Automation and computing (icac)* (pp. 1–6), IEEE.
Sanders, A., Subramanian, K. R., Redlich, T., & Wulfsberg, J. P. (2017). Industry 4.0 and lean management-synergy or contradiction? In *IFIP International Conference on Advances in Production Management Systems* (pp. 341–349). Cham: Springer.
Santos, M. Y., e Sá, J. O., Costa, C., Galvão, J., Andrade, C., Martinho, B., Costa, E., et al. (2017). A big data analytics architecture for Industry 4.0. In *World Conference on Information Systems and Technologies* (pp. 175–184). Cham: Springer.
Saucedo-Martínez, J. A., Pérez-Lara, M., Marmolejo-Saucedo, J. A., Salais-Fierro, T. E., & Vasant, P. (2017). Industry 4.0 framework for management and operations: a review. *Journal of Ambient Intelligence and Humanized Computing, 9*(3), 789–801.
Sauer, P. C., & Seuring, S. (2017). Sustainable supply chain management for minerals. *Journal of Cleaner Production, 151,* 235–249.
Savtschenko M., Schulte F., & Voß S. (2017). IT governance for cyber-physical systems: The case of Industry 4.0. In A. Marcus, W. Wang (Eds.), *Design, User Experience, and Usability: Theory, Methodology, and Management. DUXU 2017. Lecture Notes in Computer Science* (p. 10288). Cham: Springer.
Schiederig, T., Tietze, F., & Herstatt, C. (2012). Green innovation in technology and innovation management—An exploratory literature review. *R&D Management, 42,* 180–192.
Schmidt, R., Möhring, M., Härting, R.C., Reichstein, C., Neumaier, P., & Jozinović, P. (2015a). Industry 4.0-potentials for creating smart products: Empirical research results. In *Proceedings of the International Conference on Business Information Systems* (pp. 16–27). Cham: Springer.
Schmidt, B., Wallenburg, C. M., Rutkowsky, S., Einmahl, L., Petersen, I., & Klötzke F. (2015b). Digital supply chains: Increasingly critical for competitive edge.
Scholz, J., De Meyer, A., Marques, A.S., Pinho, T.M., Boaventura-Cunha, J., et al. (2018). Digital technologies for forest supply chain optimization: Existing solutions and future trends. *Environmental Management, 62*(6), 1108–1133.
Schrauf, S., & Berttram, P. (2016). Industry 4.0: How digitization makes the supply chain more efficient. Agile, and Customer-focused, PWC, White Paper.

Schröder, C. (2018). The challenges of Industry 4.0 for small and medium-sized enterprises. https://insme.wordpress.com/2018/04/26/the-challenges-of-industry-4–0-for-small-and-medium-sized-enterprises.

Schuh, G., Anderl, R., Gausemeier, J., TenHompel, M., & Wahlster, W. (2017). Industrie4.0 maturity index. Munich: Managing the Digital Transformation of Companies. HerbertUtz.

Sendlhofer, T., & Lernborg, C. M. (2018). Labour rights training 2.0: The digitalisation of knowledge for workers in global supply chains. *Journal of Cleaner Production, 179,* 616–630.

Seuring, S. (2013). A review of modeling approaches for sustainable supply chain management. *Decision Support Systems, 54*(4), 1513–1520.

Seuring, S., & Müller, M. (2008). From a literature review to a conceptual framework for sustainable supply chain management. *Journal of Cleaner Production, 16*(15), 1699–1710.

Shamim, S., Cang, S., Yu, H., & Li, Y. (2017). Examining the feasibilities of Industry 4.0 for the hospitality sector with the lens of management practice. *Energies, 10*(4), 499.

Sommer, L. (2015). Industrial revolution-Industry 4.0: Are German manufacturing SMEs the first victims of this revolution? *Journal of Industrial Engineering and Management, 8*(5), 1512–1532.

Speranza, M. G. (2018). Trends in transportation and logistics. *European Journal of Operational Research, 264*(3), 830–836.

Srivastava, S. K. (2007). Green supply-chain management: A state-of-the-art literature review. *International Journal of Management Reviews, 9,* 53–80.

Strandhagen, J. O., Vallandingham, L. R., Fragapane, G., Strandhagen, J. W., Stangeland, A. B. H., & Sharma, N. (2017). Logistics 4.0 and emerging sustainable business models. *Advances in Manufacturing, 5*(4), 359–369.

Strange, R., & Zucchella, A. (2017). Industry 4.0, global value chains and international business. *Multinational Business Review, 25*(3), 174–184, https://doi.org/10.1108/mbr-05-2017-0028.

Sultan, N. (2013), Knowledge management in the age of cloud computing and Web 2.0: Experiencing the power of disruptive innovations. *International Journal of Information Management, 33*(1), 160–165.

Sun, Y., Zhang, C., Dong, K., & Lang, M. (2018). Multiagent modelling and simulation of a physical internet enabled rail-road intermodal transport system. *Urban Rail Transit, 4*(3), 141–154.

Tan, J. S. K., Ang, A. K., Lu, L., Gan, S. W. Q., & Corral, M. G. (2017). Quality analytics in a big data supply chain: Commodity data analytics for quality engineering. In *IEEE Region 10 Annual International Conference, Proceedings/TENCON* (Vol. 7848697, pp. 3455–3463).

Theorin, A., Bengtsson, K., Provost, J., Lieder, M., Johnsson, C., Lundholm, T., et al. (2017). An event-driven manufacturing information system architecture for Industry 4.0. *International Journal of Production Research, 55*(5), 1297–1311.

Todorovic, V., Maslaric, M., Bojic, S., Jokic, M., Mircetic, D., & Nikolicic, S. (2018). Solutions for more sustainable distribution in the short food supply chains. *Sustainability (Switzerland), 10*(10), 3481.

Tombido, L. L., Louw, L., & van Eeden, J. (2018). A systematic review of 3pls' entry into reverse logistics. *South African Journal of Industrial Engineering, 29*(3 Special Edition), 235–260.

Tseng, M.–L., Tan, R. R., Chiu,A. S. F., Chien, C.–F., & Kuo, T. C. (2018). Circular economy meets industry 4.0: Can big data drive industrial symbiosis? *Resources, Conservation and Recycling, 131,* 146–147.

Turhan, Z. D., Vayvay, O., & Birgun, S. (2011). Supply chain reengineering in a paint company using axiomatic design. *The International Journal of Advanced Manufacturing Technology, 57*(5–8), 421–435.

Tzoulis, I., Andreopoulou, Z., Trigkas, M., Tsekouropoulos, G., & Voulgaridis, E. (2015). Wood trade in greece: The impact of economic crisis and the use of new technologies. *CEUR Workshop Proceedings, 1498,* 26–37.

Van Hemel, C., & Cramer, J. (2012). Barriers and stimuli for ecodesign in SMEs. *Journal of Cleaner Production, 10*(5), 439–453.

Viswanadham, N. (2002). The past, present, and future of supply-chain automation. *IEEE Robotics and Automation Magazine, 9*(2), 48–56.

Vogel-Heuser, B., & Hess, D. (2016). Guest editorial Industry 4.0—Prerequisites and visions. *IEEE Transactions on Automation Science and Engineering, 13*(2), 411–413.

Waldeck, N. E. (2014). *Advanced manufacturing technologies and workforce development*. Oxon, NY: Taylor & Francis.

Wang, S., Wan, J., Li, D., & Zhang, C. (2016). Implementing smart factory of industrie 4.0: An outlook. *International Journal of Distributed Sensor Networks, 12*(1), 3159805.

Wittstruck, D., & Teuteberg, F. (2011). Development and simulation of a balanced scorecard for sustainable supply chain management—A system dynamics approach. *Network, 7,* 368.

Wu, J., Chen, Z., & Ji, X. (2018) Sustainable trade promotion decisions under demand disruption in manufacturer-retailer supply chains. *Annals of Operations Research,* 1–29.

Xu, J. (2014). *Managing digital enterprise: Ten essential topics* (pp. 1–199). Paris: Atlantis Press, ISBN: 978-94-6239-094-2.

Xu, X. (2012). From cloud computing to cloud manufacturing. *Robotics and Computer-Integrated Manufacturing, 28*(1), 75–86.

Yan, M.-R., Chien, K.-M., & Yang, T.-N. (2016). Green component procurement collaboration for improving supply chain management in the high technology industries: A case study from the systems perspective. *Sustainability, 8,* 105.

Yee J. T., & Oh S. C. (2013). Background and introduction. In *Technology Integration to Business* (pp. 1–22). London: Springer. https://doi.org/10.1007/978-1-4471-4390-1_1.

Yu, H., & Solvang, W. D. (2017). Enhancing the competitiveness of manufacturers through small-scale intelligent manufacturing system (SIMS): A supply chain perspective. In *2017 6th International Conference on Industrial Technology and Management, ICITM 2017* (Vol. 7917904, pp. 101–107).

Yue, X., Cai, H., Yan, H., Zou, C., & Zhou, K. (2015). Cloud-assisted industrial cyber-physical systems: An insight. *Microprocessors and Microsystems, 39*(8), 1262–1270.

Zeng, H., Chen, X., Xiao, X., & Zhou, Z. (2017). Institutional pressures, sustainable supply chain management, and circular economy capability: Empirical evidence from Chinese eco-industrial park firms. *Journal of Cleaner Production, 155,* 54–65.

Zhong, R., Xu, X., & Wang, L. (2017). Food supply chain management: Systems, implementations, and future research. *Industrial Management and Data Systems, 117*(9), 2085–2114.

Zhou, K., Liu, T., & Zhou, L. (2015). Industry 4.0: Towards future industrial opportunities and challenges. In *12thIEEE International Conference on Fuzzy Systems and Knowledge Discovery (FSKD)* (pp. 2147–2152). http://dx.doi.org/10.1109/FSKD.2015.7382284.

Zhu, Q., Geng, Y., & Lai, K. H. (2011). Environmental supply chain cooperation and its effect on the circular economy practice-performance relationship among Chinese manufacturers. *Journal of Industrial Ecology, 15*(3), 405–419.

Zhu, L., Lian, W.-S., & Wang, F. (2013). BIM digital technology in the construction industry in the field of application. *Advanced Materials Research, 143–144,* 1375–1379.

Part I
Management of Sustainability/Industry 4.0 on Supply Chains

Chapter 2
A Framework for the Application of Industry 4.0 in Logistics and Supply Chains

Ramakrishnan Ramanathan, Elly Philpott, Usha Ramanathan, and Yanqing Duan

Abstract This research aims to identify and understand the contemporary practice of using Business Analytics (BA) in improving the performance of logistics companies by conducting exploratory case studies. We present seven case studies using a within-case and cross-case analysis of the practice of BA use in UK logistics firms. We position our analysis under major BA application areas identified in previous third-party logistics surveys. Based on an in-depth analysis, we present a Value-Adding Input-Output (VAIO) framework to support an understanding of the use of Business Analytics in logistics companies. One of the main findings is the recognition of four antecedents (skills, systems, technology, and trust issues) before deriving value from business analytics investments. When the antecedents are in place, it is possible for logistics companies to derive value by engaging in BA application areas. The value dimensions ultimately help logistics firms to be competitive in the market place. The framework supports the applicability of the Resource-Based View of a firm for BA use in logistics. The framework developed in this chapter provides a practical basis for logistics companies to derive value from their investments in Business Analytics. The Value-Adding Process Framework is a new framework suggested in this chapter.

Keywords Business analytics · Logistics · Resource-Based View · Value-Adding framework · UK

1 Introduction

Industry 4.0 is a topic of this current generation for business and research alike. A strong connection among Industry 4.0, lean manufacturing, and supply chains is evident from the literature (Ben-Daya et al. 2017; Buer et al. 2018; Tortorella and Fettermann 2018). However, adoption of Industry 4.0 is not rapid in the operations

---

R. Ramanathan · E. Philpott · Y. Duan
University of Bedfordshire, Luton LU1 3JU, UK

U. Ramanathan (✉)
Nottingham Trent University, Nottingham NG1 4FQ, UK
e-mail: usha.ramanathan@ntu.ac.uk

U. Ramanathan and R. Ramanathan (eds.), *Sustainable Supply Chains: Strategies, Issues, and Models*, https://doi.org/10.1007/978-3-030-48876-5_2

management arena due to its complex nature of algorithms and technology expertise. Data is playing a great role in the success of Industry 4.0 adoption. It is worth mentioning that the growth of Industry 4.0 is also related to the adoption of big data (BD) and Business Analytics (BA) in the industry.

Business analytics refers to the collection of analytics tools that work both on big data and traditional data to generate insights and intelligence (Gandomi and Haider 2015). The renewed significance of analytics is partly driven by the availability of big data (Goes 2014; Tian et al. 2015). There is a general appreciation that BA using big data and traditional data could hold promise to give deeper insights into customers, partners, and businesses. Several research studies have highlighted that Data and BA could be useful for advancing businesses in several sectors with Industry 4.0 (Davenport 2006; Malomo and Sena 2017; Ramanathan et al. 2017). Here, Industry 4.0 represents companies with the ability to use data extensively in routine operations. Business analytics has enormous potential value in the logistics sector as this sector traditionally uses a number of analytics tools (Ayed et al. 2015; Waller and Fawcett 2013; Wang et al. 2016). The present research study aims to understand the potential of Business Analytics in the logistics sector and the factors influencing its use.

By providing an efficient flow of goods on which other commercial sectors depend, the logistics sector is a key driver of economic growth in any economy. It has been estimated that the direct contribution of this sector is 9–10% (DHL 2010), varying between 3% in plant engineering industries to 13% in mining industries (Handfield et al. 2013). There are huge volumes of logistics-related data both historical (e.g., demand data, route data, transportation management systems, warehouse management systems, and ERP) or real time (e.g., through sensors including Internet of Things (IoT), RFID, GPS, QR codes, and social networks). It has been highlighted that BA using data from these sources can provide logistics firms with improved supply chain visibility and superior competitive advantage (Cecere 2012) and competitive intelligence (Reinmoeller and Ansari 2016). However, a comprehensive understanding of how competitive advantage can be derived by logistics firms by employing BA has not been systematically studied in the literature.

In order to fill this research gap, we have conducted multiple case studies with UK logistics companies. The companies use BA for a variety of purposes and hence the challenges or benefits are varied. We understand that the UK logistics industry is one of the perfect testing fields to see Industry 4.0 adoption. By analysing the data from these case studies, we aim to develop a structured framework to understand how logistics firms are employing BA to improve their operations. In this process, we also identify the potential of the Resource-Based View (RBV) of a firm to understand the complexity of using BA in UK businesses.

2 Background of This Study

Studies on the impact of BA in Industry 4.0 organisations are still scarce but a few studies have been published in the last few years. Most of them generally point to a strong link between BA and firm performance, often anecdotally (Agarwal and Dhar 2014) but empirical evidence is scarce (Akter et al. 2016). Ramanathan et al. (2017) have found that several organisational resources (technology, human resources, top management commitment, environment, etc.) have to be aligned to business needs for realising potential benefits from BA investments. They also highlighted important roles played by trust, privacy, etc., in this context. Côrte-Real et al. (2017) have shown that business analytics provides business value for European firms. Using a systematic literature review, Wamba et al. (2015) have identified that BA can create value in multiple ways. However, it is not clear from the existing literature how BA and Industry 4.0 will help logistics firms in the presence of big data and open information exchange. This is the focus of this research paper.

Business analytics is increasingly playing a key role in logistics and supply chain management. Using the literature review, Wang et al. (2016) have found that BA has been applied to make strategic decisions (sourcing, network design, and product development) and operational decisions (demand planning, procurement, and inventory) in logistics and supply chain management. Specifically, in the field of logistics, their literature review has identified opportunities using RFID tags, mobile devices, and EDI transactions as sources of data for BA. For analytics, they identified transportation/distribution problems, network flow problems, and crew/equipment routing problems. A study by DHL (2013) has identified optimisation, understanding customers, risk management, network analysis, and efficient fleet management as five key areas for the role of big data and analytics in logistics. The Internet of Things (IoT) has a number of potential applications in logistics including the use of RFID in inventory management, tracking using GPS, self-driving vehicles, and product traceability (Xu et al. 2014; Whitmore et al. 2015). Based on interactions with industrial experts, Hofmann and Rusch (2017) have predicted that IoT will have a very wide range of applications in the field of logistics. A regular annual survey conducted with Third Party Logistics (3PL) providers (3PL study 2014, 2017) found that business analytics tools had been employed in logistics in specific application areas listed in Table 1.

Most of the studies on BA seem to be providing a generic overview of the impact of data analytics on firms in multiple functional areas or sectors. There are very few studies that focus on specific segments. For example, Erevelles et al. (2016) have studied the impacts of analytics in marketing. Ramanathan et al. (2017) have focused on retail operations. Wang et al. (2017) studied the healthcare sector. Recent studies reviewed applications of BA in logistics and supply chains: Arunachalam et al. (2017) for supply chains and Wang et al. (2016) for logistics and supply chains. These two studies on logistics are based on the literature review, while the other two publications (DHL 2013; 3PL study 2017) are not academic research articles. Some other studies have employed quantitative studies to link BA to primarily supply chains (Chen et al. 2013; Gunasekaran et al. 2017). There is a need to extend the understanding gained from these literature reviews by linking to actual applications of BA in the logistics industry. Our study is the first to provide an understanding of the applications of BA in the logistics industry by engaging the practitioners in the field via multiple case studies.

Table 1 Business Analytics application areas in logistics

1	Transportation management (planning and execution)
2	Supply chain planning
3	Network modelling and optimisation
4	Advanced analytics and data mining
5	Generating visibility (order shipment, inventory, etc.)
6	Customer order management
7	Electronic data interchange (EDI)
8	Supply chain event management
9	Warehouse/distribution centre management
10	Global trade management
11	E-commerce functionalities and other related tools

Source 3PL study (2014, 2017)

Thus, based on the literature review, it is evident that there is an insufficient understanding of how logistics companies can utilise BA effectively for improving business performance. Hence, the primary objective of the current study is to explore how BA is being used in UK-based logistics companies. We consider specifically logistics companies to see how these companies perceive using BA helps them in improving their businesses. Accordingly, we developed the following research questions, to be answered through the case analysis.

1. How are logistics companies in the UK using BA in their businesses?
2. How is the use of BA helping these companies in improving their business (with a specific focus on logistics operations)?
3. What are the preconditions that must exist for companies to leverage value from BA better?

The rest of the paper is organised as follows. The methodology adopted in this study is discussed in the next section. In line with the choice of the qualitative methodology, a description of our case companies is provided in Sect. 4. A detailed cross-case analysis is presented in Sect. 5. The results of our research are presented in Sect. 6, while the last section provides conclusions.

3 Exploring BA in UK Logistics Companies

Given that there were a few conceptual models relating to the adoption of BA in Industry 4.0, especially for logistics business performance, we used exploratory research to gather rich data that could be analysed in-case and by cross-case methods. Due to the exploratory nature of these research questions, we have used a case-study-based qualitative research approach (Yin 2012). Case research included interviews, observational study, and site visits.

3.1 Three-Stage Approach

In line with the research questions, the main interview questions are: "*how are your businesses using BA in the operations*?", "*what are your perceptions of the relationship between the use of BA and business performance?*", and "*what are the enablers and inhibitors for the successful BA utilisation in your company?*".

We followed three stages in our qualitative approach. In the first stage, we identified leading logistics companies in the UK with interests in BA based on an analysis of publicly available information (website, reports, etc.) and contacted 14 such companies explaining the purpose of our research project with a request for an in-depth semi-structured interview. Seven companies expressed interest at this stage. All of these companies have logistics as part of their business.

In the second stage, we interviewed a key logistics management contact from each of the seven case companies. We followed university guidelines on interview practice and offered confidentiality agreements to allow discussion of commercially sensitive issues more freely. Each interview was conducted by two academics to ensure consistency and reliability of data transcription later on. Interviews took place mostly on company premises and took 1–2 h. The interviews followed the broad areas outlined in Table 1 and we designed semi-structured interviews with open-ended questions to obtain maximum data from respondents. Table 2 describes the case companies in terms of sector and operations, and the domain of BA use and the interview participants. All interviews were transcribed and checked, and interviewees were invited to check their interview transcripts to confirm the accuracy of the transcription.

Stage three involved a wider analysis of interview transcripts. At this stage, we liaised with the companies where clarifications or confirmations of specific aspects of their operations were needed. This activity created a good understanding of how data was used for better performance in each case and provided assurance that the cases were both robust and reliable. During analysis, we triangulated interview data with other relevant data from external sources (Yin 2012).

Table 2 Details of our case companies

Case company	Who is involved in case study?—Data provider
C1: Aerospace engine manufacturer and distributor	Ex-operations manager (recently retired)
C2: Third party logistics (3PL) providers operating globally	Global head of logistics
C3: Global Logistics Company	Ex-operations manager (recently retired)
C4: Global Logistics Company	Vice president of logistics
C5: Retail chain with physical stores and online sales—own their own logistics	Operations director
C6: Logistics company for B2B	Managing director
C7: Client-centric logistics and inventory company	Managing director

3.2 Description of Case Companies and the Role of BA

The description of our case companies presented below has been obtained mainly from interviews. The data was further supplemented by additional information from the company and also from the public domain.

Case Company C1—Manufacturing and Service

Company C1 designs, developments, manufactures, and services integrated power systems for use in air, on land, and at sea. The company has reported a revenue of £14,588M (2014) with £942M in profit. It has 54,100 employees worldwide.

Company C1 equips its engines with sensors that collect data from different components, systems, or subsystems. The use of BA is mainly for process improvement, production, and logistics of product pickup and delivery. Our interviewee was a recently retired operations manager in the company, with 35 years of experience, most of which is in managing logistics operations.

Case Companies C2–C4—Logistics Services

Company C2 is one of the world's largest Third Party Logistics (3PL) providers. C2 Group operates a global network with some 500 offices in more than 70 countries, and it works with partner companies in further 90 countries. C2 employs over 16000 people worldwide. Our interview was carried out with the Global Head of Continuous Improvement in Logistics. Our interviewee had previous experience at two other global logistics companies.

Company C3 is a Logistics Consultancy. Our interviewee was the MD of the company. The discussion focused primarily on his experience with the major 3PL from where he had recently retired after 16 years of service as the Global Account Director. According to him, the 3PL company was always looking at data to improve the current state of the art in its operations, logistics, change engineering, and continuous improvement.

Company C4 is a company in the top five logistics companies in the world and is involved in multiple activities in the field of logistics—sea freight, air freight, road and rail logistics, contract logistics, integrated logistics/lead logistics, real estate, and insurance brokers. The company has a global network of more than 63,000 employees at nearly 1,000 locations in over 100 countries.

In C4, data is mainly used to achieve the objectives of efficient operation and profitable growth. Our interviewee was the Vice President managing logistics for clients, with 10 years of experience in the company.

Case Companies C5–C7—Logistics-Related Services

Company C5 is a private limited company providing logistics services in the UK with a turnover of £170m (2014). Registered in 1964, C5 is a family-owned business operating by mail order, online, and in nearly 70 stores nationwide. It employs in excess of 1500 people. Our interview was carried out with the Operations Director who had 10 years of service with the company. Prior to working for C5, he had worked for major UK retailers and a printing company. His total experience in the Logistics Industry was 30 years.

Company C6 is a private limited logistics company incorporated in 1993. Our interviewee was the MD of the company, who with his business partner had grown the venture into a £1.5m business employing 22 drivers and 5 office staff. Although it has a number of manufacturing and retail clients listed on its website, its main work was for a large airside company that operates maintenance and operations for large air carriers. Through this relationship, the company is responsible for moving parts between airports for a major low-cost airline. Main BA use in C6 is linked to its online activities and the use of mobile telephony for tracking and monitoring. In total, the MD has 21 years of experience in the industry.

C7 is a private limited logistics company that has been in operation for 25 years. It employs over 200 people. It services the fulfilment and logistics needs of over 85 clients. It specialises in Storage and Distribution, Consumer Sampling, Kitting, Collation, Pre-Pack, E-commerce and Mail Order, Promotional Response Handling, Contact centre services, Handling and Fulfilment, Competition Handling, Loyalty Scheme Management, and Multichannel fulfilment. Our interviewee was the managing director of the company, with more than 25 years of experience.

All of the above-mentioned cases were further analysed on the themes that emerged from application areas of BA and value dimensions.

4 Cross-Case Analysis

The cross-case analysis is built on cross-cutting themes identified from the interviews. It starts with observations on the business analytics application areas listed in Table 1, but it also looks at other relevant factors such as building value and antecedents. Due to space restrictions, we provide only important quote(s) for each of the themes below.

4.1 Business Analytics Application Areas in Case Companies

Using Table 1 on a set of major business analytics application areas relevant for logistics companies as a basis, the following discussion compares the experience of our cases and highlights corroboration and contradiction where this was found.

4.1.1 Transportation Management (Planning and Execution)

The capability to plan transportation is an integral component of any logistics company. Planning transportation management includes the choice of appropriate mode/vehicles for a given purpose, understanding of potential traffic congestion points, rerouting, understanding the availability of parking spaces, etc., to ensure smooth flow of vehicles during logistics operations (Caicedo et al. 2012; Jiang et al. 2013; Koshak et al. 2013). Four (C1, C2, C4, and C5) of our seven case companies are using data for managing and planning their transportation requirements.

> So we started to change our efficiency. For example, we, on a weekly basis, changed vehicles that were running on, what we call milk runs, i.e., vehicles in our fleet going to the same vendor every day, to what we call "man in the van", which was sub tier contractors, less than full load, where you just employ someone to nip in and pick up. That was based on continually reviewing the demand versus what was actually happening in terms of supply, and working out which was the more cost-effective route. No one could have done that without data. (C1)

While only four of our cases have highlighted that they used BA for planning transportation, in contrast, all of them said that they are using BA for executing transportation. While transportation planning requires a more strategic approach, executing transportation management is generally at operational and tactical levels. For example, BA has been used to identify cost-effective ways of achieving same-day delivery and developing agile omnichannel networks (3PL study 2014). BA can help logistics providers achieve reductions in truck miles and saving fuel (Fuhr and Pociask 2007). Our cases highlight that BA is used more widely at operational levels for executing transportation plans. For example, C5 has specifically highlighted using BA for filling trailers, deciding the frequency of delivery and the time of releasing orders to stores. C2 installed {third-party Proprietary modelling software}, enabling analysts to simulate "what-if" scenarios, and create and visualise models of supply chains, taking into account factors such as special handling requirements for the goods being distributed, transport and logistics costs, delivery times, distances travelled, agreed service levels, and inventory management strategies.

4.1.2 Supply Chain Planning

Supply chain planning is required to coordinate various business functions (such as marketing, purchasing, and logistics) starting from procuring raw materials to distributing finished goods to the final consumer (Gupta and Maranas 2003; Meyr 2004). Varieties of data, in-house and public, are used in producing effective supply chain plans. Five of our case companies, C1, C2, C3, C4, and C5, have said that they used BA for supply chain planning. As a specific example, supply chain optimisation using BA enabled C2 to manage its own transport network more efficiently, and reduce transport and logistics costs for its customers. As a third-party logistics provider, C2 claimed to have helped a leading global electronics company to analyse delays in the flow of materials between its European suppliers. It found that the structure of the existing transport network led to inefficiencies in the consolidation process, and increased the costs of managing the network. It went on to claim that, on C2's recommendation, strategic flows of materials between suppliers were established, European consolidation centre was relocated, and a combination of road and rail for goods transport was employed, reducing costs by 21% and cutting CO_2 emissions by 34%.

4.1.3 Network Modelling and Optimisation

For logistics operations, a number of network models and their solution procedures are important in order to improve the effectiveness and efficiency of their operations. Some very relevant network applications for logistics include the shortest path problem for a distribution activity, vehicle routing, and multi-commodity flow problems (Bertsekas 1998). Five (C1–C5) of our seven case companies highlighted that they use network modelling and optimisation in their operations regularly.

> The first objective of the whole thing was the operational effectiveness area and improving efficiency. And that definitely…that was what the data was used for in the first place.… We were setting up a model from day one with a view of improving efficiency and that's exactly why we used it. (C1)

4.1.4 Advanced Analytics and Data Mining

Advanced analytics includes visualisation, forecasting, optimisation, and other BA tools, while data mining deals with automatic extraction of patterns, associations, and structures from data (Bose 2009). Many companies have realised the power of BA using advanced analytic tools to develop business insights (Barton and Court 2012). For example, Hamden and Rogers (2008) have found using data analytics that warehouses that are small, that use bulk storage, and with larger aisles are generally more efficient compared to their counterparts that are large, use selective racking and with narrow aisles, respectively. Due to complexities involved with advanced analytics, only four (C1, C2, C3, and C4) of our seven case companies have highlighted that they were using advanced analytics tools. The following quotes highlighting the emphasis on advanced analytics have been derived from the annual report of C2: "*The ability to collect, process, analyse and present information quickly and accurately enables Company B to maximize revenue, reduce costs and provide a world-class service to the customers. …*".

4.1.5 Generating Visibility

Visibility relates to the ability of supply chain partners to be able to access (e.g., track and trace via GPS) and share information related to the operations of other partners (Caridi et al. 2014). This information sharing is supposed to help all partners in better planning, better decision-making, and hence in their overall performance (in terms of cost, quality, flexibility, and time). Three of our respondents (C5, C6, and C7) have been using BA for improving visibility to end-customers, while six (C1, C2, C4, C5, C6, and C7) of our respondents reported that they use BA for improving visibility for their clients. C3 did explain the problems they faced in their warehouse because they did not have the technology to improve visibility. For example, a part had to be reordered even though the part had already been received but coded wrongly affecting visibility.

4.1.6 Customer Order Management

Customer order management involves activities such as order processing procedure, initiating production, and establishment of appropriate in-house activities such as inventory/warehouse management/transportation, and the mechanism to deal with customer complaints in an effective way (Chen et al. 2014). Our case companies have recognised the potential of BA in streamlining customer order management, and six of them (C1, C2, C4, C5, C6, and C7) are using BA for this purpose.

4.1.7 Electronic Data Interchange (EDI)

Electronic Data Interchange (EDI) has been traditionally defined as the transmission of standardised business transactions from computer to computer among business partners using appropriate industry protocols and standards (Ferguson et al. 1990). Since logistics as a function requires coordination of the flow of information as well as materials, EDI plays a critical role in the efficient functioning of logistics firms. The creation, development, and use of Data for BA are thus facilitated by the presence of modern EDI systems in firms. Six of our respondents (C1, C2, C4, C5, C6, and C7) are using EDI in their firms to facilitate efficient use of BA. According to one of our case companies (C1), the Enterprise Resource Planning, a sub-form of EDI, was a driver for the use of BA. Our interviewee recalled how he had needed to work more closely with the factory layout specialist because

> …how the material arrived at the factory and how it went through the factory and out of the other end became about more than how you lay out the factory. He {the factory layout guy} became I think, as far as I could see, there may be other people in other functions, but that guy became the champion of Big Data and he was internally grown, not externally.

4.1.8 Supply Chain Event Management

An interesting use of BA is to help identify bottlenecks and avoid potential disruption to forthcoming events and thus minimise adverse impact on businesses. This idea is supported via Supply Chain Event Management (SCEM), which can be viewed as a management concept, a software application, and a software component (Otto 2003). SCEM aims to identify deviations from production/transport schedules and minimise impact early before they begin to affect operations efficiency and customer satisfaction (Stadtler and Kilger 2002). An example of SCEM is the traditional statistical process control that identifies potential issues in production processes (Bersimis et al. 2014). Five of our respondents (C1, C2, C3, C4, and C7) claimed they were using BA for SCEM in their organisations.

4.1.9 Warehouse/Distribution Centre Management

There are huge opportunities for BA in improving the efficiency of operations in warehouses and distribution centres (Chen et al. 2013). A number of predictive analytics tools, such as mixed integer programming (Holzapfel et al. 2016), simulated annealing (Matusiak et al. 2014), and Data Envelopment analysis (Hamdan and Rogers 2008) have been employed to study warehouse efficiency. Five of our case companies (C1, C2, C4, C5, and C7) have employed BA in organising operations in their warehouses and distribution centres. Specifically, C2 described how the company was using data from two angles, warehouse location and warehouse number.

> I guess we do {use data} from two angles, there is the pure freight movements, which is the best way to get something from A to B? What's the best way to combine it? ~~Blah blah blah~~. Increasingly we are looking at where should the warehouse be? So from a logistics perspective it's modelling whether we should have one warehouse two warehouse three warehouses, and this is the project that we mentioned of {Project name}, where is the best place to put the inventory?

4.1.10 Global Trade Management

As companies globalise their operations, they face newer challenges such as variable regulatory regimes in various countries, channel fragmentation, poor logistics infrastructure, increased risk of supply chain disruption, newer cultures, new competition, etc. (Handfield et al. 2013). This increased complexity of multinational trade usually complicates logistics operations in outsourcing (Ramanathan 2017; Wu et al. 2013). However, BA can help improve decision-making, and hence facilitates global trade. Three of our case companies (C1, C2, and C4) have indicated that they use BA for helping to make decisions on managing their global trade. As a specific example, C2 mentioned that it was evaluating possibilities, with the help of BA, to establish additional satellite centres in Asia. This would allow the company to further leverage its ability to navigate Asia's fragmented trade and regulatory landscape, capitalise on regional demand, and be closer to relevant decision makers. C2 further highlighted that its staff used simulations to identify possible weaknesses and to measure the effects that changes in parameters have on the various scenarios. The new insights enabled C2's customers to base important business decisions on reliable data and to maximise the end-to-end efficiency of their supply chains.

4.1.11 E-Commerce Functionalities and Other Related Tools

E-commerce functionalities include e-commerce-based systems for online booking and order tracking (Ramanathan et al. 2016), the use of RFID tags and bar codes

(Ramanathan et al. 2012), use of collaboration tools (Ramanathan and Gunasekaran 2014), and the more recent use of Internet of Things (IoT) sensors (Yu and Wang 2016). Our case companies have used these capabilities in varying degrees: web portals for booking and order tracking (C1, C2, C4, C6, and C7), barcoding (C1, C2, C3, and C4), RFID (C1, C2, and C4), collaboration tools (C2 and C4), and yard management tools (C4).

Table 3 summarises the company's adoption of BA application areas and factors behind non-adoption of BA application areas. Comparison among companies shows a strong emphasis on visibility in Industry 4.0 setups along with EDI use to manage customer orders. All case companies used BA applications for e-commerce to survive the competition. Transportation management, supply chain planning, and global trade management have not been considered important elements of BA applications by global logistic companies as they have experienced in-house talents. One of the interviewees mentioned that it was not easy to convince all global partners to use smart operations in logistics while some partners prefer to use their experience.

4.2 *Business Analytics Value Dimensions*

The role of BA in deriving value from data is apparent from the definition of 5Vs (volume, velocity, variety, veracity, and value) associated with big data (Opresnik and Taisch 2015). Our case companies have emphasised that they are using BA in order to derive value in their businesses. The concept of value has been defined as the potential to cut costs or enhance revenues (Akter et al. 2016; LaValle et al. 2011).

In line with the conceptions of value in the literature, our case companies felt that they were able to derive value with BA in the forms of reduced delivery time, improved geographic coverage, increased resource utilisation, and improved forecasting accuracy.

4.2.1 Improved Operational Efficiency—Reduced Cost

The literature has stressed the need to improve the operations efficiency of logistics service providers (e.g., Hamden and Rogers 2008; Wang et al. 2016). This is equally true with our case companies. Cost-cutting opportunities with BA were cited by almost all our case companies (C1–C6). The following is an example quote from one respondent.

> Far more efficient, it's quicker to transfer the data, we minimise errors, that take place by processing the data cleanly…too make sure they have a clean set of data that's transmitted through to their systems…so far it has had approximately £1.5million benefit to the business in annualised turnover. (C6)

Table 3 Business analytics application areas practised by the case companies with insights on using or not using them

	Transportation management	Supply chain planning	Network modelling and optimisation	Advanced analytics and data mining	Generating visibility	Customer order management	Electronic Data Interchange (EDI)	Supply chain event management	Warehouse/distribution centre management	Global trade management	E-commerce functionalities and other related tools
Companies using BA application	C1, C2, C4, C5	C1–C5	C1–C5	C1–C4	C1–C2, C4–C7	C1–C2, C4–C7	C1–C2, C4–C7	C1–C4, C7	C1, C2, C4, C5, C7	C1, C2, C4	All cases
Key purpose of BA application	Efficiency; Day-to-day transportation	Procurement and distribution	Operational effectiveness	Quick and accurate Maximise revenue	Visibility to clients and end-customers	Basic need	Driver for all smart operations	Identify deviation from original plan	Modelling	Decision-making	Tracking
Companies not using BA application	C3, C6, C7	C6, C7	C6, C7	C5–C7	C3	C3	C3	C5, C6	C3, C6	C3, C5–C7	–
Reasons for not using BA application	Experience is valued than BA	Experience is valued than BA	Investment issues/not necessary/less knowledge	Non-availability of high skilled analytics	Planning to use BA to avoid warehouse problems	Not felt its importance yet	Progressing to use EDI	Current SC event management team is effective	Experience is valued than BA	Not necessary	–

4.2.2 Improved Reliability/Accuracy

Reliability features strongly as one of the most important requirements for efficient logistics in the literature (Cullinane and Toy 2000; Dullaert and Zamparini 2013). The role of cutting costs is by improving reliability, and forecasting accuracy has been stressed by three (C1, C2, and C5) of our case companies.

> we know what's selling, we know what's not selling, we know what's staying in stock; which is not moving quickly, so we think that potentially there is at least enough opportunity to forecast their inventory. We do. But a logistics company never thinks of doing that, you know our job is usually is to try and fill the space…we would be able to go back to the customer …and say don't worry about inventory planning because we will do it all for you. (C2)

4.2.3 Improved Customer Experience

Logistics operations should be conducted efficiently for improved customer service (Wang et al. 2016). Our case companies were not an exception to this trend. Five (C1, C2, C4, C5, and C6) of our case firms have highlighted improved customer experience with BA to enhance their revenues. For them, BA has helped them to derive value by helping them learn more about their customers.

4.2.4 Improved Marketing Efficiency

Logistics plays an important role in supporting marketing activities that would enhance customer experience (Rinehart et al. 1989). For example, BA could be helpful in pinpointing specific issues with customers and reacting faster. This aspect of improving marketing efficiency has been cited as one of the strong points of BA in the literature (e.g., Wang et al. 2016). Improved marketing efficiency as a way of cost reduction has been cited by two (C5 and C7) firms. For example, one respondent (C3) recently introduced a system designed to give an end-to-end service to a specific industry, while mitigating risk in an increasingly lean environment. The system is designed to react quickly to fluctuating demand and supply.

4.2.5 New Business Models (C2, C3, C6)—Agility

Business analytics, as in any other functional discipline, provides opportunities in logistics to rethink existing ways of conducting business and developing new business models with significant growth opportunities (e.g., Wang et al. 2016). Three (C2, C3, and C6) of our respondents have highlighted this potential and developed new business models with BA.

C3 used BA to identify business opportunities for consolidation in thinly populated markets.

> I spent quite a lot of time in the Nordic countries, where you got lot land mass not many people are living there, so product that is going into this country is expensive to distribute. …. For new business models you would be saying for example, you are delivering to city A in Norway, so we will consolidate at city B and deliver onward to you once a week may be if that service level is acceptable. …. (C3)

4.2.6 Reacting to Wider Economic and Environmental Changes

A very important advantage of BA is that companies are able to make data-driven decisions in response to newer economic changes (Wang et al. 2016). Our respondents have specifically highlighted leveraging this opportunity using BA. For example, three firms (C2, C3, and C5) have improved their capabilities to react to wider economic changes by analysing BA and hence managed to enhance revenues. Two of them (C1 and C2) have cited using BA to understand and respond quickly to environmental changes (weather and tax regimes).

4.2.7 Linking Value Dimensions to BA Application Areas

Further discussions with the case companies revealed that business analytics application areas in their organisations have helped to achieve one or more of the above value dimensions. For example, a case company (C1) highlighted that their BA-based capabilities (optimisation, transportation management, traceability, global trade management, and customer order management) have helped them not only in improving their operational efficiency but also in developing newer supply chain designs and improving the reliability of their operations. Another one (C2) highlighted that its BA-related capabilities (forecasting, optimisation, visibility, transportation planning/management, traceability, and global coverage) have helped it improve the efficiency of operations, customer management, visualisation, and the agility to adapt to wider economic/environmental changes. Another company (C5) explained how it employed the idea of heat maps (Heijden and Garn 2013) to identify new shop locations for ensuring business sustainability. One more case company (C6) explained the use of SCEM for improving their resilience to disruption and enjoy better integration with clients.

4.3 *Antecedents to Effective BA Utilisation*

A number of previous studies have argued that any big investment such as that for business analytics faces significant issues prior to implementing the project (e.g., Alharthi et al. 2017; Hashem et al. 2015; LaValle et al. 2011). For example, Ramanathan et al. (2017) have highlighted that incompatibility among multiple IT platforms would be a potential barrier to deriving benefits from BA. Alharthi et al.

(2017) have highlighted barriers related to technology infrastructure, privacy, human skills, and organisational vision.

Our interviewees have also highlighted a number of antecedents to successful BA utilisation in their organisations.

4.3.1 Analytics and Business Skills

All of our case companies felt that human resources with necessary IT skills were the most important limitation for their ability to exploit BA.

> I would say there's an issue with lack of skills available to do all of the IT things that everybody wants to be able to do. So there's a lack of skilled people who can do all of this data analytics and particularly to build the databases that underpin it. Erm… And that's one barrier to it. (C2)

4.3.2 Data and Systems

In addition to skills, issues related to data and systems have also been highlighted. The availability of accurate data in an appropriate form often limits their ability to improve reliability (C1, C3, and C4).

Data in multiple formats was recognised as an antecedent.

> for a quite a lot of customers we are producing the data in a different way. We have to manipulate that data to tell us what we want to know. (C3)

> …The problem comes when specific formats are required and doing this is human-intensive. … This is further complicated when each point in the supply chain is in effect its own node. In single supply chains where there is one customer and one supplier benefits of the increased collection of data would possibly be financially feasible but with such complexity, to invest in collecting more data for analysis would be too large a burden on profit margins. (C7)

4.3.3 Technology

Technology issues were highlighted as antecedents. This included issues such as bandwidth, battery life for asset tracking, and lack of mobile phone-based capabilities (C6). The need to be familiar with emerging and new technologies affecting the logistics industry was also highlighted as shown in the following quote.

> Our customers are thinking about making big changes which as a logistics company that opens up the opportunity for us, and the other side is definitely the technological developments, so we spent quite some money last year and the year before, just on a research project saying what are the new technologies that are going to come into our industry and affect our industry? And a lot of those were around data, so it was cloud computing, big data, RFID, Internet of things, and obviously when somebody comes back and says, these are the four things hitting your industry, you need to do something about it… (C2)

4.3.4 Trust, Power, and Legal Issues

In addition, our interviewees have also stressed the need for taking cautious gradual approach in BA investments (C2), legal/behavioural issues such as data access, power, trust, and privacy (C2, C4, C5), standardisation of data formats (C7), and issues with linking with legacy IT systems (C7). There were also views that smaller firms have limited capabilities to exploit BA (C2). These could be viewed as antecedents to achieving these capabilities. The following quote from C2 on trust and power relations on the use of BA is interesting.

> For sure. It's a trust and a power thing as well, whether they see us as a competitor, whether they see us as a partner, and of course we deal with, if you think the top technology companies, we deal with them all. And they start giving information to us that we then use for one of their competition then we're in a dangerous game there. So, there's a big power question around who's got that data and what I can do with it. (C2)

5 The Value-Adding Input-Output Framework

5.1 The Framework

Analysis and discussions in Sect. 4 enabled us to develop a Value-Adding Input-Output framework (VAIO) for using BA in Logistics (Fig. 1). This framework has four elements—antecedents, BA application areas, value dimensions, and competitive advantage. Individual elements of the framework have been discussed in detail

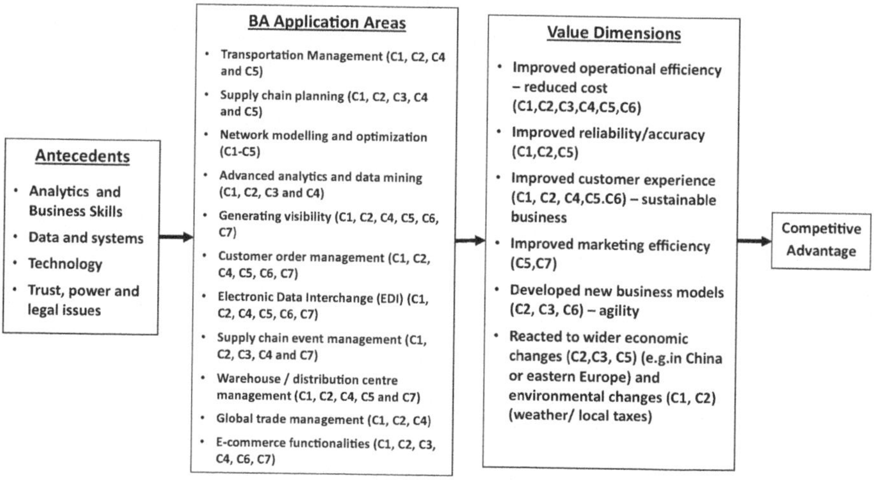

Fig. 1 A Value-Adding Input-Output (VAIO) framework for understanding the use of Business Analytics in logistics

in previous sections with the exception of competitive advantage, which will be discussed further here. The following example quote from C2 on BA highlights its ability to achieve competitive advantage.

> I guess where it's a little bit come from is, we've now started to ask the question, how can we differentiate from {Other company1}, {Other company 2}, {Other company 3}? So we have had lots of discussions about what we could do different, how we could do things differently? And this one kept popping up, we didn't really know what we could do differently, but we think now that with an angle, we can say that we have the best company at analysing your information {to the} customer. So now that's been decided that it could be a potential differentiator, now it's a case of, resources available, now go show us how you're going to do it. (C2)

The VAIO framework helps understanding how logistics firms could derive value from their investments in business analytics. Before embarking on a BA project, it is important to ensure that antecedents are in place to launch the project. The availability of adequate human resource skills (in analytics capability and also with the necessary business acumen) is an important antecedent that would make or break a BA project. Recognition of data of varying types and quality is an important requirement to plan right the kind of hardware/software investments. A continuous understanding of emerging new technologies is another antecedent. The need to understand the legal, power, and trust issues is also important.

When the antecedents are adequately put in place, it is possible to derive value from BA investments for logistics companies. Our case companies have used their BA investments to gain further insights in transportation management, supply chains, network optimisation, data mining, visibility generation, and other relevant areas, so that the potential benefits are realised not only to their own companies but also to their suppliers and clients.

A superior understanding of these logistics application areas has supported in deriving value to these companies. Value has been derived in the form of improved operational efficiency, improved reliability of logistics provision, better understanding of customers, higher agility in meeting customer expectations, and better ability to react to changes in the business environment. There is a recognition that these values have provided the case companies a competitive advantage.

We term this framework as an input-output framework because the inputs on the left-hand side are necessary to achieve the outputs on the right-hand side. For example, adequate considerations for antecedents are necessary before the application areas are attempted. Values can be derived from BA investments only when application areas are systematically employed. Deriving value is necessary in order to achieve a competitive advantage.

The VAIO framework builds on a similar analysis from the literature. Most of the previous studies focused on one or two elements of the VAIO framework. For example, there are studies that focused on antecedents and value of BA (Chen et al. 2013), and application areas and value (DHL study 2013). However, our study explores the relations among all the four elements together.

We believe that the framework stresses the importance of innovation-oriented value creation in logistics (Marchet et al. 2017) that highlights the importance of shared IT systems (TMS, WMS, etc.). It also echoes the findings of Chen et al. (2013) on the antecedents and impacts of BA in creating value. Our study echoes another position paper on the same subject DHL (DHL 2013). The report by DHL highlighted that BA has been used in various sectors (not focusing exclusively on logistics) to gain competitive advantage using at least three value dimensions: improving operational efficiency, improving customer experience, and developing new business models for Industry 4.0 companies. Our framework is a special representation of the information system value chain of Abbasi et al. (2016) for the specific context of logistics.

5.2 Supporting Theory—The Resource-Based View of a Firm

Though we did not derive our VAIO framework based on a specific theory, the link of the framework to the theory of Resource-Based View has emerged as a prominent theory with a very significant agreement with many case companies on the BA concepts identified. The Resource-Based View (RBV) of a firm is a strategic management theory that helps in understanding how organisations are using their bundles of resources and capabilities to achieve an edge over competition in the market place (Barney 1991). Resources may have different characteristics in terms of being valuable (V), rare (R), inimitable (I), and proper organisations (O) (Barney et al. 2001). They are tangible or intangible assets. Capabilities are resources that are non-transferable and are shaped by the business environment so as to impart a competitive edge to the organisation (Hitt et al. 2015). Especially at this era of Industry 4.0, BA will help identifying the non-transferable skills within the supply chains and help decision makers to plan their operations effectively.

The VAIO framework highlights that BA assets are utilised in our case companies in specific ways. These BA assets not only include technological investments but also others such as human resources aimed at analytics. These BA assets are a set of resources, which when bundled with other resources of an organisation, provide insights into various IT application areas (e.g., optimisation) giving inimitable capabilities. These capabilities provide value to the firm and ultimately help the firm to achieve a competitive advantage. Thus, the VAIO framework supports the tenets of RBV in the context of the use of BA in the logistics sector with smart operations.

Our research is not new in linking RBV in the BA and business analytics literature. RBV has been generally supported by the literature as a suitable theoretical underpinning to understand the impact of BA in firms (e.g., Abbasi et al. 2016; Akter et al. 2016; Wamba et al. 2017; Opresnik and Taisch 2015; Gunasekaran et al. 2017). Hazen et al. (2016) have included RBV as one of the eight organisational theories generally cited for understanding the role of BA in supply chains. Dubey et al. (2017) have used RBV to link BA capabilities to social and environmental performance in firms. Thus, our research has provided further insights into supporting the applicability of RBV for business analytics, especially for the logistics industry in the era of Industry 4.0 with smart operations and capability of Business Analytics.

6 Conclusions

The primary objective of our study was to explore how Business Analytics (BA) is being used in UK-based logistics companies. By focusing on one particular sector, we believe we generated some sector-specific insights on the use of BA in logistics. Based on case study research with seven logistics companies in the UK, we proposed a Value-Adding Input-Output framework to understand how logistics firms could derive value and stay ahead of the competition with the help of BA investments. The framework supports the applicability of the Resource-Based View of a firm for BA in logistics. Thus, our study has implications both for practice and theory.

While doing so, we believe that our study has made some contributions to the literature. The Value-Adding Input-Output framework has been proposed in this paper for supporting the use of BA in the logistics sector. The framework stresses the need for having the key antecedents—human resources with right analytics and business skills, right infrastructure and systems to suit the quality and quantity of available data, the technological aspect, and understanding of trust, power, and legal issues—for a BA project in logistics to succeed. The framework also highlights various IT-related application areas that would help make a BA project in logistics successful. The framework thus provides a more focused and in-depth guidance for the sector-specific research and practice agenda. The other contribution of our study is that the framework is evidence-based and linked to the relevant theory—RBV.

Our framework is built on insights derived from seven firms, but involvement from more firms would help improving the wider applicability of our results. Further research can extend the VAIO framework in other application domains (e.g., manufacturing, health, etc.). A larger scale testing of the framework using a quantitative survey could also be attempted in future studies. In spite of these limitations, we believe that the VAIO framework proposed in this study will help extend our understanding of how business analytics can be advantageously utilised in the logistics sector.

Acknowledgments The authors would like to thank British Academy for funding to carry out this research.

References

3PL study. (2014). 2014 THIRD-PARTY LOGISTICS STUDY: The state of logistics outsourcing—Results and findings of the 18th Annual Study, C. John Langley, Jr., Ph.D., and Capgemini.

3PL study. (2017). 2017 THIRD-PARTY LOGISTICS STUDY: The state of logistics outsourcing—Results and findings of the 21st Annual Study, C. John Langley, Jr., Ph.D., and Capgemini.

Abbasi, A., Sarker, S., & Chiang, R. H. (2016). Big data research in information systems: Toward an inclusive research agenda. *Journal of the Association for Information Systems, 17*(2), 1–32.

Agarwal, R., & Dhar, V. (2014). Big data, data science, and analytics: The opportunity and challenge for IS research. *Information Systems Research, 25*(3), 443–448.

Akter, S., Wamba, S. F., Gunasekaran, A., Dubey, R., & Childe, S. J. (2016). How to improve firm performance using big data analytics capability and business strategy alignment? *International Journal of Production Economics, 182,* 113–131.

Alharthi, A., Krotov, V., & Bowman, M. (2017). Addressing barriers to big data. *Business Horizons, 60*(3), 285–292.

Arunachalam, D., Kumar, N., & Kawalek, J. P. (2017). Understanding big data analytics capabilities in supply chain management: Unravelling the issues, challenges and implications for practice. *Transportation Research Part E: Logistics and Transportation Review* (In Press).

Ayed, A. B., Halima, M. B., & Alimi, A. M. (2015, May). Big data analytics for logistics and transportation. In *2015 4th International Conference on Advanced Logistics and Transport (ICALT)* (pp. 311–316), IEEE.

Ben-Daya, M., Hassini, E., & Bahroun, Z. (2017). Internet of things and supply chain management: A literature review. *International Journal of Production Research*, 1–24.

Buer, S. V., Strandhagen, J. O., & Chan, F. T. (2018). The link between Industry 4.0 and lean manufacturing: Mapping current research and establishing a research agenda. *International Journal of Production Research, 56*(8), 2924–2940.

Barney, J. (1991). Firm resources and sustained competitive advantage. *Journal of Management, 17*(1), 99–120.

Barney, J., Wright, M., & Ketchen, D. J. (2001). The resource-based view of the firm: Ten years after 1991. *Journal of Management, 27,* 625–641.

Barton, D., & Court, D. (2012). Making advanced analytics work for you: A practical guide to capitalizing on big data. *Harvard Business Review, 90*(10), 78–83.

Bersimis, S., Koutras, M. V., & Maravelakis, P. E. (2014). A compound control chart for monitoring and controlling high quality processes. *European Journal of Operational Research, 233*(3), 595–603.

Bertsekas, D. P. (1998). *Network optimization: Continuous and discrete models*. Belmont: Athena Scientific.

Bose, R. (2009). Advanced analytics: Opportunities and challenges. *Industrial Management & Data Systems, 109*(2), 155–172.

Caicedo, F., Blazquez, C., & Miranda, P. (2012). Prediction of parking space availability in real time. *Expert Systems with Applications, 39*(8), 7281–7290.

Caridi, M., Moretto, A., Perego, A., & Tumino, A. (2014). The benefits of supply chain visibility: A value assessment model. *International Journal of Production Economics, 151,* 1–19.

Cecere, L. (2012). *Big data: Go big or go home*, Supply Chain Insights LLC. Retrieved March 21, 13, from http://supplychaininsights.com.

Chen, J. C., Cheng, C. H., Huang, P. B., Wang, K. J., Huang, C. J., & Ting, T. C. (2013). Warehouse management with lean and RFID application: A case study. *The International Journal of Advanced Manufacturing Technology, 69*(1–4), 531–542.

Chen, M. C., Hsu, C. L., Hsu, C. M., & Lee, Y. Y. (2014). Ensuring the quality of e-shopping specialty foods through efficient logistics service. *Trends in Food Science & Technology, 35*(1), 69–82.

Côrte-Real, N., Oliveira, T., & Ruivo, P. (2017). Assessing business value of big data analytics in European firms. *Journal of Business Research, 70,* 379–390.

Cullinane, K., & Toy, N. (2000). Identifying influential attributes in freight route/mode choice decisions: A content analysis. *Transportation Research Part E: Logistics and Transportation Review, 36*(1), 41–53.

Davenport, T. (2006). Competing on analytics. *Harvard Business Review, 84*(5), 150–151.

DHL. (2010). *Delivering tomorrow: Towards sustainable logistics*, Deutsche Post AG, Headquarters, Bonn, Germany. Retrieved October 23, 2017, from http://www.dhl.co.uk/content/dam/downloads/g0/logistics/green_logistics_sustainable_logistics_study_en.pdf.

DHL. (2013). *Big data in logistics: A DHL perspective on how to move beyond the hype*. Troisdorf, Germany: DHL Customer Solutions & Innovation.

Dubey, R., Gunasekaran, A., Childe, S. J., Papadopoulos, T., Luo, Z., Wamba, S. F., & Roubaud, D. (2017). Can big data and predictive analytics improve social and environmental sustainability? *Technological Forecasting and Social Change* (in press).

Dullaert, W., & Zamparini, L. (2013). The impact of lead time reliability in freight transport: A logistics assessment of transport economics findings. *Transportation Research Part E: Logistics and Transportation Review, 49*(1), 190–200.

Erevelles, S., Fukawa, N., & Swayne, L. (2016). Big data consumer analytics and the transformation of marketing. *Journal of Business Research, 69*(2), 897–904.

Ferguson, D. M., Hill, N. C., & Hansen, J. V. (1990). Electronic data interchange: Foundations and survey evidence on current use. *Journal of Information Systems, 4*(2), 81–91.

Fuhr, J. P., & Pociask, S. B. (2007). Broadband services: Economic and environmental benefits. *American Consumer Institute.*

Goes, P. B. (2014). Editor's comments: Big data and IS research. *MIS Quarterly, 38*(3), iii–viii.

Gunasekaran, A., Papadopoulos, T., Dubey, R., Wamba, S. F., Childe, S. J., Hazen, B., et al. (2017). Big data and predictive analytics for supply chain and organizational performance. *Journal of Business Research, 70,* 308–317.

Gandomi, A., & Haider, M. (2015). Beyond the hype: Big data concepts, methods, and analytics. *International Journal of Information Management, 35*(2), 137–144.

Gupta, A., & Maranas, C. D. (2003). Managing demand uncertainty in supply chain planning. *Computers & Chemical Engineering, 27*(8), 1219–1227.

Hamdan, A., & Rogers, K. J. (2008). Evaluating the efficiency of 3PL logistics operations. *International Journal of Production Economics, 113*(1), 235–244.

Handfield, R., Straube, F., Pfohl, H. C., & Wieland, A. (2013). *Embracing global logistics complexity to drive market advantage.* BVL International: DVV Media Group GmbH.

Hashem, I. A. T., Yaqoob, L., Anuar, N. B., Mokhtar, S., Gani, A., & Khan, S. U. (2015). The rise of "big data" on cloud computing: Review and open research issues. *Information Systems, 47,* 98–115.

Hazen, B. T., Skipper, J. B., Ezell, J. D., & Boone, C. A. (2016). Big data and predictive analytics for supply chain sustainability: A theory-driven research agenda. *Computers & Industrial Engineering, 101,* 592–598.

Heijden, H. V., & Garn, W. (2013). Profitability in the car industry: New measures for estimating targets and target directions. *European Journal of Operational Research, 225*(3), 420–428.

Hitt, M. A., Xu, K., & Carnes, C. M. (2015). Resource based theory in operations management research. *Journal of Operations Management, 41,* 77–94.

Hofmann, E., & Rüsch, M. (2017). Industry 4.0 and the current status as well as future prospects on logistics. *Computers in Industry, 89,* 23–34.

Holzapfel, A., Kuhn, H., & Sternbeck, M. G. (2016). Product allocation to different types of distribution center in retail logistics networks. *European Journal of Operational Research, 264*(3), 948–966.

Jiang, S., Fiore, G. A., Yang, Y., Ferreira, Jr, J., Frazzoli, E., & González, M. C. (2013). A review of urban computing for mobile phone traces: Current methods, challenges and opportunities. In *Proceedings of the 2nd ACM SIGKDD international workshop on Urban Computing* (p. 2). ACM.

Koshak, N., Nour, A., & Center, K. G. I. (2013). Integrating RFID and GIS to support urban transportation management and planning of Hajj. In *The 13th International Conference on Computers in Urban Planning and Urban Management.*

LaValle, S., Lesser, E., Shockley, R., Hopkins, M. S., & Kruschwitz, N. (2011). Big data, analytics and the path from insights to value. *MIT Sloan Management Review, 52*(2), 21–32.

Malomo, F., & Sena, V. (2017). Data intelligence for local government? Assessing the benefits and barriers to use of big data in the public sector. *Policy & Internet, 9*(1), 7–27.

Marchet, G., Melacini, M., Perotti, S., Sassi, C., & Tappia, E. (2017). Value creation models in the 3PL industry: What 3PL providers do to cope with shipper requirements. *International Journal of Physical Distribution & Logistics Management, 47*(6), 472–494.

Matusiak, M., de Koster, R., Kroon, L., & Saarinen, J. (2014). A fast simulated annealing method for batching precedence-constrained customer orders in a warehouse. *European Journal of Operational Research, 236*(3), 968–977.

Meyr, H. (2004). Supply chain planning in the German automotive industry. *OR Spectrum, 26*(4), 447–470.

Opresnik, D., & Taisch, M. (2015). The value of big data in servitization. *International Journal of Production Economics, 165,* 174–184.

Otto, A. (2003). Supply chain event management: Three perspectives. *International Journal of Logistics Management, 14*(2), 1–13.

Ramanathan, R. (2017). Energy and environmental issues in international business. *Journal of Contemporary Management and Development Studies, 4*(1), 44–60.

Ramanathan, R., Bentley, Y., & Ko, L. W. L. (2012). Investigation of the status of RFID applications in the UK logistics sector. *Logistics & Transport Focus, the official magazine of the Chartered Institute of Logistics and Transport (UK), 14*(11), pp. 45–49.

Ramanathan, R., Karpuzcu, T., & Ramanathan, U. (2016). Impact of e-commerce in B2B physical distribution: Diffusion of innovations perspective. In I. N. Lee (Ed.), *Encyclopedia of E-Commerce Development, Implementation, and Management* (pp. 2256–2270), IGI Global, Hershey, PA 17033, USA, Chapter 162.

Ramanathan, R., Philpott, E., Duan, Y., & Cao, G. (2017). Adoption of business analytics and impact on performance: A qualitative study in retail. *Production Planning and Control, 28*(11–12), 985–998.

Ramanathan, U., & Gunasekaran, A. (2014). Supply chain collaboration: Impact of success in long-term partnerships. *International Journal of Production Economics, 147,* 252–257.

Reinmoeller, P., & Ansari, S. (2016). The persistence of a stigmatized practice: A study of competitive intelligence. *British Journal of Management, 27*(1), 116–142.

Rinehart, L. M., Cooper, M. B., & Wagenheim, G. D. (1989). Furthering the integration of marketing and logistics through customer service in the channel. *Journal of the Academy of Marketing Science, 17*(1), 63–71.

Stadtler, H., & Kilger, C. (2002). *Supply chain management and advanced planning: Concepts, models*. Springer, Berlin: Software and Case Studies.

Tian, X., Han, R., Wang, L., Lu, G., & Zhan, J. (2015). Latency finance. *The Journal of Finance and Data Science, 1*(1), 33–41.

Tortorella, G. L., & Fettermann, D. (2018). Implementation of Industry 4.0 and lean production in Brazilian manufacturing companies. *International Journal of Production Research, 56*(8), 2975–2987.

Wamba, S. F., Akter, S., Edwards, A., Chopin, G., & Gnanzou, D. (2015). How 'big data' can make big impact: Findings from a systematic review and a longitudinal case study. *International Journal of Production Economics, 165,* 234–246.

Wamba, S. F., Gunasekaran, A., Akter, S., Ren, S. J. F., Dubey, R., & Childe, S. J. (2017). Big data analytics and firm performance: Effects of dynamic capabilities. *Journal of Business Research, 70,* 356–365.

Wang, G., Gunasekaran, A., Ngai, E. W., & Papadopoulos, T. (2016). Big data analytics in logistics and supply chain management: Certain investigations for research and applications. *International Journal of Production Economics, 176,* 98–110.

Waller, M. A., & Fawcett, S. E. (2013). Data science, predictive analytics, and big data: A revolution that will transform supply chain design and management. *Journal of Business Logistics, 34*(2), 77–84.

Wang, Y., Kung, L., & Byrd, T. A. (2017). Big data analytics: Understanding its capabilities and potential benefits for healthcare organizations. *Technological Forecasting and Social Change, 126,* 3–13.

Whitmore, A., Agarwal, A., & Da Xu, L. (2015). The internet of things—A survey of topics and trends. *Information Systems Frontiers, 17*(2), 261–274.

Wu, Y. C. J., Huang, S. K., Goh, M., & Hsieh, Y. J. (2013). Global logistics management curriculum: Perspective from practitioners in Taiwan. *Supply Chain Management: An International Journal, 18*(4), 376–388.

Xu, L. D., He, W., & Li, S. (2014). Internet of things in industries: A survey. *IEEE Transactions on Industrial Informatics, 10*(4), 2233–2243.

Yin, R. K. (2012). *Applications of case study research* (3rd ed.). Thousand Oaks, USA: Sage Publications.

Yu, Q., & Wang, K. (2016). Applications of IoT in production logistics: Opportunities and challenges. *WIT Transactions on Engineering Sciences, 113,* 233–240.

Chapter 3
Mathematical Models for Sustainable Inventory and Production Plans for Component Remanufacturing Problems of OEM with Break-Even Period Determination

S. Malolan and M. Mathirajan

Abstract To make products and processes sustainable, Original Equipment Manufacturers (OEM) adopt remanufacturing practices. Component Remanufacturing (CR) problems arise for any OEM that produce durable products and acquires returns. Returns are of three categories: Commercial Returns, End-of-Use Returns and End-of-Life Returns for remanufacturing components, which are equivalent in all aspects as compared to the manufactured and/or purchased components. The final product assembled using manufactured and/or purchased and/or remanufactured components by OEM are also equivalent in all aspects to satisfy the demand. The remanufacturing of components from the return will be carried out by installing required capacities for both dismantling the returns and remanufacturing the various components (called together as Reverse Flow Capacities (RFC)). Moreover, the final products' demand can be satisfied by the same product and/or by Backordering (BO) and/or by Product Substitution (PS). With these problem characteristics of CR, this study focuses on three research problems. Accordingly, the *first problem* of focus is on *the "Multi-Products-CR (MP-CR) system with RFC&Break-Even Period problem"*, where the strategic decisions of RFC installations and the break-even period for the capital investments are determined. In *the second problem*, demand management by BO along with the real-life procurement restrictions are incorporated into the MP-CR System and *are termed as "MP-CR System with BO problem"*. Finally, the additional and important problem characteristics on PS mechanism are incorporated in *the third problem*, and this is the *"MP-CR System with BO&PS problem"*. A systematic 5-step methodology, in which the MP-CR system is modelled as an Integer Linear Programme, is proposed to address the first problem. For the second and third research problems, Integer Non-Linear Programming models are proposed. All the

S. Malolan
Department of Management Studies, National Institute of Technology Tiruchirappalli, 620 015 Tiruchirappalli, India
e-mail: malolan@nitt.edu

M. Mathirajan (✉)
Department of Management Studies, Indian Institute of Science, 560 012 Bangalore, India
e-mail: msdmathi@iisc.ac.in

U. Ramanathan and R. Ramanathan (eds.), *Sustainable Supply Chains: Strategies, Issues, and Models*, https://doi.org/10.1007/978-3-030-48876-5_3

proposed models determine the optimal inventory and production decisions for the MP-CR system. A suitable numerical example, based on data from a real-life Indian Automobile OEM, is developed for representing each of the three research problems and for demonstrating the workability of the proposed models. Finally, managerial insights are presented from the solutions obtained.

Keywords Remanufacturing · Break-even period · Inventory · Production plans · Integer programming

1 Introduction

In the modern era, waste management emphasizes on "value recovery". Today most manufacturing sectors, particularly the automotive industry, are increasingly encouraged to recover value from used products (Doppelt and Nelson 2001). The used products are called "returns". The most popular practice for value recovery from returns, by Original Equipment Manufacturers (OEM) in general, is "Remanufacturing". This is primarily because remanufacturing is an industrial process that restores items to "good-as-new conditions". Moreover, it has been established that remanufacturing is profitable (Maslennikova and Foley 2000; Lund and Hauser 2012; Sustainability Accounting Standards Board (2014)) and versatile to apply on a variety of products and their components (Lund 1998). In addition, remanufacturing has the social benefit of reducing landfill quantities (Li et al. 2017).

However, when an OEM incorporates remanufacturing practices, several operational challenges arise (Parkinson and Thompson 2003). Many studies have addressed different combinations of the operational challenges for problems, where the entire return is remanufactured. Such studies are generally categorized as "product remanufacturing". In product remanufacturing, the returns are examined for its remanufacturing capabilities. This is based on the number of components in the return that are worn-out and need replacement with new components to bring the return product to "good-as-new" working condition. The new components are those obtained from the conventional sources: component manufacturing and/or component purchasing. If too many components need replacement in the return product, then the entire return is disposed.

On the contrary, a more profitable new business proposal for any OEM is to remanufacturing components from the returns. This type of remanufacturing can be applied to both dismantled return products and return components for bringing back all the components to "good-as-new" condition. Such a process of remanufacturing components from returns is referred to as "Component Remanufacturing (CR)" in this study. This is possible as remanufacturing is cheaper than the conventional sources of component manufacturing and component purchasing (Shumon et al. 2011; Mondal and Mukherjee 2012; Jung et al. 2014). There are two salient differences between product remanufacturing and component remanufacturing:

1. In component remanufacturing, specific worn-out components alone can be disposed. In product remanufacturing, disposal is always at a whole return level.
2. A clear distinction of dismantling and assembly activates is possible in component remanufacturing, whereas the same is not the case in product remanufacturing (as usually remanufacturing and dismantling occur simultaneously).

In this study, end-to-end problems concerning the operational decisions and two strategic decisions for the incorporation of component remanufacturing, which is an additional source of components, by OEM are considered. Here, the operational decisions are the various production decisions and inventory decisions. The strategic decisions are (a) determination of ideal capacities to install for the various recovery operations and (b) break-even period for the capital investments made for these installations. Subsequently, the behaviour of the operational decisions is studied when "Backordering" and a "Proposed Product Substitution" concept is incorporated. The research problems considered in this study address the decisions for OEM that assemble discrete multiple products using multiple unique and common components. The required components for producing multiple products are obtained by either manufacturing them from raw materials or purchasing them in ready-to-use condition from external vendors or remanufacturing them in-house by the OEM from acquired returns.

With these backgrounds, the subsequent sections are organized in the following sequence. The detailed problem of focus along with assumptions is presented in the next section. The newness of the research problems considered in this study is established through a literature review in Sect. 3. Section 4 proposes mathematical model for each of the three research problems considered in this study. In Sect. 5, the real-life data collected, by observation from an Indian Automobile OEM, to represent each of the three research problems is presented. The detail on the validation/workability of the proposed mathematical models and managerial insights obtained from the solutions are also discussed in the sub-sections of Sect. 5. The study concludes in Sect. 6.

2 Problem Description

The existing production environment of most OEM assembles multiple products using multiple components. The multiple products can be different products or variants of the same product. In this study, "Multi-Products (MP)" refers to different products and their variants. All the MP are assembled in an existing single common capacitated assembly line. At present, the components used in assembly activities for producing the MP are obtained either from manufacturing and/or purchasing sources. The manufacturing sources produce components by performing manufacturing activities in-house on raw materials procured from external suppliers. The manufactured components are produced using existing dedicated component manufacturing line available for each component. The raw materials used in manufacturing of components can be common or unique among the various components. In addition to manufacturing components, at present, the components can be purchased in ready-to-use condition from external vendors. To exploit the scale of economics and to cater to demand fluctuations, OEM has dedicated inventories for all MP, components and raw materials. This existing production environment is called, in this study, as the "Multi-Product-Component Manufacturing (MP-CM) system".

In the existing MP-CM system, the incorporation of component remanufacturing would imply the additional source for components. The remanufacturing components from the remanufacturing source ensure that they are in "good-as-new condition" and equivalent to the current manufactured and/or purchased components. This incorporation of remanufactured components would involve the establishment of a reverse flow in addition to the existing forward flow. Particularly, in this study, to establish a reverse flow, as a first mechanism, acquisition of returns will be carried out. The returns are acquired by paying a monitory incentive to the customers, referred in literature as "Return Acquisition Price (RAP)". RAP would depend on the category of the return. In this study, three categories of returns are considered: Commercial returns, End-Of-Use (EOU) returns and End-Of-Life (EOL) returns. This categorization of returns is predominantly based on the usage time, although it could also be based on combinations of other characteristics which varies from product to product. In general, commercial returns have the least use time and EOL have the maximum usage time. Obviously, the RAP is maximum for the commercial returns and least for the EOL. Moreover, over the period, components are worn-out in most durable products. Therefore, the number of possible remanufacturable components that can be obtained from commercial returns are more as compared to EOU returns.

Once returns are acquired, they can be stored in dedicated return-inventories. The returns from these inventories are dismantled, on a common dismantling line for all return products, to obtain the return components. The non-remanufacturable return components are disposed. The remanufacturable return components are sent to an exclusive remanufacturing line where they are remanufactured. These remanufactured components are then stored in the same inventories where the manufactured and purchased components are stored. With these, for an OEM, there is a new production environment where components are obtained from three sources (manufacturing, purchasing and remanufacturing) and this is referred in this study as "Multi-Product-Component Remanufacturing (MP-CR)" system. A schematic diagram representing the MP-CR system, described here, is presented in Fig. 1.

In the MP-CR system considered in this study, we propose to address three integrated decision problems. The ***first*** problem addresses a strategic issue involving the determination of the optimal dismantling capacities and remanufacturing capacities, which needs to be setup. These dismantling capacities and remanufacturing capacities are together called, in this study, as "Reverse Flow Capacities (RFC)". The RFC are determined considering the capital cost for setting up RFC and the regular running operational costs involved in the planning horizon. The various operational costs considered are the cost for {assembling the products, manufacturing components, purchasing components, remanufacturing components, dismantling returns, procuring raw materials, storing products, storing components, storing

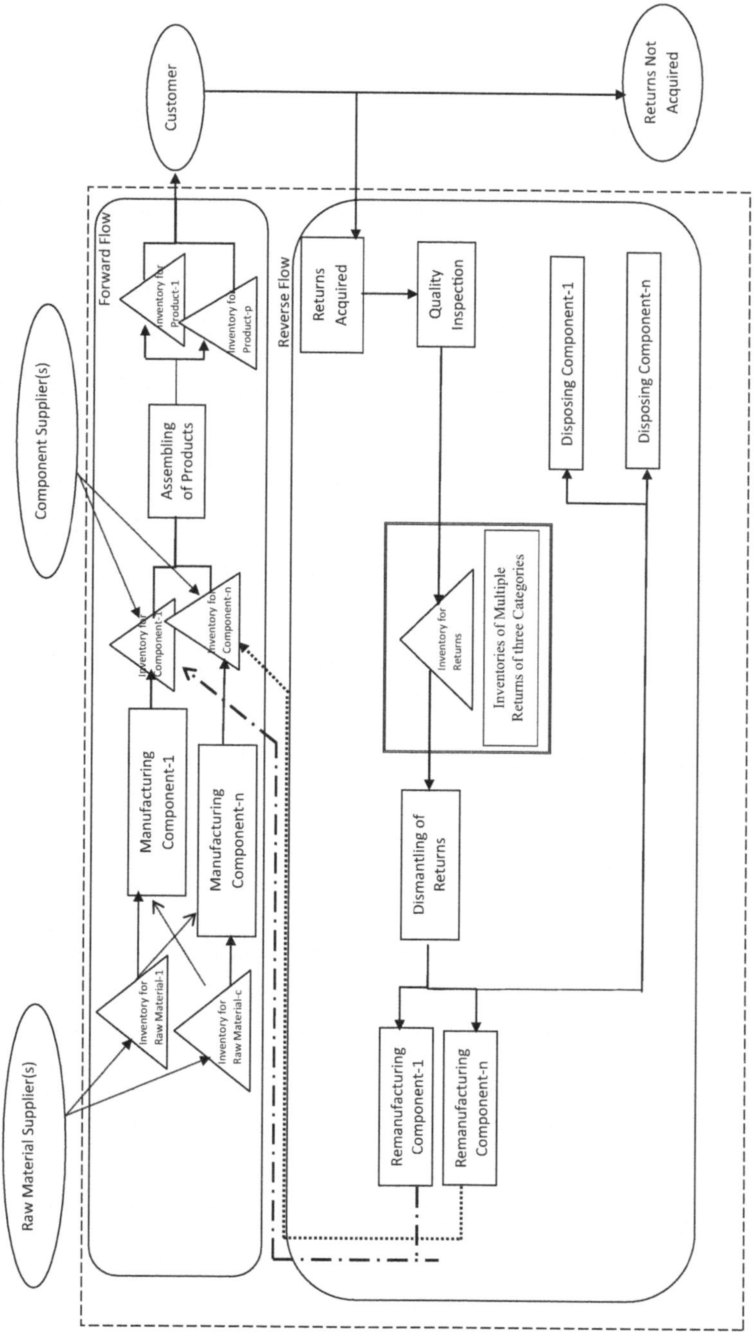

Fig. 1 A schematic representation of the MP-CR system considered in this study

returns, storing raw materials and disposing non-remanufacturable components}. This first research problem is referred as "MP-CR system with RFC&BEP" problem, in this study.

Once the ideal RFC is determined (along with break-even period to recover back the investment of the fixed cost in introducing the ideal RFC) for installation, the study extends the focus in the second problem. Accordingly, the ***second*** problem focuses on the operational concern of incorporating backordered demands for the product. This is an important aspect as many durable products, particularly the products related to the automobile sector, will be delivered against the orders placed. Such scenarios can be considered as backordered demands. The study refers to this second problem as "MP-CR system with Backordering (BO)" problem.

Finally, the ***third*** problem considered in this study is incorporating a proposed product substitution concept to the MP-CR system with BO problem. The proposed product substitution suggests the substitution of a low-priced product with a higher priced product, provided the higher priced product is made with maximum number of remanufactured components. Here, the emphasis is on the term "maximum" which means that not all components need to be utilized from the available remanufacturing component source only. This is because, some components could be single-use components which may get destroyed while dismantling the return. Therefore, by emphasizing "maximum", it is imposed that only such single-use components should be obtained from the conventional sources (that is from manufacturing and/or purchasing sources), and all other components must be from the remanufacturing source for a higher priced product to substitute a lower priced product. It is believed that this proposed product substitution will yield positive results for the OEM as it establishes an alternative method of satisfying demand. This final problem, in this study, is called the "MP-CR system with BO and Product Substitution (PS)" problem.

2.1 Assumptions for the MP-CR System

The following assumptions are considered while developing models to address each of the three problems described under the MP-CR system.

1. The demand for the multi-products is independent, deterministic, known and time varying.
2. The forward flow (manufacturing components, purchasing components, producing multiple products by assembling components) activities have fixed and known capacities.
3. All the multi-products are assembled on a single assembly line. Similarly, all the multi-returns are dismantled on a single dismantling line.
4. There are separate manufacturing/remanufacturing lines for all manufactured/remanufactured components.
5. The lead times/setup times for all the activities are assumed to be negligible (zero).

6. The cost of assembling any product and manufacturing any component is fixed and known.
7. The cost of dismantling any return across all three return categories is fixed and known and is significantly less than the cost of assembling that product.
8. The cost of remanufacturing any component obtained from any return category is fixed and known.
9. For all components, the respective purchasing cost, manufacturing cost (which is lesser than purchasing cost), the remanufacturing cost (which is lesser than manufacturing cost), is fixed and known.
10. The various manufactured components are manufactured using unique as well as common raw materials.
11. The procurement cost for all raw materials is fixed and known.
12. The inventory holding cost for the various inventories associated with products, components, raw materials and returns is fixed and known.
13. The return rate for the multi-products, representing a portion of the demand for that product in that period, across the three return categories is deterministic, know and fixed.
14. The return acquisition price for the multi-products across the three return categories is fixed and known.
15. There is no cost involved in checking and inspecting any of the multiple returns.
16. The remanufactured components are good-as-new and due to this, the multi-products which are assembled using them are good-as-new. Further, the remanufactured components are exclusively used only in the assembly of multi-products, and not for service/spare parts requirements.
17. The disposal rates for a component obtained from a return category are fixed and known. Furthermore, the disposal costs for a component obtained from any return category are fixed and known.
18. The disposal capacity, for all components, is infinite (un-capacitated).

3 Literature Review

Remanufacturing as a business practice has been in existence for over 6 decades (Parkinson and Thompson 2003). Thus, remanufacturing and closed-loop supply chain, in general, has attracted significant research interest and several operational challenges, both theoretical and practical, have been envisaged and addressed. The studies by Krikke et al. (2013), Souza (2013) and Govindan et al. (2015) have reviewed the earlier studies addressing various challenges in general. It is noted that the challenges have been addressed for *production* which can be continuous (Ervasti et al. 2016) or discrete, and for *products* which are durable, non-durable (Welle 2011; Zhang and Wen 2014) or *technology* (Rathore et al. 2011; Sarath et al. 2015). The earlier studies that have addressed discrete production of durable products can be grouped based on the values of the parameters considered as either stochastic (Sun et al. 2013; Francie et al. 2015; Mawandiya et al. 2018) or deterministic.

Furthermore, as suggested by Govindan et al. (2015) various studies on deterministic closed-loop supply chain, adopting remanufacturing in discrete production of durable products, can be broadly grouped based on the focus as Designing and Planning, Survey, Price and Coordination, Production Planning, Inventory Management, etc. For this study, only those published literature that have explored research problems focused on the Production Planning and Inventory Management are considered. Over the past, several issues pertaining to production planning and inventory management for remanufacturing have been addressed and one can refer Akcali and Cetinkaya (2011) for a detailed review in this space.

For the MP-CR system considered in this study, the OEM performs the remanufacturing in-house. Thus, among the literature that address discrete production of durable products, the ones where the OEM adopts in-house remanufacturing are only analysed in depth. However, the studies by Azadeh and Mazaheri (2013), Li et al. (2014a), Papachristos and Adamides (2014) and Zheng and Wu (2016) may act as a good starting point for readers interested in third-party remanufacturing. From the analysis of the literature, it is observed that though several research studies have focused on a manufacturing–remanufacturing production environment, most of the studies consider remanufacturing of products (Tang and Teunter 2006; Teunter et al. 2008; Zhang et al. 2011; Li et al. 2014a; Torkaman et al. 2017, 2018).

To the best of our knowledge, only five studies have considered remanufacturing components and have adopted them as an additional source of getting components for making final products. Accordingly, the study by Geyer and Van Wassenhove (2005) addresses a multi-product-component remanufacturing problem. In this specific study, they considered separate demands for the products made with new components and for the products made with new and remanufactured components. Here, the impact of two constraints on reusing end-of-life components in product manufacturing is addressed. The first constraint addresses the non-usability of all returned end-of-life returns for remanufacturing. The second constraint focuses on the lack of demand for the remanufactured product, when excess return products are available. For both the constraints, the study determines the (a) number of returned components for remanufacturing and for using in the products and (b) number of components to be disposed.

Kim et al. (2006) address a remanufacturing environment where the manufacturer obtains components, for producing a product. The components are obtained either by newly procuring from external suppliers or by refurbishing at a refurbishing/remanufacturing site by a sub-contractor. The study proposes an (0–1) integer linear programming model for the problem. The objective of the study is maximizing the total cost savings by optimally determining the number of components to be obtained from the two sources. Chen and Abrishami (2014) address a hybrid manufacturing–remanufacturing line where components required are manufactured or remanufactured to satisfy separate demands using shared resources. The remanufacturing operation is performed on the returns, which are of three categories. The objective is to make the optimal production plans by determining the number of components to be manufactured, the number of returns to be dismantled, the number of components to be remanufactured, the number of return product to be acquired and the inventory level of manufactured/remanufactured components and returns at the end of each period to reduce the total cost.

Jung et al. (2014) address the component remanufacturing problem at the manufacturing site. In this study two products are made from two distinct components which are purchased. Particularly, they develop inventory and production plans for a two-component deterministic remanufacturing disposal system. Here, the just-in-time inventory policy is considered for the serviceable item inventory. No other inventories are present. The returns obtained are dismantled by third parties to obtain the components. The components are then sent to the manufacturer for remanufacturing and are used to produce the product. The excess components required are purchased from external suppliers.

Zahraei and Teo (2018) focus on a two-stage system. Here, the upstream stage recovers components from used products. These components are supplied to the downstream stage where assembly is performed. The assembly also has components supplied from other external sources. The stages are setup with inventories to store returns, components obtained from upstream remanufacturing source, components obtained from external sources and the assembled products. Moreover, the upstream remanufacturing and downstream assembly operations are capacitated. For such a two-stage system, the study determines the various stock levels for the inventories to minimize cost, while preventing stock outs. The inventory level depends on the production variability, which in turn depends on the uncertainty in demand and arrival of returns. With these, this study presented a mathematical model for the two-stage system and studied the effect of different production smoothing options to mitigate the uncertainties.

The problem configurations considered in the above five closely related studies on component remanufacturing are summarized in Table 1. The last three rows in Table 1 give three different problem configurations considered in this study. From the analysis of the literature review as well as from Table 1 the research problem with three different problem configurations considered in this study has the following research gaps:

1. There is no literature that considers the production and inventory plans for multiple products in an integrated level when component remanufacturing is initiated by an OEM.
2. Very limited research literature considered the cost of acquiring the returns and no research has considered return categorization.
3. The determination of optimal reverse flow capacities (dismantling capacity for returns and remanufacturing capacities for components), when an OEM install component remanufacturing, is not addressed by any previous research study.
4. It appears that there is no previous literature in the area of component remanufacturing, which has "backordering" and "downward substitution" mechanisms for satisfying demand.

The main objective of this study is to fill all these research gaps.

Table 1 Summary of literature on component remanufacturing

Study	Production planning											
	# of products	# of components	Common components	Capacitated manufacturing	Capacitated remanufacturing	Capital cost for setting up capacities	Lead time	Disposal	Return categories	Return acquisition	Backordering	Substitution
Geyer and Van Wassenhove (2005)	Multiple	Multiple	Yes					X				
Kim et al. (2006)	Multiple	Multiple	Yes	X	X			X				
Chen and Abrishami (2014)	Multiple	Multiple	Yes	X	X		X			X		
Jung et al. (2014)	Two	Two	No					X				
Zahraei and Teo (2018)	Multiple	Multiple	Yes	X	X							
MP-CR system with RFC & BEP Problem	Multiple	Multiple	Yes	X	X	X		X	X	X		
MP-CR system with BO Problem	Multiple	Multiple	Yes	X	X			X	X	X	X	
MP-CR system with BO&PS Problem	Multiple	Multiple	Yes	X	X			X	X	X	X	X

Inventories considered				Objective	Solution methodology
Raw materials	Components	Products	Returns		
				Maximization of yields	Mathematical model
	X	X		Maximizing cost savings	Binary programming model
X	X		X	Minimizing cost	Mixed integer linear programming model

(continued)

Table 1 (continued)

Inventories considered				Objective	Solution methodology
Raw materials	Components	Products	Returns		
				Minimizing cost	Linear programming
	X	X	X	Minimizing cost	Mathematical model
X	X	X	X	Minimizing cost	Linear programming
X	X	X	X	Minimizing cost	Non-Linear programming
X	X	X	X	Minimizing cost	Non-Linear programming

4 Development of Mathematical Models for the Problems of the MP-CR Systems

To address the three problems of the MP-CR system, considered in this study, a mathematical modelling approach is adopted. To address the first problem of determining the RFC for the MP-CR system, a systematic methodology is proposed. In the proposed systematic methodology, the MP-CR system is modelled mathematically. The methodology uses the proposed mathematical model and determines (a) the optimal RFC (b) the optimal operational decisions, with respect to the optimal RFC obtained, pertaining to the operation costs and (c) the break-even period to cover the capital cost incurred to setup the required optimal RFC. For the second problem, the proposed mathematical model for the first problem is extended to additionally consider the costs of incorporating backordering (after the RFC are setup). However, for the second problem the objective is modified to maximize profit rather than minimize costs. For the final (third) problem, the model for the MP-CR system with BO problem is further extended to accommodate the PS and to capture the behaviour of the system in terms of operational characteristics and profitability. With these briefs, to address each of the three research problems considered in this study, the complete list of notations considered is presented first in the following sub-section.

4.1 Notations

All the notations used in the development of the mathematical models for each of the three problems under the MP-CR system are presented in this section.

Sets	
a	Number of periods in the planning horizon and range 1, 2, … *m*
b	Number of components used in product assembly and range 1, 2, …, *n*
c	Number of raw materials used in component manufacturing and range 1, 2, …, *s*
d	Number of Return categories (3 return categories, with return category number $d = 1$ – Commercial Returns $d = 2$ – End-Of-Use Returns $d = 3$ – End-Of-Life Returns)
e, f	Number of products and range 1, 2, …, *p*
Uncontrollable variables	
X_{ea}	Demand for product-e in period-a
AAC	Available assembly capacity
CAP_e	Cost to Assemble Product-e
SCP_e	(Inventory) Storing Cost for Product-e
UCP_{be}	Units of Component-b needed to assemble single unit of Product-e

(continued)

(continued)

Sets	
$MCAC_b$	Manufacturing Capacity Available for Component-b
$IRMCC_b$	Installed Remanufacturing Capacity for Component-b
CMC_b	Cost to Manufacture Component-b
CPC_b	Cost to Purchase Component-b
$CRMC_{bd}$	Cost to Remanufacture Component-b obtained from return category-d
CSC_b	(Inventory) Cost for Storing Component-b
$URawMMC_{cb}$	Units of Raw Material-c needed to Manufacture a single unit of Component-b
$CPRawM_c$	Cost for Procuring Raw Material-c
$CSRawM_c$	(Inventory) Cost for Storing Raw Material-c
$RRRP_{ed}$	Return Rate for Return Product-e of category-d
UR_{eda}	Units of Returns-e of category-d received in period-a
APR_{ed}	Acquisition Price for Return-e of category-d
SCR_{ed}	(Inventory) Storing Cost for Return-e of category-d
$IDisCR$	Installed Dismantling Capacity for Returns
$CDisR_e$	Cost to Dismantling Return-e
$RDRC_{bd}$	Rate of Disposal for Return Component-b obtained from category-d
$CDRC_b$	Cost to Dispose Return Component-b
IQP_{e0}	Inventory Quantity for Product-e that is carried from the last period of previous planning horizon
IQC_{b0}	Inventory Quantity for Component-b that is carried from the last period of previous planning horizon
$IQRawM_{c0}$	Inventory Quantity for Raw Material-c that is carried from the last period of previous planning horizon
IQR_{ed0}	Inventory Quantity for Return-e of category-d that is carried from the last period of the previous planning horizon
$CCURMCC_b$	Capital Cost per Unit of Remanufacturing Capacity installed for Component-b
$CCUDisC$	Capital Cost per unit of Dismantling Capacity installed
PSP_e	Price for Selling Product-e
CBP_e	Cost for Backordering Product-e
BQP_{e0}	Backordered Quantity for Product-e in the last period of the previous planning horizon
$UPLC_b$	Upper Purchase Limit for Component-b
$LPLC_b$	Lower Purchase Limit for component-b (if it is purchased)
$UPLRawM_c$	Upper Procurement Limit for Raw Material-c
$LPLRawM_c$	Lower Procurement Limit for Raw Material-c (if it is purchased)
$CISP_{ef}$	Cost Incurred for Substituting Product-e with product-f ($f < e$)
Decision variables	
AQP_{ea}	Assembly Quantity for product-e in period-a/ Assembly Quantity for (Type-One) Product-e in period-a

(continued)

(continued)

Sets	
BQP_{ea}	Backordered Quantity for Product-e in period-a
IQP_{ea}	Inventory Quantity for Product-e at the end of period-a
MQC_{ba}	Manufacture Quantity for Component-b in period-a
PQC_{ba}	Purchase Quantity for Component-b in period-a
$RMQC_{bda}$	Remanufactured Quantity for Component-b from category-d in period-a
$DQRC_{bda}$	Disposal Quantity for Return Component-b from category-d in period-a
IQC_{ba}	Inventory Quantity for Component-b at the end of period-a
$PQRawM_{ca}$	Purchase Quantity for Raw Material-c in period-a
$IQRawM_{ca}$	Inventory Quantity for Raw Material-c at the end of period-a
$DisQR_{eda}$	Dismantling Quantity for Return-e of category-d in period-a
IQR_{eda}	Inventory Quantity for Return-e of category-d at the end of period-a
$EIRMCC_{ba}$	Excess Installed Remanufacturing Capacity for Component-b in period-a
EIDisC_a	Excess Installed Dismantling Capacity in period-a
$TEIRMC_b$	Total Excess Installed Remanufacturing Capacity for component-b in the entire planning horizon
$AQTTP_{ea}$	Assembly Quantity for Type-Two Product-e in period-a
SQP_{efa}	Substitution Quantity for product-e by Product-f in period-a
$IQRMC_{ba}$	Inventory Quantity for Remanufactured Component-b in period-a

4.2 *Methodology to Determine Ideal Reverse Flow Capacities (RFC) for the MP-CR System*

For introducing an efficient MP-CR system in the existing MP-CM system, it is imperative to determine the ideal capacities for the reverse flow activities for introducing the required remanufacturing process. It is well known in the operational management world that capacity installation is an important strategic decision as significant capital investment is involved. Therefore, in this section, the following step-by-step methodology is proposed to simultaneously determine (a) the ideal Reverse Flow Capacities (RFC) to install and (b) the corresponding Break-Even Period (BEP), where the capital investment made for installing the RFC facility is gained by an OEM through the cost-benefit of introducing the remanufacturing of components, for the determined ideal RFC:

Step 1: A mathematical model is developed for the proposed MP-CR system (Sect. 4.2.1). The objective of the proposed model is to determine the optimal total cost for the system. The costs considered are those related to the operational decisions: inventory and production decisions, in the planning horizon.

Step 2: For the given numerical problem, the proposed mathematical model is generated and solved for various system scenarios. Each of the system scenarios depicts a possible RFC installation, which is generated by varying simultaneously and systematically the capacity installed for (i) dismantling of returns and (ii) remanufacturing of components from the returns. The various system scenarios are generated for the given numerical problem by assuming the RFC installations in terms of the percentage capacity of (a) assembly capacity available and (b) manufacturing capacity available for the respective components. The respective total cost is determined for each of the system scenarios.

Step 3: A relation, in the form of graph, is developed between the obtained total cost for each of the system scenarios and the various RFC installations. From this graph, the point where the total cost becomes least is the Total optimal Cost for the MP-CR system (called as TC-MP-CR system) for the given numerical problem, where the optimal RFC occurs corresponding to both capacities for dismantling and remanufacturing operations.

Step 4: To determine the Break-even Period (BEP) for recovering the capital cost to be invested for installing the RFC by gaining the cost difference between manufacturing and remanufacturing components, a comparison with the existing MP-CM system is needed. To do this the total optimal cost for the MP-CM system (called as TC-MP-CM system) for the given numerical problem needs to be determined. To obtain TC-MP-CM, we need to model the existing MP-CM system. Accordingly, in this study, the mathematical model proposed for the newly proposed MP-CR system is appropriately modified by removing the returns and remanufacturing components related variables and constraints. Accordingly the model for MP-CM system is developed and presented in Sect. 4.2.2.

Step 5: Using the TC-MP-CM and TC-MP-CR the BEP is determined by directly substituting in the following BEP formula (Mathirajan et al. 2011)

$$\mathbf{BEP} = \frac{\text{Capital cost incurred for setting up RFC}}{(\text{TC} - \text{MP} - \text{CM}) - (\text{TC} - \text{MP} - \text{CR})} \times \text{number of periods in the planning horizon} \tag{A}$$

4.2.1 Proposed Mathematical Model for MP-CR System to Determine Optimal RFC

To develop a mathematical model for determining optimal RFC the following assumptions are made, in addition to the assumptions presented in Sect. 2.1 related to the MP-CR system:

1. Demand in a period must be satisfied in the same period by the same product (that is, backordering and substitution of the multi-products are not permitted).

2. The reverse flow capacities (capacities for dismantling and remanufacturing) are not yet installed and are unknown. However, the remanufacturing capacity that can be setup for a component is a fixed percentage (same for all components) of that components' manufacturing capacity. Similarly, the dismantling capacity that can be setup for a return is a percentage proportion of the assembly capacity of the corresponding product.
3. A capital cost is incurred for setting up the reverse flow capacities. This capital cost is assumed to be known [which is assumed to be a percentage proportion of the per unit cost of the forward flow (assembly and manufacturing) operations, respectively].
4. There is no quantity restriction for purchasing either raw materials or components.

With the above assumptions, the objective of the proposed model is to minimize the total cost. The total cost is the summation of costs incurred for operational decisions (that is, inventory and production decisions) of the MP-CR system and the capital costs incurred for setting up RFC. Accordingly, the verbal formulation of this objective function is, minimize {(Assembly + Inventory) costs for multi-product + (Manufacturing + Purchasing + Remanufacturing + Inventory) costs for various components + (Acquisition + Dismantling + Inventory) costs for multiple returns + (Return disposal costs for various components) + (Purchase + Inventory) costs for various raw materials + (The capital cost incurred in installing excess reverse flow capacities)}.

As the objective is to minimize the total cost, with respect to the capital investment decisions, it would be to minimize the capital cost lost by excessive installation of RFC. As the RFC has two capacities, related to dismantling and remanufacturing, it is computed as the summation of capital costs invested for the excessive installation of both. Accordingly, the capital cost for excessive dismantling capacities installed for the planning horizon is the summation multiplication of (the excess dismantling capacity installed in each of the planning period horizon and the cost involved in installing per unit of the dismantling capacity). For the remanufacturing operation, as capacities are installed for each individual component, it is the total excess remanufacturing capacity installed for each component in the entire planning horizon multiplied by the capital cost involved in installing per unit of remanufacturing capacity for that component.

With the above verbal explanation on the objective function for the proposed new MP-CR systems, the required objective function is mathematically represented as given below in (1-A):

$$\text{Minimize}: \left\{ \sum_{a=1}^{m} \sum_{e=1}^{p} AQP_{ea} * CAP_e + \sum_{a=1}^{m} \sum_{e=1}^{p} IQP_{ea} * SCP_e \right\}$$

$$+ \left\{ \sum_{a=1}^{m} \sum_{b=1}^{n} MQC_{ba} * CMC_b \right.$$

$$+\sum_{a=1}^{m}\sum_{b=1}^{n} PQC_{ba} * CPC_b + \sum_{a=1}^{m}\sum_{b=1}^{n}\sum_{d=1}^{3} RMQC_{bda} * CRMC_{bd} + \sum_{a=1}^{m}\sum_{b=1}^{n} IQC_{ba} * CSC_b \Bigg\}$$

$$+\sum_{a=1}^{m}\sum_{e=1}^{p}\sum_{d=1}^{3} UR_{eda} * APR_{ed} + \sum_{a=1}^{m}\sum_{e=1}^{p}\sum_{d=1}^{3} DisQR_{eda} * CDisR_e$$

$$+\sum_{a=1}^{m}\sum_{e=1}^{p}\sum_{d=1}^{3} IQR_{eda} * SCR_{ed} + \left\{\sum_{a=1}^{m}\sum_{b=1}^{n}\sum_{d=1}^{3} DQRC_{bda} * CDRC_b\right\}$$

$$+\left\{\sum_{a=1}^{m}\sum_{c=1}^{s} PQRawM_{ca} * CPRawM_c + \sum_{a=1}^{m}\sum_{c=1}^{s} IQRawM_{ca} * CSRawM_c\right\}$$

$$+\left\{\sum_{a=1}^{m} EIDisC_a * CCUDisC + \sum_{b=1}^{n} TEIRMCC_b * CCURMCC_b\right\} \quad (1\text{-A})$$

To optimize the above objective function of the MP-CR system, there are various operational level requirements that must be met. These are captured in the form of constraints, and the same are given in constraints (2)–(20). Furthermore, these constraints are grouped and presented related to products (constraints 2–5), components (constraints 6 and 7), raw materials (constraint 8), returns (constraints 9–16), and excess RFC installation computations (constraints 17–19). Finally, the constraint number (20) is the non-negativity requirement.

Products Demand Constraint: The availability on the number of each multi-products (assembled in that period and/or stored in the product inventory) to be greater than the demand for each of the multi-products across all periods in the planning horizon is established in this constraint. Furthermore, for the first period of the current planning horizon, the inventory for the product-e carried from last period of the previous planning horizon (IQP_{e0}) is assumed to be "zero".

$$AQP_{ea} + IQP_{ea-1} \geq X_{ea} \quad \forall e \in [1, \ldots, p] \text{ and } a \in [1, \ldots, m] \quad (2)$$

Product Ending Inventor Constraint: The inventory quantity of product-e that is carried from one period to another is computed as the difference between the available quantity for product-e in a period and the demand for that product-e in the same period. In any period, the available quantity for product-e is the summation of inventory quantity of product-e carried from the previous period and the assembly quantity for product-e in the current period.

$$IQP_{ea} = AQP_{ea} + IQP_{ea-1} - X_{ea} \quad \forall e \in [1, \ldots, p] \text{ and } a \in [1, \ldots, m] \quad (3)$$

Products Assembly Capacity Constraint: This makes sure that sum of all assembly quantities for all products in a period is less than the available assembly capacity.

$$AAC \geq \sum_{a=1}^{m}\sum_{e=1}^{p} AQP_{ea} \quad (4)$$

Assembly Flow Balance Constraint: To assemble products, the components must be available. Therefore, in each period of the planning horizon the available quantity for each component (used in the multi-product assembly) must be greater than all the multi-products assembled using that component. For any period, availability of a component is based on whether it is present in the inventory carried from the previous period, and/or it is obtained from any of the three sources: manufacture, remanufacture and purchase in that period.

$$MQC_{ba} + PQC_{ba} + IQC_{ba-1} + \sum_{d=1}^{3} (RMQC_{bda}) \geq \sum_{e=1}^{p} (AQP_{ea} * UCP_{be}) \quad \forall b \in [1, \ldots, n], \text{and } a \in [1, \ldots, m] \tag{5}$$

Component Manufacturing Capacity Constraint: This constraint ensures that the total manufactured quantity for a component does not exceed the available manufacturing capacity for that component in any given period.

$$MCAC_b \geq MQC_{ba} \qquad \forall b \in [1, \ldots, n] \text{ and } a \in [1, \ldots, m] \tag{6}$$

Component Manufacturing Flow Balance Constraint: Similar to constraint (5), here the availability of raw materials to manufacture a particular component is enforced. Here, a raw material is available if it is carried from the previous period inventory or is procured from the external suppliers in the current period.

$$PQRawM_{ca} + IQRawM_{ca-1} \geq \left(\sum_{b=1}^{n} MQC_{ba} * URawMMC_{cb} \right) \quad \forall c \in [1, \ldots, s] \text{ and } a \in [1, \ldots, m] \tag{7}$$

Raw Materials Ending Inventory Constraint: The inventory quantity for a raw material is the quantity of raw materials procured in a period plus the inventory quantity of raw materials carried from the previous period differenced with the quantity of raw materials consumed in the manufacturing of components in the current period. In addition, for the first period, the quantity carried from the last period of the previous planning horizon ($IQRawM_{r0}$) is considered as "zero".

$$IQRawM_{ca} = PQRawM_{ca} + IQRawM_{ca-1} - \sum_{b=1}^{n} (MQC_{ba} * URawMMC_{cb}) \quad \forall c \in [1, \ldots, s] \text{ and } a \in [1, \ldots, m] \tag{8}$$

Component Ending Inventory Constraint: This constraint acts as the bridge that links the forward and reverse flows of the MP-CR system. In a period, a component's inventory is replenished by the manufacturing, purchasing, remanufacturing activities or by the quantity carried from the previous period. The depletion of the inventory is the quantity of components used in the assembly of the products in the same period. The difference between the replenishment and depletion quantities gives the inventory quantity of component stored and carried to the next period.

$$IQC_{ba} = MQC_{ba} + PQC_{ba} + IQC_{ba-1} + \sum_{d=1}^{3} (RMQC_{bda}) - \left(\sum_{e=1}^{p} UCP_{be} * AQP_{ea} \right) \quad \forall b \in [1, \ldots, n],\ \text{and}\, a \in [1, \ldots, m] \tag{9}$$

Component Remanufacturing Capacity Constraint: The total quantity of remanufactured components across all three return categories cannot exceed the installed capacity for remanufacturing.

$$IRMCC_b \geq \sum_{d=1}^{3} RMQC_{bda} \qquad \forall b \in [1, \ldots, n]\, \text{and}\, a \in [1, \ldots, m] \tag{10}$$

Components Remanufacturing Flow Balance Constraint: The quantity of components remanufactured in a period must be equal to the number of remanufacturable components obtained from the dismantled returns. This is an equality constraint as inventories are available only for returns and not for dismantled return components.

$$RMQC_{bda} = \left(\sum_{e=1}^{p} DisQR_{eda} * UCP_{be} \right) - DQRC_{bda} \quad \forall b \in [1, \ldots, n],\, d \in [1, 2, 3]\, \text{and}\, a \in [1, \ldots, m] \tag{11}$$

Returns Dismantling Capacity Constraint: This constraint assures that the dismantled quantity of returns in a period cannot exceed the installed dismantling capacity.

$$IDisCR \geq \sum_{e=1}^{p} \sum_{d=1}^{3} DisQR_{eda} \qquad \forall a \in [1, \ldots, m] \tag{12}$$

Returns Dismantling Flow Balance Constraint: The dismantled quantity of returns in a period can be utmost the total quantity available for dismantling. The total quantity available for dismantling is the summation of returns acquired in a period and those carried from the previous period.

$$UR_{eda} + IQR_{eda-1} \geq DisQR_{eda} \quad \forall e \in \left[1, \ldots, p\right],\, d \in [1, 2, 3]\, \text{and}\, a \in [1, \ldots, m] \tag{13}$$

Return Rate Constraint: The quantity of returns acquired in a period for a product is the multiplication of return rate for the product and the demand for that product in that period (based on the assumption 13, presented in Sect. 2.1). Nevertheless, any deterministic return rate computation technique can be adopted to obtain the quantity.

$$UR_{eda} = RRRP_{ed} * X_{ea} \qquad \forall e \in [1, \ldots, p], d \in [1, 2, 3] \text{ and } a \in [1, \ldots, m] \tag{14}$$

Returns Ending Inventory Constraint: Based on the units of returns received in a period, the inventory quantity carried from the previous period and the dismantle quantity in the current period, inventory quantity for a return is computed as given below.

$$IQR_{eda} = IQR_{eda-1} + UR_{eda} - DisQR_{eda} \quad \forall e \in [1, \ldots, p], d \in [1, 2, 3] \text{ and } a \in [1, \ldots, m] \tag{15}$$

Return Component Disposal Rate Constraint: The fact that all return components cannot be remanufactured is established in this constraint. Here, the disposal rate specifies the minimum disposal requirement for a component obtained from a return category. Thus, this constraint is "greater than or equal to" one, imposing a minimum disposal. Therefore, it is possible for the system to dismantle returns just to obtain a few components and dispose the rest if need be. Accordingly, for each period of the planning horizon the minimum disposal quantity is determined as the multiplication of the disposal rate for a component obtained from a return category and the total number of components obtained from the same return category,

$$DQRC_{bda} \geq \left(\sum_{e=1}^{p} DisQR_{eda} * UCP_{de} \right) * RDRC_{bd} \quad \forall b \in [1, \ldots, n], d \in [1, 2, 3] \text{ and } a \in [1, \ldots, m] \tag{16}$$

Excess Installed Remanufacturing Capacity Constraint: The excess installed remanufacturing capacity for a component is the difference between the installed remanufacturing capacity for a component and the total quantity of remanufactured components, across all three return categories, obtained in that period.

$$EIRMCC_{ba} = IRMCC_{b} - \sum_{d=1}^{3} RMQC_{bda} \quad \forall b \in [1, \ldots, n] \text{ and } a \in [1, \ldots, m] \tag{17}$$

Total Excess Installed Remanufacturing Capacity Constraint: The excess installed remanufacturing capacity for every component in each period (computed in 17) when summed up over the entire planning horizon is the total excess remanufacturing capacity installed for component-b.

$$TEIRMCC_{b} = \sum_{a=1}^{m} EIRMCC_{ba} \qquad \forall b \in [1, \ldots, n] \tag{18}$$

Excess Installed Dismantling Capacity Constraint: The unutilized dismantling capacity (across all return categories in all periods) is the excess dismantling capacity installed in a period.

$$EIDisC_a = IDisCR - \sum_{e=1}^{p}\sum_{d=1}^{3} DisQR_{eda} \forall\, a \in [1, \ldots, m] \tag{19}$$

Integer and Non-Negativity Constraint: As all the decisions of the model need to be integer, we enforce it through this constraint.

$$\begin{aligned} &AQP_{ea}, IQP_{ea}, MQC_{ba}, PQC_{ba}, RMQC_{bda}, IQC_{ba}, UR_{eda}, DisQR_{eda}, IQR_{eda}, PQRawM_{ca}, \\ &IQRawM_{ca}, DQRC_{bda}, EIRMCC_{ba}, TEIRMCC_b, EIDisC_a \geq 0 \text{ and integers} \\ &\forall a \in [1, \ldots, m], b \in [1, \ldots, n], c \in [1, \ldots, s], d \in [1, 2, 3] \text{ and } e \in [1, \ldots, p] \end{aligned} \tag{20}$$

The proposed mathematical model for MP-CR system, considered in this study, has the objective function and all the constraints as linear, and all the decision variables are integers. Therefore, the proposed model is referred as Integer Linear Programming (ILP) model.

4.2.2 Proposed Mathematical Model for the MP-CM System

The MP-CM system follows the major processes of (a) procuring raw materials to manufacture components, (b) purchasing components, which are either cost wise advantageous instead of manufacturing in-house or not possible to manufacture and (c) assembling the components (obtained from manufacturing and purchasing) to obtain the final products. That is, the products are assembled using components obtained from conventional sources of in-house manufacturing or external purchase from suppliers. The raw materials needed for the in-house manufacturing operations are purchased from out-sourced suppliers. There are no returns in the system, and hence there is no remanufacturing source. Therefore, to formulate the mathematical model for the MP-CM system, the proposed mathematical model for the MP-CR system is condensed by removing all the decision variables, constraints and cost-related details related to returns and remanufacture of components. For the sake of brevity, the complete proposed mathematical model for the existing MP-CM system (referred as Baseline-ILP (B-ILP) model) is presented in the Annexure 1.

4.3 Development of Mathematical Model for the MP-CR System with Backordering (BO)

The second research problem defined in this study related to consideration of back-ordering in the MP-CR system, which gives the system flexibility to satisfy the fluctuations in demand. The various assumptions, in addition to those in Sect. 2.1, for modelling the MP-CR system with BO problem are as follows:

1. Backordering of products is permitted. However, product substitution is not permitted.
2. The RFC are already setup at predetermined ideal capacity levels and the break-even period for the capital cost for its setup has already been reached.
3. There are quantity restrictions for the various purchased components and procured raw materials.

The presence of backordering will skew the results to always backorder products, as it is the cheapest. However, this is not practically allowable. Therefore, for overcoming this effect of backordering a profit maximization objective is considered. In addition, there are some constraints that need to satisfy with respect to "If-Condition" for incorporating the backordering strategy. With these backgrounds, the concept of backordering is incorporated in the proposed model presented here for MP-CR system to obtain various operational decisions. Particularly, the model developed to address the second problem would get into an Integer Non-Linear Program (INLP) with maximization objective. Accordingly, the objective function for the second research problem is

$$\text{Maximize Total Profit} = \text{Total Revenue} - \text{Total Cost} \tag{1-B}$$

The total revenue for the MP-CR system with BO problem is the revenue from selling the products (selling price of a product multiplied with total number of products sold at that selling price). That is, the total number of products sold at a selling price in a period is the summation of demand in that period and the number of that product backordered in the previous period, differenced with the number of that product backordered in that period. From there, the total revenue is mathematically represented in equation (C).

$$\text{Total Revenue} = \sum_{a=1}^{m} \sum_{e=1}^{p} (X_{ea} + BQP_{ea-1} - BQP_{ea}) * PSP_e \tag{C}$$

The total cost for the MP-CR system with BO problem is the sum of cost obtained using (1-A) and the cost for backordering products as given in (D)

$$\begin{aligned}
\text{Total Cost} = & \left\{ \sum_{a=1}^{m} \sum_{e=1}^{p} AQP_{ea} * CAP_e + \sum_{a=1}^{m} \sum_{e=1}^{p} IQP_{ea} * SCP_e \right\} \\
& + \left\{ \sum_{a=1}^{m} \sum_{b=1}^{n} MQC_{ba} * CMC_b \right. \\
& + \sum_{a=1}^{m} \sum_{b=1}^{n} PQC_{ba} * CPC_b
\end{aligned}$$

$$+\sum_{a=1}^{m}\sum_{b=1}^{n}\sum_{d=1}^{3} RMQC_{bda} * CRMC_{bd}$$

$$+\sum_{a=1}^{m}\sum_{b=1}^{n} IQC_{ba} * CSC_{b}\}$$

$$+\sum_{a=1}^{m}\sum_{e=1}^{p}\sum_{d=1}^{3} UR_{eda} * APR_{ed}$$

$$+\sum_{a=1}^{m}\sum_{e=1}^{p}\sum_{d=1}^{3} DisQR_{eda} * CDisR_{e}$$

$$\left.+\sum_{a=1}^{m}\sum_{e=1}^{p}\sum_{d=1}^{3} IQR_{eda} * SCR_{ed}\right\}$$

$$+\left\{\sum_{a=1}^{m}\sum_{b=1}^{n}\sum_{d=1}^{3} DQRC_{bda} * CDRC_{b}\right\}$$

$$+\left\{\sum_{a=1}^{m}\sum_{c=1}^{s} PQRawM_{ca} * CPRawM_{c}\right.$$

$$\left.+\sum_{a=1}^{m}\sum_{c=1}^{s} IQRawM_{ca} * CSRawM_{c}\right\}$$

$$+\left\{\sum_{a=1}^{m}\sum_{e=1}^{p} \boldsymbol{BQP_{ea} * CBP_{e}}\right\} \tag{D}$$

Like the ILP model in Sect. 4.2.1, the INLP model is also operationally constrained. Most of the constraints for the INLP model are same as the ILP model. However, there are some modifications to the existing constraints, and some additional constraints pertaining to the BO are required to represent the MP-CR system with BO. These are discussed as follows:

4.3.1 Modified Constraints

Modified Product Demand Constraint: A product's total demand in a period is given as the demand in that period plus the quantity backordered from the previous period. This total demand needs to be met either from current assembly or from the inventory carried from the previous period or must be backordered. For representing this, the constraint (2) is modified as follows:

$$X_{ea} + BQP_{ea-1} \leq AQP_{ea} + IQP_{ea-1} + BQP_{ea}$$
$$\forall e \in [1, \ldots, p] \text{ and } a \in [1, \ldots, m] \tag{2}$$

Modified Product Ending Inventory Level Constraints: Based on the quantity of product available to satisfy the total demand for a product, there are two cases for the ending inventory. Based on this, the product inventory constraint (3) is modified to (3A) and (3B) as follows:

Case 1: ***The demand is lesser than or equal to the quantity of product available in a period***: In this case, the demand is less than the quantity of product available for satisfying it. Therefore, there is a possibility to store products in inventory. Accordingly, the inventory quantity is the difference between the quantity of product available and the total demand for that product in that period.

$$\begin{gathered} IQP_{ea} = AQP_{ea} + IQP_{ea-1} - (X_{ea} + BQP_{ea-1}) \\ \text{if } AQP_{ea} + IQP_{ea-1} \geq X_{ea} + BQP_{ea-1} \\ \forall e \in [1, \ldots, p] \text{ and } a \in [1, \ldots, m] \end{gathered} \tag{3A}$$

Case 2: ***The demand is greater than the quantity of product available***: Due to large demand, there is no excess product available to store in the inventory and with that the ending inventory is set to zero.

$$\begin{gathered} IQP_{ea} = o \\ \text{if } AQP_{ea} + IQP_{ea-1} < X_{ea} + BQP_{ea-1} \\ \forall e \in [1, \ldots, p] \text{ and } a \in [1, \ldots, m] \end{gathered} \tag{3B}$$

Modified Integer and Non-Negativity Constraints: The new decision variable "BQP_{ea}" is added and this constraint makes all the decision variables in the model as integers.

$$\begin{gathered} AQP_{ea}, IQP_{ea}, MQC_{ba}, PQC_{ba}, RMQC_{bda}, IQC_{ba}, UR_{eda}, DisQR_{eda}, IQR_{eda}, PQRawM_{ca}, \\ IQRawM_{ca}, DQRC_{bda}, \boldsymbol{BQP_{ea}} \geq 0 \text{ and integers} \\ \forall e \in [1, \ldots, p], b \in [1, \ldots, n], c \in [1, \ldots, s], d \in [1, 2, 3] \text{ and } a \in [1, \ldots, m] \end{gathered} \tag{20}$$

4.3.2 Additional Constraints

Product Backordering Constraints: The two cases specified for the product inventory is applicable to the backordering constraints. Accordingly, constraints (21A) and (21B) are as follows:

Case 1: ***The quantity of product available is greater than or equal to the total demand in that period***: The backorder quantity for the product is zero, as sufficient availability of products to satisfy the demand.

$$\begin{gathered} BQP_{ea} = 0 \\ \text{if } AQP_{ea} + IQP_{ea-1} \geq X_{ea} + BQP_{ea-1} \\ \forall e \in [1, \ldots, p] \text{ and } a \in [1, \ldots, m] \end{gathered} \tag{21A}$$

Case 2: ***The quantity of product available is less than the total demand in that period:*** Due to insufficient availability of the product, a portion of the total demand must be backordered. The backordered quantity is the difference between the total demand in a period and the total quantity of product available in that period.

$$\begin{array}{c} BQP_{ea} = X_{ea} + BQP_{ea-1} - (AQP_{ea} + IQP_{ea-1}) \\ \text{if } AQP_{ea} + IQP_{ea-1} < X_{ea} + BQP_{ea-1} \\ \forall e \in \left[1, \ldots, p\right] \text{ and } a \in [1, \ldots, m] \end{array} \tag{21B}$$

Procurement Restrictions Constraints: To represent the procurement constraints of the MP-CR system with BO problem, the constraints (22A), (22B), (22C) and (22D) are added to impose the limits on the quantity of items to be obtained from external sources. Accordingly, (22A) and (22B) enforce the lower limit on the procurement quantities for components and raw materials. These two constraints are deliberately made to have "IF conditions" as the enforcements need to kick-in only if the item needs to be obtained from the external sources. If they are not obtained, the lower limits are zero.

$$PQC_{ba} \geq LPLC_b \quad \text{if} \quad PQC_{ba} > 0 \qquad \forall b \in [1, \ldots, n] \text{ and } a \in [1, \ldots, m] \tag{22A}$$

$$PQRawM_{ca} \geq LPLRawM_c \quad \text{if} \quad PQRawM_{ca} > 0 \quad \forall c \in [1, \ldots, s] \text{ and } a \in [1, \ldots, m] \tag{22B}$$

Similarly, (22C) and (22D) enforce the upper limits on the procurement quantities for components and raw materials, respectively.

$$PQC_{ba} \leq UPLC_b \qquad \forall b \in [1, \ldots, n] \text{ and } a \in [1, \ldots, m] \tag{22C}$$

$$PQRawM_{ca} \leq UPLRawM_c \quad \forall c \in [1, \ldots, s] \text{ and } a \in [1, \ldots, m] \tag{22D}$$

4.3.3 Removed Constraints

To complete the INLP model for MP-CR system with BO problem, the constraints (17), (18) and (19) are removed from the ILP model presented in Sect. 4.2.1. In addition to the modified objective function, modified constraints, additional constraints, and removed constraints presented here, all other constraints pertaining to the MP-CR system presented in Sect. 4.2.1 are required for MP-CR system with BO problem. This completes the proposed INLP model for the MP-CR system with BO problem.

4.4 Development of Mathematical Model for the MP-CR System with BO&PS Problem

The third and final problem of this study considers the inclusion of the proposed Product Substitution (PS) mechanism for the MP-CR system with BO problem. The proposed PS emphasizes on assembling high-priced products with maximum remanufactured components and using such products for substitution of demand for lower priced products. Therefore, every product can be of two types, the first is the assembled type-one product (called as AT1P), and the second is the assembled type-two product (called as AT2P). Here, the AT2P is assembled using maximum remanufactured components. TheAT2P is same in all aspects as the AT1P product, as the remanufactured components are "good-as-new". Therefore, a downward substitution in which a higher priced product satisfies the demand for a lower price product is logically acceptable. It is noted that without loss of generality any criteria, such as price, usability, life, and preference, or their combinations can be used to determine which product substitutes another product. In this section, the model presented in Sect. 4.3, for MP-CR system with BO problem, is appropriately extended to additionally consider the requirement for PS. Therefore, the proposed model in this section is referred to as the Extended-INLP (E-INLP) model for MP-CR system with BO&PS problem. To propose the required model, the first assumption mentioned in Sect. 4.3 is modified as follows:

1. Backordering of products is permitted. Moreover, a high-priced AT2P can be used to substitute and satisfy the demand of all its low-priced products by incurring a pre-specified substitution cost.

Furthermore, to complete the development of the proposed model for MP-CR system with BO&PS problem, only the required modified objective function, modified constraints and the required additional constraints are discussed as follows.

4.4.1 Modified Objective Function

The objective function of the MP-CR system with BO&PS problem must additionally consider the cost for the substitution, particularly in total cost Eq. (D). However, the total revenue equation (C) defined in the proposed INLP Model for MP-CR system with BO problem is not changed. The justification for this is that the additional sale of all higher price products, that is used in substitution, is still selling at the same price as the lower price product. Therefore, it is more appropriate to captured this by modifying the quantity of product that is backordered (which will be discussed later when the backordering constraint is modified). However, for easy readability, revenue part of the objective function (C) is re-iterated here as well.

$$\text{Total Revenue} = \sum_{a=1}^{m}\sum_{e=1}^{p}(X_{ea} + BQP_{ea-1} - BQP_{ea}) * PSP_e \tag{C}$$

The total cost equation given under (D) to generate the possible sales revenue is modified here to additionally consider the **cost incurred due to the proposed product substitution**. Furthermore, as the assembly of AT2P are also happening on the same line incurring the same cost as the AT1P, **the quantity of AT2P assembled is added to the cost function when the assembly cost is computed**. Accordingly, the modified mathematical equation for computing the total cost is as given below as (MD).

$$\begin{aligned}
\text{Total Cost} = {} & \left(\sum_{a=1}^{m}\sum_{e=1}^{p}(AQP_{ea} + \boldsymbol{AQTTP_{ea}}\right) \boldsymbol{*CAP_e} + \sum_{a=1}^{m}\sum_{e=1}^{p} IQP_{ea} * SCP_e \\
& + \sum_{a=1}^{m}\sum_{e=1}^{p} BQP_{ea} * CBP_e + \boldsymbol{\sum_{a=1}^{m}\sum_{e=1}^{p}\sum_{f<e} SQP_{efa} * CISP_{ef}} \\
& + \sum_{a=1}^{m}\sum_{b=1}^{n} MQC_{ba} * CMC_b \\
& + \sum_{a=1}^{m}\sum_{b=1}^{n} PQC_{ba} * CPC_b + \sum_{a=1}^{m}\sum_{b=1}^{n}\sum_{d=1}^{3} RMQC_{bda} * CRMC_{bd} \\
& + \sum_{a=1}^{m}\sum_{b=1}^{n} IQC_{ba} * CSC_b \\
& + \sum_{a=1}^{m}\sum_{e=1}^{p}\sum_{d=1}^{3} UR_{eda} * APR_{ed} + \sum_{a=1}^{m}\sum_{e=1}^{p}\sum_{d=1}^{3} DisQR_{eda} * CDisR_e \\
& + \sum_{a=1}^{m}\sum_{b=1}^{n}\sum_{d=1}^{3} DQRC_{bda} * CDRC_b + \sum_{a=1}^{m}\sum_{e=1}^{p}\sum_{d=1}^{3} IQR_{eda} * SCR_{ed} \\
& + \sum_{a=1}^{m}\sum_{c=1}^{s} PQRawM_{ca} * CPRawM_c \\
& + \sum_{a=1}^{m}\sum_{c=1}^{s} IQRawM_{ca} * CSRawM_c)
\end{aligned} \tag{MD}$$

From these, the modified objective function to maximize profit for the MP-CR system with BO&PS problem is the difference between total revenue equation (C) and modified total cost equation (MD), which is represented in (1-C).

$$\text{Total Profit} = C - MD \tag{1-C}$$

4.4.2 Modified Constraints

Modified Product Demand Constraints: The demand for a product in the MP-CR system with BO&PS problem is the demand in a period plus backordered demand from the previous period plus the quantity of the product used in substitution of all lower priced products in that period. This demand is referred to as "overall demand" henceforth. To satisfy the overall demand, the product is made available by assembling AT1P, assembling AT2P, substituting the current product with a higher priced product, getting inventory product carried from the previous period or backordering the product. The summation of all these availability options is referred from now onwards as "overall availability for a product". To represent this, the constraint (2) is modified as follows:

$$\begin{aligned} X_{ea} + BQP_{ea-1} + \sum_{f>e} SQP_{fea} \le AQP_{ea} + AQTTP_{ea} \\ + \sum_{q<e} SQP_{eqa} + IQP_{ea-1} + BQP_{ea} \\ \forall e, f, q \in [1, \ldots, p] \text{ and } a \in [1, \ldots, m] \end{aligned} \quad (2)$$

Modified Product Ending Inventory Constraints: Based on the overall demand and the overall availability for a product, there are two cases for the inventory quantity of a product. To represent each of the two cases mathematically, the constraint (3) is modified into constraint (3A) and constraint (3B), respectively, as follows:

Case 1: ***The overall availability for a product is greater than or equal to the overall demand in a period***:

$$\begin{aligned} IQP_{ea} = AQP_{ea} + AQTTP_{ea} + IQP_{ea-1} + \sum_{f<e} SQP_{efa} - \left(X_{ea} + BQP_{ea-1} + \sum_{q>e} SQP_{qea} \right) \\ \text{if} AQP_{ea} + AQTTP_{ea} + IQP_{ea-1} + \sum_{f<e} SQP_{efa} \ge X_{ea} + BQP_{ea-1} + \sum_{q>e} SQP_{qea} \\ \forall e, f, q \in [1, \ldots, p] \text{ and } a \in [1, \ldots, m] \end{aligned} \quad (3A)$$

Case 2: ***The overall availability for a product is less than the overall demand in that period***:

$$\begin{aligned} IQP_{ea} = o \\ \text{if} AQP_{ea} + AQTTP_{ea} + IQP_{ea-1} + \sum_{f<e} SQP_{efa} < X_{ea} + BQP_{ea-1} + \sum_{q>e} SQP_{qea} \\ \forall e, f, q \in [1, \ldots, p] \text{ and } a \in [1, \ldots, m] \end{aligned} \quad (3B)$$

Modified Product Assembly Capacity Constraint: The assembly of multi-products and their types happen on the same line. Hence, the constraint (4) is modified to additionally consider the assembly quantity of AT2P while computing the total assembly quantity in that period. This must be less than or equal to the available assembly capacity.

$$\sum_{e=1}^{p}(AQP_{ea} + AQTTP_{ea}) \leq AAC \qquad \forall a \in [1, \ldots m] \tag{4}$$

Modified Assembly Flow Balance Constraints: Whether a component can be remanufactured or not (i.e. disposal rate of the component) determines the assembly flow balance for the MP-CR system with BO&PS. Based on this, the following two cases arise, and to represent each of these two cases mathematically the constraint (5) is modified appropriately into constraints (5A), (5B) and (5C).

Case 1: ***A component cannot be remanufactured from any of the return categories***: When a component cannot be remanufactured from any return category, the sum of its disposal rates across the return categories is three (this is because, the disposal rate for a component obtained from a return category can be a number between 0 and 1). This implies the components are obtained only from the two conventional sources: manufacturing and/or purchasing, and the component carried in inventory from the previous period. This must be enough to assemble both the AT1P and AT2P. Mathematically, this is represented as constraint (5A), as follows:

$$\begin{aligned} &MQC_{ba} + PQC_{ba} + IQC_{ba-1} \geq \sum_{e=1}^{p}\{(AQP_{ea} + AQTTP_{ea}) * UCP_{be}\} \\ &\text{if} \sum_{d=1}^{3} RDRC_{bd} = 3, \qquad \forall b \in [1, \ldots, n] \text{ and } a \in [1, \ldots m] \end{aligned} \tag{5A}$$

Case 2: ***A component can be remanufactured from at least one of the return categories***: When the component can be remanufactured, the summation of disposal rates across the return categories is less than three. Such components can be obtained from three sources: manufacturing and/or purchasing and/or remanufacturing, and from the inventories carried from the previous period. In this case, the flow balance is twofold. One is for the AT2P and the other for AT1P. For AT2P, the remanufactured quantity of components in a period must be enough to assemble it.

$$\begin{aligned} &\sum_{d=1}^{3} RMQC_{bda} \geq \sum_{e=1}^{p}(AQTTP_{ea} * UCP_{be}) \\ &\text{if} \sum_{d=1}^{3} RDRC_{bd} < 3 \qquad \forall b \in [1, \ldots, n] \text{ and } a \in [1, \ldots m] \end{aligned} \tag{5B}$$

To assemble the AT1P, the quantity of components obtained from manufacturing and/or remanufacturing and/or purchase sources and the number of component carried in the inventory from the previous period must be sufficiently available. It is noted that the remanufactured components are added to the source of components for the AT1P only if it is in excess after being used to assemble the AT2P in a period.

$$MQC_{ba} + PQC_{ba} + IQC_{ba-1} + IQRMC_{ba} \geq \sum_{e=1}^{p} (AQP_{ea} * UCP_{be})$$
$$\text{if} \sum_{d=1}^{3} RDRC_{bd} < 3, \quad \forall b \in [1, \ldots, n] \text{ and } a \in [1, \ldots m] \tag{5C}$$

The $IQRMC_{ba}$ represents an essential inventory that will be explained next when the component inventory constraints are modified.

Modified Component Inventory Constraints: The inventory linking the forward and reverse flows in the MP-CR system with BO&PS problem is the component inventory. This inventory is replenished from the sources of manufacturing and/or remanufacturing and/or purchasing component. This inventory, like in the assembly flow balance (again based on a components' disposal rate), is represented either by the assembly of AT1P and AT2P or only by the assembly of the AT1P. To mathematically represent these two cases, the constraint (6) is modified appropriately into Constraint (6A), (6B) and (6C), as follows:

Case 1: ***A component cannot be remanufactured from any of the return categories***: For a component that cannot be remanufactured, components from the manufacturing and purchasing sources are needed for the assembly of both AT1P and AT2P. Therefore, the ending inventory is the total quantity of components obtained from manufacturing, purchasing, stored and carried from the previous period, differenced with the consuming by assembling of AT1P and AT2P.

$$IQC_{ba} = MQC_{ba} + PQC_{ba} + IQC_{ba-1} - \sum_{e=1}^{p} \{(AQP_{ea} + AQTTP_{ea}) * UCP_{be}\}$$
$$\text{if} \sum_{d=1}^{3} RDRC_{bd} = 3, \quad \forall b \in [1, \ldots, n] \text{ and } a \in [1, \ldots m] \tag{6A}$$

Case 2: ***When a component can be remanufactured from at least one of the return categories***: In this case, the components obtained from the remanufacturing source are first used to assemble AT2P. The remaining are stored in the virtual inventory $IQRMC_{ba}$ and this replenishes the component inventory. The virtual inventory is used solely for computational purpose and can be mathematically represented as given below.

$$IQRMC_{ba} = \sum_{d=1}^{3} RMQC_{bda} - \sum_{e=1}^{p} \{AQTTP_{ea} * UCP_{be}\}$$
$$\text{if} \sum_{d=1}^{3} RDRC_{bd} < 3, \quad \forall b \in [1, \ldots, n] \text{ and } a \in [1, \ldots m] \tag{6B}$$

The component inventory, in this case, is replenished by components obtained from manufacturing source, purchasing source, the inventory carried from the previous period and the virtual inventory of remanufactured components. The depletion is when AT1P are assembled. This is captured in constraint (6C).

$$IQC_{ba} = MQC_{ba} + PQC_{ba} + IQC_{ba-1} + IQRMC_{ba} - \sum_{e=1}^{p}\{AQP_{ea} * UCP_{be}\}$$
$$\text{if} \sum_{d=1}^{3} RDRC_{bd} < 3, \qquad \forall b \in [1, \ldots, n] \text{ and } a \in [1, \ldots m] \tag{6C}$$

Modified Integer and Non-Negativity Constraint: The non-negativity constraint is extended to additionally enforce the decision variables "$\mathbf{AQTTP_{ea}}$", "$\mathbf{SQP_{efa}}$" and "$\mathbf{IQRMC_{ba}}$" to be non-negative and integers.

$$\begin{gathered} AQP_{ea}, IQP_{ea}, MQC_{ba}, PQC_{ba}, RMQC_{bda}, IQC_{ba}, UR_{eda}, DisQR_{eda}, IQR_{eda}, PQRawM_{ca}, \\ IQRawM_{ca}, DQRC_{bda}, BQP_{ea}, \boldsymbol{AQTTP_{ea}}, \boldsymbol{SQP_{efa}}, \boldsymbol{IQRMC_{ba}} \geq 0 \text{ and integers} \\ \forall e, f \in [1, \ldots, p], b \in [1, \ldots, n], c \in [1, \ldots, s], d \in [1, 2, 3] \text{ and } a \in [1, \ldots, m] \end{gathered} \tag{20}$$

Modified Product Backordering Constraints: Based on the overall availability for a product and the overall demand, there are two cases for product backordering. Accordingly, the constraints (21A) and constraints (21B) are modified as follows:

Case 1: The overall availability for a product is greater than or equal to the overall demand: In this case, products are sufficiently available to satisfy the demand in that period completely. Hence, there is no need to backorder the demand.

$$\begin{gathered} BQP_{ea} = 0 \\ \text{if} AQP_{ea} + AQTTP_{ea} + \sum_{f<e} SQP_{efa} + IQP_{ea-1} \geq X_{ea} + BQP_{ea-1} + \sum_{q>e} SQP_{qea} \\ \forall e, q, f \in [1, \ldots p,] \text{ and } a \in [1, \ldots m] \end{gathered} \tag{21A}$$

Case 2: The overall availability for a product is lesser than the overall demand: Since the demand cannot be satisfied, a portion of it must be backordered. The backordered quantity is the difference between the overall demand and overall availability for a product.

$$\begin{gathered} BQP_{ea} = \left\{X_{ea} + BQP_{ea-1} + \sum_{q>e} SQP_{qea}\right\} - \left\{AQP_{ea} + AQTTP_{ea} + \sum_{f<e} SQP_{efa} + IQP_{ea-1}\right\} \\ \text{if } AQP_{ea} + AQTTP_{ea} + \sum_{f<e} SQP_{efa} + IQP_{ea-1} < X_{ea} + BQP_{ea-1} + \sum_{q>e} SQP_{qea} \\ \forall e, q, f \in [1, \ldots, p] \text{ and } a \in [1, \ldots m] \end{gathered} \tag{21B}$$

In the above formulations, the quantity of higher priced products used to substitute a lower priced product directly reduces the backorder quantity for the lower priced product. Thus, in the computation of the overall sales a higher priced product is selling at the price of the lower priced product. This avoids the need to consider a separate sale for the higher priced products that substitute and sell at the selling price of the lower priced product. Therefore, in the total revenue equation (C) the substitution sales are not separately included.

4.4.3 Additional Constraints

Product Substitution Constraints: The quantity of higher priced products that can be used to substitute a lower priced product depends on the demand for the higher priced product and the availability of the higher priced product. Hence, there are two cases: (That is, the demand for a higher priced product is the summation of demand for the higher priced product in a period and the backorder quantity for the higher priced product in the previous period.)

Case 1: The total available higher priced product is greater than or equal to the demand for the higher priced product: Availability of excess higher priced products facilitates the substitution to be possible. The quantity of higher priced products that can be used to substitute all lower priced products is equal to how many higher products products are left after their respective demands are satisfied.

$$\sum_{e<f} SQP_{fea} \leq AQP_{ea} + AQTTP_{ea} + IQP_{ea-1} + \sum_{e>w} SQP_{ewa} - X_{ea} - BQP_{ea-1}$$
$$if\ AQP_{ea} + AQTTP_{ea} + IQP_{ea-1} + \sum_{e>w} SQP_{ewa} \geq X_{ea} + BQP_{ea-1}$$
$$\forall e, f, w \in [1, \ldots, p] \text{ and } a \in [1, \ldots m] \tag{23A}$$

Case 2: The total available higher priced product is less than the demand for the higher priced product: Since the demand for a higher value product is itself greater than its availability quantity, substitution of lower priced products is not feasible. Therefore, the substitution quantities for such higher priced products are set to zero.

$$\sum_{e<f} SQP_{fea} = 0$$
$$if AQP_{ea} + AQTTP_{ea} + IQP_{ea-1} + \sum_{e>w} SQP_{ewa} < X_{ea} + BQP_{ea-1} \tag{23B}$$
$$\forall e, f, w \in [1, \ldots, p] \text{ and } a \in [1, \ldots m]$$

Downward Substitution Constraint: In order to make sure that the lower priced products do not substitute higher priced products, we need to make such substitution quantities are assigned as "zero".

$$\sum_{f<e} SQP_{fea} = 0 \qquad \forall e, f \in [1, \ldots, p] \text{ and } a \in [1, \ldots m] \tag{23C}$$

Restricted Substitution Constraint: The proposed product substation can happen only when the higher priced products are of AT2P. To enforce this requirement, the total substitution quantity for higher priced product is made to be utmost the assembled quantity of higher priced AT2P in the same period.

$$\sum_{q<e} SQP_{eqa} \leq AQTTP_{qa} \qquad \forall e, q \in [1, \ldots, p] \text{ and } a \in [1, \ldots m] \tag{23D}$$

Note that the higher priced AT2P cannot be stored in the inventories to substitute lower priced product demands at a later period. The substitution happens immediately in the same period.

The modifications and additions to the proposed INLP model presented for MP-CR with BO system will complete the required INLP model for the MP-CR system with BO&PS problem. This extended INLP model is called as E-INLP model for MP-CR system with BO&PS problem. Finally, for generating each of the three proposed mathematical models, suitable LINGO Set Code is developed. Due to the brevity of the paper size, the LINO Set Code developed for E-INLP Model for MP-CR System with BO&PS Problem is given in Annexure 2.

5 Validation of Each of the Three Proposed Models for MP-CR System

In this section, a suitable numerical example for validating each of three proposed models for MP-CR system, considered in this study, is presented based on the observation from a real-life Indian Automobile OEM (IA-OEM). This data is obtained by observation for all the forward flow-related input. As data for the reverse flow-related input is unavailable (due to non-existence of the reverse flow), they are suitably assumed. After discussing about the numerical problem, the validation of each of the proposed models is discussed.

5.1 A Numerical Example for MP-CR System Considered in the Study

A numerical example is developed to represent each of the three research problems for MP-CR system considered in this study. Accordingly, it is assumed that three different engines are assembled (E1, E2 and E3). Here, engines can be considered as products because in general each of these engines are produced by assembling various components in separate plants (generally these plants are called as engine plants) and then shipped to other plants (generally called vehicle plants) for fitment into automobiles. Considering a four-month planning horizon, the complete data for all forward flow-related parameters such as demand, price of selling, cost to assemble, storing cost, cost to backorder, cost to substitute, cost to setup per unit of dismantling capacity, and available assembly capacity for each of the engines/products are collected and presented in Table 2.

As each of the three engines are assembled using several components, for the sake of demonstration and simplicity, only 13 different components are used in the assembly process to obtain the engines/products. These 13 components are chosen in such a way that they capture both the diversity in the engines and the diversity in the

Table 2 Froward flow-related data for the final product (Engines)

Parameter	Period wise the forward flow parameter values for an engine											
	E1				E2				E3			
Planning horizon	1	2	3	4	1	2	3	4	1	2	3	4
Demand (Units/Month)	1500	1800	2200	2300	3000	4000	4000	4500	4500	4800	4800	5000
Price of selling (Rs./Unit)	18000				17520				17380			
Cost to Assemble (Rs./Unit)	1100				1000				1050			
Storing cost (Rs./Unit-Month)	165				157.5				150			
Cost to backorder (Rs./Unit)	825				787.5				750			
Cost to substitute (Rs./Unit)	Not applicable				E2 with E1 is 161				E3 with E2 is 379 E3 with E1 is 539			
Assembly capacity (Units/Month)	11600											

Table 3 Composition of components in producing engines

Component ID	Units of component required to assemble a single unit of an engine		
	E1	E2	E3
C1	1	1	1
C2	1	1	1
C3	1	2	2
C4	1	1	1
C5	2	2	2
C6	8	8	8
C7	2	2	2
C8	0	1	1
C9	0	1	1
C10	0	1	1
C11	1	0	0
C12	1	0	0
C13	1	0	1

various sources from which they can be obtained. Accordingly, the 13 components considered are indexed C1 to C13, and the composition of each engine assembled using them is presented in Table 3.

In Table 4, the various forward flow related parameters for each of the 13 components: manufacturing capacity available, cost to manufacture, cost to purchase, storing cost and upper and lower limits for purchase quantities are specified. Among the 13 considered components, those that are manufactured in-house utilize unique and common raw materials procured from external sources. For the sake of simplicity, 13 raw materials are chosen. These chosen raw materials, RM1 to RM13, capture the unique nature of the 13 components' manufacturing processes. The utilization of the raw materials in the manufacturing of components is specified in Table 5. The procurement cost and the storing cost for the 13 raw materials are given in Table 6.

Apart from the forward flow related data in Tables 2, 3, 4, 5 and 6, the reverse flow related parameters for the returned engines (returns) and the components obtained from them are assumed suitably and presented in Tables 7 and 8. Table 7 specifies the return-engine category wise: cost to dismantle, capital cost to install per unit of dismantling capacity, storing cost, acquisition price and rate of return. Further, in Table 8, for the 13 components, the assumed values related to: cost to dispose, capital cost to install per unit of remanufacturing capacity for a component and the return category-wise cost to remanufacture and return disposal rates are presented.

In addition to the data presented in Tables 2, 3, 4, 5, 6, 7 and 8, it is assumed that installation of the reverse flow (remanufacturing and dismantling) capacities can only happen in incremental steps of 10%, as compared to the respective forward flow (manufacturing and assembling) capacities. With these data, the various models

Table 4 Forward flow related data for components

Component ID	Manufacturing capacity available (Units/Month)	Cost to manufacture (Rs./Unit)	Cost to purchase (Rs./Unit)	Storing cost (Rs./Unit-Month)	Upper purchase limit (Units/order)	Lower purchase limit (Units/order)
C1	4320	600	1800	88	17280	1728
C2	4320	600	1800	88	17280	1728
C3	8640	300	900	75	34560	3456
C4	4320	300	900	75	17280	1728
C5	0	–	900	188	69120	6912
C6	0	–	450	63	138240	13824
C7	4320	500	1500	200	17280	1728
C8	4800	200	600	188	19200	1920
C9	0	–	900	188	6720	672
C10	0	–	900	188	6720	672
C11	3360	50	150	200	13440	1344
C12	0	–	225	188	6720	672
C13	3360	20	255	50	13440	1344

Table 5 Forward flow related data for raw materials

Raw Material ID	Cost to procure (Rs./Unit)	Storing cost (Rs./Unit-Month)	Upper procurement limit (Units/order)	Lower procurement limit (Units/order)
RM1	100	15	4320	432
RM2	100	15	4320	432
RM3	100	15	4320	432
RM4	50	7.5	8640	864
RM5	50	7.5	4320	432
RM6	40	6	4800	480
RM7	35	5.25	3360	336
RM8	75	30	3360	336
RM9	15	2.25	13440	1344
RM10	10	6.75	65280	6528
RM11	25	3.75	51840	5184
RM12	30	4.5	4800	480
RM13	30	4.5	3360	336

Table 6 Composition of raw materials to manufactured component

Raw material ID	Units of a raw material ID needed to manufacture one unit of a component ID												
	C1	C2	C3	C4	C5	C6	C7	C8	C9	C10	C11	C12	C13
RM1	1	0	0	0	0	0	0	0	0	0	0	0	0
RM2	0	1	0	0	0	0	0	0	0	0	0	0	0
RM3	0	0	0	0	0	0	1	0	0	0	0	0	0
RM4	0	0	1	0	0	0	0	0	0	0	0	0	0
RM5	0	0	0	1	0	0	0	0	0	0	0	0	0
RM6	0	0	0	0	0	0	0	1	0	0	0	0	0
RM7	0	0	0	0	0	0	0	0	0	0	1	0	0
RM8	0	0	0	0	0	0	0	0	0	0	0	0	1
RM9	0	0	0	0	0	0	0	0	0	0	3	0	4
RM10	0.75	0.75	0.75	0.75	0	0	0.75	0	0	0	0.1	0	0.1
RM11	2	2	2	2	0	0	2	0	0	0	0	0	0
RM12	0	0	0	0	0	0	0	1	0	0	0	0	0
RM 13	0	0	0	0	0	0	0	0	0	0	0	0	2

Table 7 Reverse flow parameters for engine returns (category wise)

Parameter	Engine return category wise the values of the parameter								
	E1			E3			E2		
	Commercial return	EOU return	EOL return	Commercial return	EOU return	EOL return	Commercial return	EOU return	EOL return
Cost to dismantling (Rs./Unit)	110			105			100		
Capital cost to install dismantling capacity (Rs./Unit)	840								
Storing cost (Rs./Unit-Month)	82.5	66	49.5	78.75	63	47.25	75	60	45
Rate of return (% of demand)	0	10	50	10	20	30	20	30	20
Acquisition price for return (Rs./Unit-category)	9153	4576	2197	8023	4012	1926	8043	4021	1930

Table 8 Assumed data for components related reverse flow parameters

Component ID	Disposal cost (Rs./Unit)	Capital cost to install remanufacturing capacity (Rs./Unit)	Cost to remanufacture a return category (Rs./Unit-category)			Rate of disposal for a return category (%/Unit-category)		
			Commercial return	EOU return	EOL return	Commercial return	EOU return	EOL return
C1	120	480	120	180	240	0.10	0.20	0.30
C2	120	480	120	180	240	0.10	0.20	0.30
C3	50	240	50	75	100	0.00	0.10	0.20
C4	50	240	50	75	100	0.00	0.10	0.20
C5	60	480	60	90	120	0.20	0.50	0.70
C6	30	0	–	–	–	1.00	1.00	1.00
C7	100	400	100	150	200	0.20	0.50	0.70
C8	40	160	40	60	80	0.00	0.00	0.00
C9	60	480	60	90	90	0.00	1.00	1.00
C10	60	480	60	90	120	0.00	1.00	1.00
C11	50	40	50	75	100	0.00	0.00	0.00
C12	45	480	45	68	90	0.00	0.50	1.00
C13	20	16	20	30	40	0.00	0.00	0.00

developed in Sect. 4 are validated in the following sections. Furthermore, in the following sections, only the salient features of the various problems are presented.

5.2 Validation of the Proposed Model for MP-CR System to Determine RFC & BEP

For the numerical example considered, dismantling and remanufacturing capacities are varied in steps of 10% up to 200%. With these, we have 400 scenarios (representing 20 remanufacturing capacity × 20 dismantling capacity) for the given numerical example. For each of the 400 scenarios of MP-CR system, the proposed ILP model for the MP-CR system, presented in Sect. 4.2.1, is generated using the LINGO Set Code developed and solved using LINGO 11.0 Solver. The optimum total cost obtained for each of the 400 scenarios is presented in Table 9. From the analysis and the interpretation of the solution obtained from the proposed ILP model, there is clear indications that the proposed ILP model represents the new MP-CR system considered in this study.

In addition, for each of the scenarios "how the optimum total cost varies with respect to the varying RFC" is captured and presented the same in Fig. 2. From Fig. 2, it is seen that the total cost shows a distinctive tend. For any installed dismantling capacity, as the installed remanufacturing capacity is increased the total cost decreases. This is because more components are available from the cheaper remanufacturing source. However, after a certain point, the installed remanufacturing capacity is in excess than the availability of the return components, which leads to increase in total cost. Therefore, based on the numerical example with 400 scenarios considered here, the combination of best dismantling and remanufacturing capacities, which is to be installed for the implementation of the MP-CR system by the IA-OEM, is 70% and 150%, respectively. This means that the dismantling capacity to be installed must be 70% of the available assembly capacity, and the remanufacturing capacity installed for a component must be 150% of the available manufacturing capacity for that component. At this RFC, the optimum total cost is Rs. 627.03 Million (Rs. 596.13 Million for operational decisions and Rs. 30.90 Million for excess capacity installation).

Further, using the input data in Tables 2, 3, 4, 5 and 6, the proposed B-ILP model for the existing MP-CM system, presented in Sect. 4.2.2, is generated using the LINGO Set Code developed and solved using a LINGO 11.0 solver. The optimum total cost for the existing MP-CM system is Rs. 669.94 Million. To validate the final step (Step 5) of the proposed methodology, the optimum computed total cost for the MP-CR system and total cost for the MP-CM systems, and RFC capacity installations are appropriately substituted into the BEP equation (A) for determining BEP. Accordingly, the BEP for the numerical problem considered in this study is as follows:

Table 9 Variation in total cost for the MP-CR system (vs) varying RFC installations for the IA-OEM data

% of Assembly capacity installed as dismantling capacity	Optimal total cost for MP-CR system at the remanufacturing capacity installed as % of the manufacturing capacity																			
	10	20	30	40	50	60	70	80	90	100	110	120	130	140	150	160	170	180	190	200
10	744.77	730.5	722.56	722.61	725.44	729.43	733.42	737.42	741.41	745.4	749.4	753.39	757.38	761.38	765.37	769.36	773.36	777.35	781.35	785.34
20	747.7	733.42	719.18	707.52	698.94	698.25	698.4	701.24	705.23	709.23	713.22	717.21	721.21	725.2	729.2	733.19	737.18	741.18	745.17	749.16
30	750.62	736.35	722.1	710.44	699.46	688.56	679.06	676.55	676.71	677.01	680.91	684.91	688.9	692.89	696.89	700.88	704.88	708.87	712.86	716.86
40	753.54	739.27	725.02	713.36	702.38	691.48	680.62	669.84	660.83	657.83	657.98	658.61	662.32	666.32	670.31	674.3	678.3	682.29	686.28	690.28
50	756.47	742.19	727.95	716.29	705.3	694.41	683.54	672.77	662.33	652.24	644.32	641.01	641.26	643.04	647.04	651.03	655.02	659.02	663.01	667
60	759.39	745.12	730.87	719.21	708.23	697.33	686.46	675.69	665.25	655.16	645.15	635.23	628.83	627.28	627.56	630.86	634.86	638.85	642.84	646.84
70	762.31	748.04	733.79	722.13	711.15	700.25	689.39	678.61	668.17	658.09	648.07	638.15	630.54	627.3	**627.03**	629.29	632.88	636.88	640.87	644.86
80	765.23	750.96	736.72	725.06	714.07	703.18	692.31	681.54	671.09	661.01	651	641.07	633.46	630.22	629.95	632.21	635.81	639.8	643.79	647.79
90	768.16	753.89	739.64	727.98	717	706.1	695.23	684.46	674.02	663.93	653.92	644	636.38	633.14	632.87	635.14	638.73	642.72	646.72	650.71
100	771.08	756.81	742.56	730.9	719.92	709.02	698.16	687.38	676.94	666.86	656.84	646.92	639.31	636.06	635.8	638.06	641.65	645.65	649.64	653.63
110	774	759.73	745.49	733.83	722.84	711.95	701.08	690.3	679.86	669.78	659.76	649.84	642.23	638.99	638.72	640.98	643.24	645.51	647.77	650.03
120	776.93	762.66	748.41	736.75	725.76	714.87	704	693.23	682.79	672.7	662.69	652.77	645.15	641.91	641.64	643.91	647.5	651.49	655.49	659.48
130	779.85	765.58	751.33	739.67	728.69	717.79	706.93	696.15	685.71	675.63	665.61	655.69	648.08	644.83	644.57	646.83	650.42	654.42	658.41	662.4
140	782.77	768.5	754.26	742.6	731.61	720.72	709.85	699.07	688.63	678.55	668.53	658.61	651	647.76	647.49	649.75	653.35	657.34	661.33	665.33
150	785.7	771.43	757.18	745.52	734.53	723.64	712.77	702	691.56	681.47	671.46	661.54	653.92	650.68	650.41	652.68	656.27	660.26	664.26	668.25
160	788.62	774.35	760.1	748.44	737.46	726.56	715.7	704.92	694.48	684.4	674.38	664.46	656.85	653.6	653.34	655.6	659.19	663.19	667.18	671.17
170	791.54	777.27	763.03	751.37	740.38	729.49	718.62	707.84	697.4	687.32	677.3	667.38	659.77	656.53	656.26	658.52	662.12	666.11	670.1	674.1
180	794.47	780.19	765.95	754.29	743.3	732.41	721.54	710.77	700.33	690.24	680.23	670.31	662.69	659.45	659.18	661.45	665.04	669.03	673.03	677.02
190	797.39	783.12	768.87	757.21	746.23	735.33	724.47	713.69	703.25	693.17	683.15	673.23	665.62	662.37	662.11	664.37	667.96	671.95	675.95	679.94
200	800.31	786.04	771.8	760.14	749.15	738.26	727.39	716.61	706.17	696.09	686.07	676.15	668.54	665.3	665.03	667.29	670.88	674.48	678.07	681.67

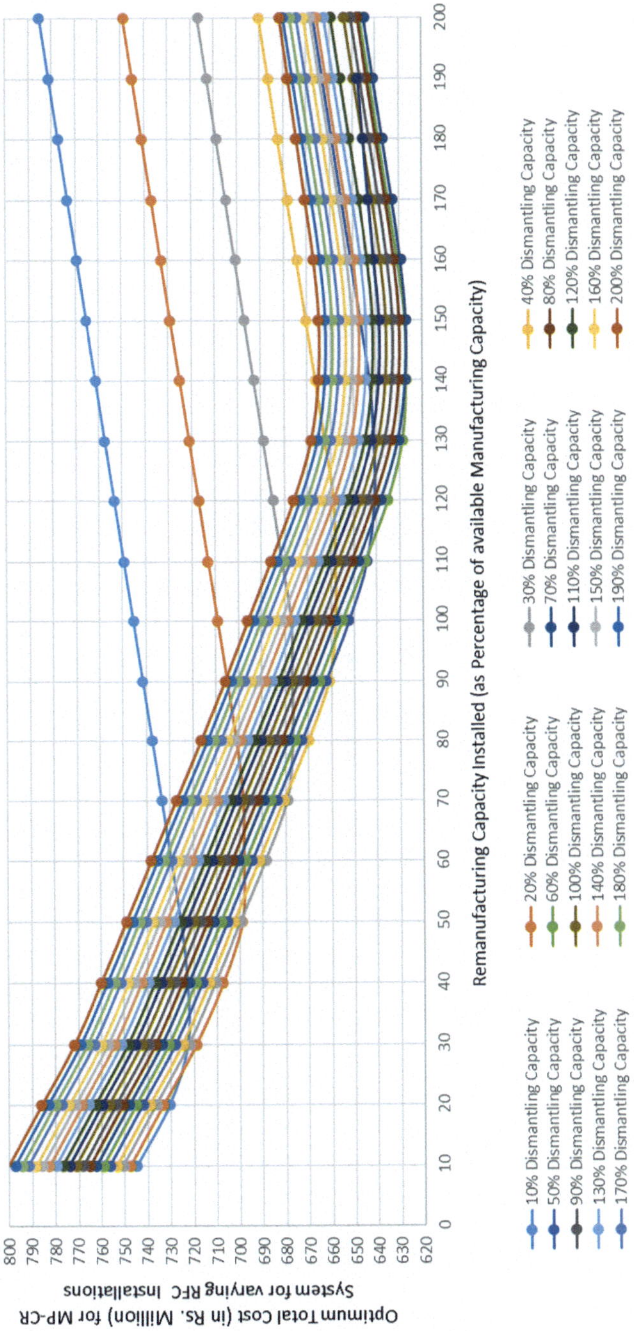

Fig. 2 Graphical representation of variation in total cost for the MP-CR system (vs) varying RFC installations for the IA-OEM data

$$BEP = \frac{(0.7 * 11600 * 840) + \sum_{b=1}^{13} (0.8 * CMC_b * 1.5 * MCC_b)}{(669.94 - 596.13) * 10^6} * 4\,\text{Month}$$

$$BEP = \frac{26.79 * 10^6}{76.81 * 10^6} * 4\,\text{Months}$$

$$BEP = \sim 0.4 * 4\,\text{Months}$$

$$\sim 1.6\,\text{Months}$$

Finally, from then BEP result for the given numerical example, it can be observed that (a) the MP-CR system is profitable to implement for the IA-OEM and (b) the break-even period is approximately 1 month and 18 days, after which the capital costs invested for setting up the RFC for incorporation of CR will be reclaimed.

5.3 *Validation of the Proposed Models for Problems 2 & 3 and Comparison with Insights*

This section demonstrates workability of the proposed INLP and E-INLP models developed in Sects. 4.3 and 4.4, for the MP-CR system with BO problem and MP-CR system with BO&PS problem, respectively. Since both the models are solved for the same input data, a comparison is presented and the insights for the management are highlighted. Accordingly, all the data, except for the demand, presented in Tables 2, 3, 4, 5, 6, 7 and 8 in Sect. 5.1 are used for validation of the proposed models. Only the demand data is modified here. This is because, for the MP-CR system and MP-CM system, it is mandatory that the total demand across all multi-product in the first period of the planning horizon must be less than or equal to the assembly capacity available. It is also required that the demand of all multi-products in the entire planning horizon needs to be less than the overall assembly capacity (summation of assembly capacity over the planning horizon). These are required conditions, as violations would result in the model being infeasible as demands cannot be shifted to other periods. However, for problem-2 and problem-3 of this study, these requirements are not necessary. Therefore, demand data is appropriately modified as given in Table 10.

For these demand data along with the numerical example, the proposed mathematical models are generated and solved using LINGO 11.0 Solver. However, the

Table 10 Modified demand to demonstrate the INLP and E-INLP models

Parameter	Period wise the values of the parameter for an engine											
	E1				E2				E3			
	1	2	3	4	1	2	3	4	1	2	3	4
Demand (Units/Month)	2050	1800	2200	2300	5000	4000	4000	4500	5000	4800	4800	5000

proposed INLP and E-INLP models are concave in nature. Therefore, the convergence to a global optimal solution is difficult, and only a local optimal solution is obtained. Nevertheless, a solution report obtained for the E-INLP model is presented in Annexure 3 – Exhibits 3.1 to 3.3 (similar solution reports can be generated for the first and second problems from the ILP and INLP models, respectively). For comparison purposes, a brief consolidated cost report for the BO problem and BO&PS problem for the MP-CR system is captured in Table 11.

From Table 11, the following managerial insights can be implied for the numerical problem considered in this study:

Table 11 Consolidated cost comparison at the solution identified for BO and BO&PS problems

Row no.	Operation aspect	Local optimal solution for	
		MP-CR system with BO problem	MP-CR system with BO&PS problem
1	Total profit (Rs. Million)	134.62	136.25
2	Total revenue (Rs. Million)	718.45	725.62
3	Total cost (Rs. Million)	583.83	589.37
4	Total sales (Units)	40,899	41,311
5	Sales from substitution in total sales (Units)	–	412
6	Total backorder quantity at the end of the planning horizon (Units)	4,551	4,139
7	Overall demand satisfied (%)	~90	~90.9
8	Increase or decrease in sales (%)		Sales increase by ~0.9
9	Total value of assembled products (both types) (in Rs. Million)	~42.61	~43.04
10	Proportion of total cost associated to component manufacturing (%)	~7.4	~7.2
11	Proportion of total cost associated to component remanufacturing (%)	~4.0	~4.0
12	Proportion of total cost associated to return dismantling (%)	~0.5	~0.5
13	Proportion of total cost associated to return component disposal (%)	~3.0	~2.9

1. The MP-CR system is more profitable when BO and PS are simultaneously implemented (1st row). This is because more demand is satisfied (7th and 8th rows in Table 11).
2. The revenue and the cost for the BO&PS problem are more as compared to the BO problem (2nd & 3rd rows in Table 11). The cost for the BO&PS problem is high because the substitution cost is additionally present. Nevertheless, the increase in revenue (~Rs. 7.17 Million more than the BO problem) is more than the increased cost (~Rs 5.54 Million more than the BO problem).
3. There is significant decrease in backordered quantity in the BO&PS problem as compared to the BO problem (6th row in Table 11). The decreased backorder is the demand that is satisfied through substitution (5th row in Table 11). This clearly indicates that the proposed PS is a good mechanism to increase flexibility to satisfy demand and generate more revenue efficiently.
4. Having higher sales (4th row in Table 11), the BO&PS problem is more profitable as lesser proportion of components is obtained from manufacturing (10th row in Table 11). Moreover, lesser proportion of return components are disposed (13th row in Table 11) even when the same proportion of returns are dismantled (12th row in Table 11). This implies that increased utilization is from the cheapest remanufacturing source.
5. As the utilization from remanufacturing increases (as indicated previously in point 4), there is a significant reduction in waste generation. This can be seen if we compare the number of components disposed in the CR problems and the existing MP-CM system, where all returns are disposed. Therefore, the proposed CR problem leads to decreased generation of landfills.
6. To reduce disposal costs and increase utilization of remanufactured components, E1 engines are assembled and these substitute the demand for E2 engines wherever possible. The substitution happens when enough excess assembly capacity is available for the higher priced E1, after the self-demand for the same is satisfied.
7. On deeper analysis of the solution report (in Annexure 3: Exhibit 3.3) it can be established that components C9 and C10 are scarce resources (reasoning for this is left as exercise to the interested reader), and hence E3, which shares these components with the higher priced E2, cannot be assembled to satisfy its own demand. Hence, E2 engines which have higher selling price are assembled using the common components obtained from E3 returns. Wherever substitution of E2 with E1 happens, E3 is assembled using the excess available C9 and C10 components. When substitution is not possible, E3 is backordered (in Annexure 3: Exhibit 3.1 row number 3).

From the above discussions, based on the consolidated report in Table 11, it can be concluded that INLP and E-INLP model capture all the important operational aspects of the respective problems of focus. Moreover, complete solution reports can be generated for in depth analysis to establish that the models completely capture all the considered inventory and production decisions, thereby validating them.

6 Summary and Conclusion

In summary, the study addresses important strategic and operational challenges when remanufacturing is incorporated as a mechanism to reduce waste by an OEM. More precisely, the problems addressed in this study focuses on a ***new business proposal*** by setting up a reverse flow of returns into an OEM and determination of associated decisions of capital investments required for setting up the Reverse Flow Capacities (RFC), Break-Even Period (BEP) for the investments and the corresponding inventory and production decisions. The setting up of the reverse flow is to establish a cheaper source of getting components from remanufacturing, in addition to the current sources of components from manufacturing and purchasing sources. With this, there is a system where components are obtained from three sources referred to as Component Remanufacturing (CR) system, and as they are used for assembling Multi-Products (MP) it is called the "MP-CR System".

For the new and proposed MP-CR system, in this study three research problems are addressed. The first is the strategic decision problem of determining the ideal RFC to setup and the BEP to reclaim the investments incurred in setting up the RFC. The second problem addresses the option of having the flexibility to satisfy the demand by the concept of Backordering (BO). Finally, a Product Substitution (PS) mechanism is proposed to make the system even more profitable.

For all three research problems, mathematical models are developed. The proposed mathematical models are demonstrated using a suitable numerical example, which is developed based on the observation from a real-life Indian Automobile OEM (IA-OEM). From the demonstration of the workability of the proposed mathematical models, the following managerial insights are inferred:

- The MP-CR system is profitable to incorporate, in comparison to the existing MP-CM system.
- Incorporating the proposed BO and PS together makes the MP-CR system very attractive as it gives the system flexibility in satisfying demand, wherein more revenue is generated by maximizing the utilization of common resources based on the product compositions.
- The incorporation of remanufacturing, in general, leads to reduced waste generation as more components are obtained from the remanufacturing source, and the corresponding quantities are reduced while manufacturing or purchasing the components. This also directly translates to reduction in landfill generation.
- The scarce resources, if any, can be effectively identified from the analysis of the solution obtained from the proposed models.

Though a holistic study of the MP-CR system is addressed here, right from the strategic decisions to operational decisions, the work is limited in scope as the considered CR system (where parameters: return rates and acquisition prices) is deterministic. Moreover, the study is based on the underlying premise that remanufacturing of the components is equivalent to manufacturing components (that is, remanufactured component is good-as-new). This premise has been widely debated in the fraternity that broadly research remanufacturing. Finally, many industry experts believe that dismantling is more expensive than assembling. Therefore, all these limitations are directions for the future research where stochastic CR systems, with separate demand for products assembled using components from conventional sources and from remanufacturing source can be studied to determine the break-even points for various dismantling-assembling cost structures.

Annexure 1: Proposed B-ILP Model for MP-CM System

Objective Function:

$$\text{Minimize}: \left\{\sum_{a=1}^{m}\sum_{e=1}^{p} AQP_{ea} * CAP_e + \sum_{a=1}^{m}\sum_{e=1}^{p} IQP_{ea} * SCP_e\right\}$$
$$+ \left\{\sum_{a=1}^{m}\sum_{b=1}^{n} MQC_{ba} * CMC_b\right.$$
$$\left. + \sum_{a=1}^{m}\sum_{b=1}^{n} PQC_{ba} * CPC_b + \sum_{a=1}^{m}\sum_{b=1}^{n} IQC_{ba} * CSC_b\right\}$$
$$+ \left\{\sum_{a=1}^{m}\sum_{c=1}^{s} PQRawM_{ca} * CPRawM_c\right.$$
$$\left. + \sum_{a=1}^{m}\sum_{c=1}^{s} IQRawM_{ca} * CSRawM_c\right\}$$

Products Demand Constraint:

$$AQP_{ea} + IQP_{ea-1} \geq X_{ea} \forall e \in [1, \ldots, p] \text{ and } a \in [1, \ldots, m] \quad (2)$$

Product Ending Inventor Constraint:

$$IQP_{ea} = AQP_{ea} + IQP_{ea-1} - X_{ea} \forall e \in [1, \ldots, p] \text{ and } a \in [1, \ldots, m] \quad (3)$$

Products Assembly Capacity Constraint:

$$AAC \geq \sum_{a=1}^{m} \sum_{e=1}^{p} AQP_{ea} \tag{4}$$

Assembly Flow Balance Constraint:

$$\begin{gathered} MQC_{ba} + PQC_{ba} + IQC_{ba-1} \geq \sum_{e=1}^{p} (AQP_{ea} * UCP_{be}) \\ \forall b \in [1, \ldots, n], \text{ and } a \in [1, \ldots, m] \end{gathered} \tag{5}$$

Component Manufacturing Capacity Constraint:

$$MCAC_b \geq MQC_{ba} \forall\, b \in [1, \ldots, n] \text{ and } a \in [1, \ldots, m] \tag{6}$$

Component Manufacturing Flow Balance Constraint:

$$\begin{gathered} PQRawM_{ca} + IQRawM_{ca-1} \geq \left(\sum_{b=1}^{n} MQC_{ba} * URawMMCM_{cb} \right) \\ \forall\, c \in [1, \ldots, s] \text{ and } a \in [1, \ldots, m] \end{gathered} \tag{7}$$

Raw Materials Ending Inventory Constraint:

$$\begin{gathered} IQRawM_{ca} = PQRawM_{ca} + IQRawM_{ca-1} - \sum_{b=1}^{n} (MQC_{ba} * URawMMC_{cb}) \\ \forall c \in [1, \ldots s] \text{ and } a \in [1, \ldots, m] \end{gathered} \tag{8}$$

Component Ending Inventory Constraint:

$$\begin{gathered} IQC_{ba} = MQC_{ba} + PQC_{ba} + IQC_{ba-1} - \left(\sum_{e=1}^{p} UCP_{be} * AQP_{ea} \right) \\ \forall\, b \in [1, \ldots, n], \text{ and } a \in [1, \ldots, m] \end{gathered} \tag{9}$$

Integer and Non-Negativity Constraints:

$$\begin{gathered} AQP_{ea}, IQP_{ea}, MQC_{ba}, PQC_{ba}, IQC_{ba}, PQRawM_{ca}, IQRawM_{ca} \geq 0 \text{ and integers} \\ \forall\, a \in [1, \ldots, m], b \in [1, \ldots, n], c \in [1, \ldots, s] \text{ and } e \in [1, \ldots, p] \end{gathered} \tag{20}$$

Annexure 2: A LINGO Set Code to Generate the E-INLP Model for MP-CR System with BO&PS Problem

```
SETS:
a/1..4/  : AAC, IDisCR;
b/1..13/ : CMC, MCAC, CPC, IRMCC, CDRC ,CSC, UPLC, LPLC;
e/1..3/  : CAP, CDisR, SCP, CBP, PSP;
d/1..3/  : ;
c/1..13/ : CPRawM, CSRawM, UPLRawM, LPLRawM;
cb(c,b)  : URawMMC;
ca(c,a)  : IQRawM, PQRawM;
ea(e,a)  : X, AQP, IQP, BQP,AQTTP;
ee(e,e)  : CISP;
eea(e,e,a): SQP;
be(b,e)  : UCP;
ba(b,a)  : IQC, MQC, PQC, IQRMC;
bd(b,d)  : CRMC, RDRC;
ed(e,d)  : APR, RRRP, SCR;
bda(b,d,a): RMQC, DQRC;
eda(e,d,a): DisQR, IQR,UR;
ENDSETS

!OBJECTIVE FUUCPTION;
[OBJROW]
Min = @SUM(ea(l,j)|j#GT#1:PSP(l)*(X(l,j) + BQP(l,j-1) - BQP(l,j))) + @SUM(ea(l,j)|j#EQ#1:PSP(l) *(X(l,j) -
BQP(l,j))) - (@SUM(ea(l,j):CAP(l)*(AQP(l,j))+ AQTTP(l,j)) +@SUM(eea(l,q,j): SQP(l,q,j)*CISP(l,q)) +
@SUM(ea(l,j):SCP(l)*IQP(l,j)) + @SUM(ba(i,j):CMC(i)*MQC(i,j)) + @SUM(ba(i,j):CPC(i)*PQC(i,j)) +
@SUM(bda(i,k,j):CRMC(i,k)*RMQC(i,k,j)) + @SUM(ba(i,j):CSC(i)*IQC(i,j)) +
@SUM(ca(o,j):CSRawM(o)*IQRawM(o,j)) + @SUM(ca(o,j):CPRawM(o)*PQRawM(o,j))+ @SUM(eda(l,k,j):
(CDisR(l)*DisQR(l,k,j)) + (SCR(l,k)*IQR(l,k,j)) + (APR(l,k)*UR(l,k,j))) +
@SUM(bda(i,k,j):CDRC(i)*DQRC(i,k,j))+ @SUM(ea(l,j):CBP(l)*BQP(l,j)));

!Product Demand Constraint;
@FOR(a(j)|j#EQ#1: @FOR(e(l): X(l,j) + @SUM(e(w)|w#GT#l: SQP(w,l,j)) <= AQP(l,j) + AQTTP(l,j) + BQP(l,j)
+ @SUM(e(q)|q#LT#l: SQP(l,q,j))));

@FOR(a(j)|j#GT#1: @FOR(e(l): X(l,j) + BQP(l,j-1) + @SUM(e(w)|w#GT#l: SQP(w,l,j)) <= AQP(l,j) +
AQTTP(l,j) + BQP(l,j) + IQP(l,j-1) + @SUM(e(q)|q#LT#l: SQP(l,q,j))));

!Product Inventory Level Constraint;
@FOR(a(j)|j#GT#1: @FOR(e(l): IQP(l,j) = @IF(AQP(l,j) + IQP(l,j-1) + AQTTP(l,j) + @SUM(e(q)|q#LT#l:
SQP(l,q,j)) #GT# X(l,j) + BQP(l,j-1) + @SUM(e(w)|w#GT#l: SQP(w,l,j)),
AQP(l,j) + IQP(l,j-1) + AQTTP(l,j) + @SUM(e(q)|q#LT#l: SQP(l,q,j)) - X(l,j) - BQP(l,j-1) - @SUM(e(w)|w#GT#l:
SQP(w,l,j)), 0)));

@FOR(a(j)|j#EQ#1: @FOR(e(l): IQP(l,j) = @IF(AQP(l,j) + AQTTP(l,j) + @SUM(e(q)|q#LT#l: SQP(l,q,j)) #GT#
X(l,j) + @SUM(e(w)|w#GT#l: SQP(w,l,j)),
AQP(l,j) + AQTTP(l,j) + @SUM(e(q)|q#LT#l: SQP(l,q,j)) - X(l,j) - @SUM(e(w)|w#GT#l: SQP(w,l,j)), 0)));


!Product Assembly Capacity Constraint;
@FOR(a(j): AAC(j) >= @SUM(e(l): AQP(l,j)+AQTTP(l,j)));
```

!Assembly Flow Balance;
@FOR(a(j)|j#GT#1: @FOR(b(i): MQC(i,j) + PQC(i,j) + IQRMC(i,j) + IQC(i,j-1) >= @IF(@SUM(d(k):RDRC(i,k)) #EQ# 3, @SUM(e(l): (AQP(l,j) + AQTTP(l,j))*UCP(i,l)), @SUM(e(l): AQP(l,j)*UCP(i,l)))));

@FOR(a(j)|j#EQ#1: @FOR(b(i): MQC(i,j) + PQC(i,j) + IQRMC(i,j) >= @IF(@SUM(d(k):RDRC(i,k)) #EQ# 3, @SUM(e(l): (AQP(l,j) + AQTTP(l,j))*UCP(i,l)), @SUM(e(l): AQP(l,j)*UCP(i,l)))));

@FOR(a(j): @FOR(b(i): @SUM(d(k): RMQC(i,k,j)) >= @IF(@SUM(d(k): RDRC(i,k)) #EQ# 3, 0, @SUM(e(l): AQTTP(l,j)*UCP(i,l)))));

!Component Inventory Level Constraints;
@FOR(a(j)|j#GT#1: @FOR(b(i): IQC(i,j) = (@IF(@SUM(d(k): RDRC(i,k)) #EQ# 3,
MQC(i,j) + PQC(i,j) + IQRMC(i,j) + IQC(i,j-1) - @SUM(e(l): UCP(i,l)*(AQP(l,j) + AQTTP(l,j))),
MQC(i,j) + PQC(i,j) + IQRMC(i,j) + IQC(i,j-1) - @SUM(e(l): UCP(i,l)* AQP(l,j))))));

@FOR(a(j)|j#EQ#1: @FOR(b(i): IQC(i,j) = (@IF(@SUM(d(k): RDRC(i,k)) #EQ# 3,
MQC(i,j) + PQC(i,j) + IQRMC(i,j) - @SUM(e(l): UCP(i,l)*(AQP(l,j) + AQTTP(l,j))),
MQC(i,j) + PQC(i,j) + IQRMC(i,j) - @SUM(e(l): UCP(i,l)* AQP(l,j))))));

@FOR(a(j): @FOR(b(i): IQRMC(i,j) = @IF(@SUM(d(k): RDRC(i,k)) #EQ# 3, 0, @SUM(d(k):RMQC(i,k,j)) - @SUM(e(l): UCP(i,l)*AQTTP(l,j)))));

!Component Manufacturing Capacity Constraint;
@FOR(a(j): @FOR(b(i): MQC(i,j)<= MCAC(i)));

!Manufacturing Flow Balance;
@FOR(a(j)|j#GT#1: @FOR(c(s): @SUM(b(i):URawMMC(s,i)*MQC(i,j)) <= PQRawM(s,j) + IQRawM(s,j-1)));
@FOR(a(j)|j#EQ#1: @FOR(c(s): @SUM(b(i):URawMMC(s,i)*MQC(i,j)) <= PQRawM(s,j)));

!Raw Material Inventory Level Constraint;
@FOR(a(j)|j#GT#1: @FOR(c(s): IQRawM(s,j) = PQRawM(s,j) - @SUM(b(i): URawMMC(s,i)*MQC(i,j)) + IQRawM(s,j-1)));
@FOR(a(j)|j#EQ#1: @FOR(c(s): IQRawM(s,j) = PQRawM(s,j) - @SUM(b(i): URawMMC(s,i)*MQC(i,j))));

!Return Rate Constraint;
@FOR(a(j)|j#GT#1: @FOR(e(l): @FOR(d(k): UR(l,k,j) = RRRP(l,k)*X(l,j))));
@FOR(a(j)|j#EQ#1: @FOR(e(l): @FOR(d(k): UR(l,k,j) = 0)));

!Returns Dismantling Flow Balance Constraint;
@FOR(a(j)|j#GT#1: @FOR(e(l): @FOR(d(k): IQR(l,k,j-1) + UR(l,k,j) >= DisQR(l,k,j))));
@FOR(a(j)|j#EQ#1: @FOR(e(l): @FOR(d(k): UR(l,k,j) >= DisQR(l,k,j))));

!Return Dismantling Capacity;
@FOR(a(j): @SUM(ed(l,k): DisQR(l,k,j)) <= IDisCR(j));

!Returns Inventory Level Constraint;
@FOR(a(j)|j#GT#1: @FOR(e(l): @FOR(d(k): IQR(l,k,j) = UR(l,k,j) - DisQR(l,k,j) + IQR(l,k,j-1))));
@FOR(a(j)|j#EQ#1: @FOR(e(l): @FOR(d(k): IQR(l,k,j) = UR(l,k,j) - DisQR(l,k,j))));

!Component Remanufacturing Capacity Constraint;
@FOR(a(j): @FOR(b(i): IRMCC(i) >= @SUM(d(k):RMQC(i,k,j))));

!Component Remanufacturing Flow Balance;
@FOR(a(j): @FOR(b(i): @FOR(d(k): RMQC(i,k,j) = @SUM(e(l): DisQR(l,k,j)*UCP(i,l)) - DQRC(i,k,j))));

```
!Returns Disposal Constraints;
@FOR(a(j): @FOR(b(i): @FOR(d(k)|k#EQ#1: DQRC(i,k,j) >= @SUM(e(l):UCP(i,l)*DisQR(l,k,j))*RDRC(i,k))));
@FOR(a(j): @FOR(b(i): @FOR(d(k)|k#EQ#2: DQRC(i,k,j) >= @SUM(e(l):UCP(i,l)*DisQR(l,k,j))*RDRC(i,k))));
@FOR(a(j): @FOR(b(i): @FOR(d(k)|k#EQ#3: DQRC(i,k,j) >= @SUM(e(l):UCP(i,l)*DisQR(l,k,j))*RDRC(i,k))));

!Integer and Non Negativity Constraints;
@FOR(ea(z,j):@GIN(AQP(z,j)));
@FOR(ea(z,j):@GIN(AQTTP(z,j)));
@FOR(ea(z,j):@GIN(BQP(z,j)));
@FOR(ea(z,j):@GIN(IQP(z,j)));
@FOR(ba(i,j):@GIN(IQC(i,j)));
@FOR(ba(i,j):@GIN(MQC(i,j)));
@FOR(ba(i,j):@GIN(PQC(i,j)));
@FOR(ca(r,j):@GIN(IQRawM(r,j)));
@FOR(ca(r,j):@GIN(PQRawM(r,j)));
@FOR(eda(l,k,j):@GIN(UR(l,k,j)));
@FOR(eda(l,k,j):@GIN(IQR(l,k,j)));
@FOR(eda(l,k,j):@GIN(DisQR(l,k,j)));
@FOR(eda(l,k,j):@GIN(DQRC(l,k,j)));
@FOR(eda(l,k,j):@GIN(RMQC(l,k,j)));
@FOR(eea(l,q,j):@GIN(SQP(l,q,j)));
@FOR(ba(i,j):@GIN(IQRMC(i,j)));

!Backordering Constraint;
@FOR(a(j)|j#GT#1: @FOR(e(l): BQP(l,j) = @IF(AQP(l,j) + IQP(l,j-1) + AQTTP(l,j) + @SUM(e(q)|q#LT#l:
SQP(l,q,j)) #GE# X(l,j) + BQP(l,j-1) + @SUM(e(w)|w#GT#l: SQP(w,l,j)),
0, X(l,j) + BQP(l,j-1) - AQP(l,j) -AQTTP(l,j) - IQP(l,j-1) - @SUM(e(q)|q#LT#l: SQP(l,q,j)))));

@FOR(a(j)|j#EQ#1: @FOR(e(l): BQP(l,j) = @IF(AQP(l,j) + AQTTP(l,j) + @SUM(e(q)|q#LT#l: SQP(l,q,j)) #GE#
X(l,j) + @SUM(e(w)|w#GT#l: SQP(w,l,j)), 0, X(l,j) - AQP(l,j) -AQTTP(l,j) - @SUM(e(q)|q#LT#l: SQP(l,q,j)))));

!Procurement Restriction Constraints;
@FOR(a(j): @FOR(b(i): PQC(i,j) <= UPLC(i)));
@FOR(a(j): @FOR(b(i): PQC(i,j) >= @IF(PQC(i,j) #GT# 0, LPLC(i), 0)));

@FOR(a(j): @FOR(c(i): PQRawM(i,j) <= UPLRawM(i)));
@FOR(a(j): @FOR(c(i): PQRawM(i,j) >= @IF(PQRawM(i,j) #GT# 0, LPLRawM(i), 0)));

!Product Substitution Constraints;
@FOR(a(j)|j#GT#1: @FOR(e(q): @SUM(e(l)|l#GT#q: SQP(l,q,j)) <= @IF( AQP(q,j) + AQTTP(q,j) + IQP(q,j-1) +
@SUM(e(w)|w#LT#q:SQP(q,w,j)) #GE# X(q,j) + BQP(q,j-1),
AQP(q,j) + AQTTP(q,j) + IQP(q,j-1) + @SUM(e(w)|w#LT#q: SQP(q,w,j)) - X(q,j) - BQP(q,j-1), 0)));

@FOR(a(j)|j#EQ#1: @FOR(e(q): @SUM(e(l)|l#GT#q: SQP(l,q,j)) <= @IF( AQP(q,j) + AQTTP(q,j) +
@SUM(e(w)|w#LT#q:SQP(q,w,j)) #GE# X(q,j),
AQP(q,j) + AQTTP(q,j) + @SUM(e(w)|w#LT#q: SQP(q,w,j)) - X(q,j), 0)));

!Downward Substitution Constraint;
@FOR(a(j): @FOR(e(l): @SUM(e(q)|q#LT#l: SQP(q,l,j))= 0));

!Restricted Substitution Constraint;
@FOR(a(j): @FOR(e(l): @SUM(e(w)|w#GT#l: SQP(w,l,j)) <= AQTTP(l,j)));
```

Annexure 3: Local Optimal Solution Report of the MP-CR System with BO&PS Problem for the IA-OEM Data

See Exhibits 3.1, 3.2 and 3.3.

Exhibit 3.1 Period wise optimal quantities for the multiple engines/multiple engine returns, and optimal costs from the proposed E-INLP model

Row no	Engine/return-engine-related decisions	Period wise optimal quantities for the multiple engines												Optimal cost (in Rupees Million)
		E-1				E-2				E-3				
		1	2	3	4	1	2	3	4	1	2	3	4	
1	Assembly quantity for type-one engine	2047	1798	2200	2300	2940	4225	413	6874	5000	2354	5691	861	38.260
2	Assembly quantity for type-two engine	0	192	110	115	0	1159	25	1450	0	3	1553	1	4.781
3	Backorder quantity for engine	3	0	0	0	2060	489	3941	2	0	2443	0	4137	10.050
4	Substituted quantity for engine	0	0	0	0	0	187	110	115	0	0	0	0	0.066
5	Inventory quantity for engines	0	0	0	0	0	0	0	0	0	0	1	0	0.000
Period wise optimum quantities for the multiple engine-return category														
6	Units of commercial engines returns received	0	0	0	0	500	400	400	450	1000	960	960	1000	45.569
7	Dismantling quantity for commercial engine returns	0	0	0	0	500	379	420	451	979	783	1158	1000	0.576
8	Inventory quantity for commercial engine returns	0	0	0	0	0	21	1	0	21	198	0	0	0.018
9	Units of EOU engine returns received	205	180	220	230	1000	800	800	900	1500	1440	1440	1500	41.506
10	Dismantling quantity for EOU engine returns	1	384	220	230	1000	795	805	900	1344	1596	1440	1500	1.047
11	Inventory quantity for EOU engine returns	204	0	0	0	0	5	0	0	156	0	0	0	0.023
12	Units of EOL engine returns received	1025	900	1100	1150	1500	1200	1200	1350	1000	960	960	1000	26.850
13	Dismantling quantity for EOL engine returns	1025	900	1100	1149	1500	1200	1199	1350	995	965	960	1000	1.403
14	Inventory quantity for EOL engine returns	0	0	0	1	0	0	1	1	5	0	0	0	0.000
Total optimal operational cost related to multiple engines and multiple engine returns for the IA-OEM in the planning horizon														170.15
Revenue from sales														(Rupees Million)

(continued)

Exhibit 3.1 (continued)

Row no	Engine/return-engine-related decisions	Period wise optimal quantities for the multiple engines												Optimal cost (in Rupees Million)
		E-1				E-2				E-3				
		1	2	3	4	1	2	3	4	1	2	3	4	
Total sales of engines in a period of the planning horizon		2047	1803	2200	2300	2940	5571	548	8439	5000	2357	7243	863	725.612

Exhibit 3.2 Period wise optimal quantities for the components, and optimal costs from the proposed E-INLP model

Component related decisions	Planning period	Optimal quantities for component in a period													Optimal cost (Rupees Million)
		C1	C2	C3	C4	C5	C6	C7	C8	C9	C10	C11	C12	C13	
Manufacture quantity	1	4320	4320	5939	3713	0	0	4320	1622	0	0	1217	0	1553	10.656
	2	4318	4319	6315	3560	0	0	4320	2023	0	0	513	0	7	10.735
	3	4320	4012	5982	4318	0	0	4320	1702	0	0	989	0	2861	10.696
	4	4015	4320	5191	4320	0	0	4320	2983	0	0	65	0	0	10.614
Purchase quantity	1	0	0	0	0	13494	79896	9174	0	6720	6719	0	2048	1344	74.757
	2	0	0	0	1728	12989	77848	8669	0	6719	6720	0	1797	0	73.7798
	3	0	1749	0	0	13504	79936	9184	0	6720	6720	0	2200	1344	77.983
	4	1728	0	3456	0	16722	92808	12402	0	6720	6720	1344	2300	0	94.452
Remanufacture quantity from commercial returns	1	1331	1329	2958	1479	2366	0	2366	1479	1479	1479	0	0	979	1.176
	2	1045	1045	2324	1162	1859	0	1859	1162	1162	1162	0	0	783	0.924
	3	1420	1401	3156	1578	2491	0	2491	1578	1578	1578	0	0	1158	1.249
	4	1305	1305	2902	1451	2321	0	2321	1451	1451	1451	0	0	1000	1.154
Disposal quantity from commercial returns	1	148	150	0	0	592	11832	592	0	0	0	0	0	0	0.485
	2	117	117	0	0	465	9296	465	0	0	0	0	0	0	0.381
	3	158	177	0	0	665	12624	665	0	0	0	0	0	0	0.525
	4	146	146	0	0	581	11608	581	0	0	0	0	0	0	0.476
Remanufacture quantity from EOU returns	1	1875	1876	4220	2110	2344	0	2345	2344	0	0	1	0	1345	1.894
	2	2220	2220	4649	2497	2775	0	2775	2391	0	0	384	192	1980	2.246
	3	1971	1972	4239	2218	2461	0	2461	2245	0	0	220	110	1660	1.993
	4	2104	2104	4527	2367	2630	0	2630	2400	0	0	230	115	1040	2.106

(continued)

Exhibit 3.2 (continued)

Component related decisions	Planning period	Optimal quantities for component in a period													Optimal cost (Rupees Million)
		C1	C2	C3	C4	C5	C6	C7	C8	C9	C10	C11	C12	C13	
Disposal quantity from EOU returns	1	470	469	469	235	2346	18760	2345	0	2344	2344	0	1	0	1.367
	2	555	555	517	278	2775	22200	2775	0	2391	2391	0	192	0	1.579
	3	494	493	471	247	2469	19720	2469	0	2245	2245	0	110	0	1.415
	4	526	526	503	263	2630	21040	2630	0	2400	2400	0	115	690	1.524
Remanufacture quantity from EOL returns	1	2464	2464	4810	2816	1770	0	1769	2495	0	0	1025	0	2009	2.894
	2	2145	2145	4184	725	1839	0	1839	2165	0	0	900	0	1865	2.447
	3	2281	2281	4329	2607	1528	0	1528	2159	0	0	1098	0	2060	2.642
	4	2449	2449	4679	2662	1529	0	1529	2350	0	0	776	0	1237	2.714
Disposal quantity from EOL returns	1	1056	1056	1205	704	5270	28160	5271	0	2495	2495	0	1025	11	2.383
	2	920	920	1046	2340	4291	24520	4291	0	2165	2165	0	900	0	2.113
	3	978	978	1089	652	4990	26072	4990	0	2159	2159	2	1100	0	2.211
	4	1050	1050	1170	837	5469	27992	5469	0	2350	2350	373	1149	912	2.438
Inventory quantity	1	3	2	0	131	0	0	0	0	259	258	196	1	183	0.156
	2	0	0	0	72	0	0	0	0	399	399	3	0	471	0.179
	3	0	1423	32	801	0	0	0	2	1015	1015	0	0	0	0.568
	4	0	0	0	0	0	0	0	0	0	0	0	0	0	0
Total optimal operational cost related to components for the IA-OEM in the planning horizon															404.913

Exhibit 3.3 Period wise optimal quantities for the raw materials, and optimal costs from the proposed E-INLP model

Component related decisions	Planning period	Optimal quantities for component in a period													Optimal cost (Rupees Million)
		C1	C2	C3	C4	C5	C6	C7	C8	C9	C10	C11	C12	C13	
Purchase quantity	1	4320	4320	4320	5939	3713	3645	1730	1560	9863	17237	45224	1622	3106	3.695
	2	4318	4319	4320	6315	4183	0	0	0	27733	17175	45664	3725	3310	3.386
	3	4320	4012	4320	5982	4320	1761	1054	2861	13440	17612	45907	0	3360	3.728
	4	4015	4320	4320	5191	3695	2924	0	0	0	19316	51224	2983	0	3.39
Inventory quantity	1	0	0	0	0	0	2023	513	7	0	1	0	0	0	0.015
	2	0	0	0	0	623	0	0	0	1166	0	0	1702	3296	0.03
	3	0	0	0	0	625	59	65	0	195	13	3	0	934	0.01
	4	0	0	0	0	0	0	0	0	0	2698	6895	0	934	0.048
Total optimal operational cost related to raw materials for the IA-OEM in the planning horizon															14.302

References

Akcali, E., & Cetinkaya, S. (2011). Quantitative models for inventory and production planning in closed loop supply chains. *International Journal of Production Research, 49,* 2373–2470.

Azadeh, A., & Mazaheri, E. (2013). Optimal production planning of a dedicated remanufacturing process by an integrated simulation based optimization approach. *International Journal of Logistics Systems and Management, 16*(1), 67–84.

Chen, M., & Abrishami, P. (2014). A mathematical model for production planning in hybrid manufacturing-remanufacturing systems. *International Journal of Advanced Manufacturing Technology, 71,* 1187–1196.

Doppelt, B., & Nelson, H. (2001). Extended producer responsibility and product take-back: Applications for the pacific northwest *97207*(503).

Ervasti, I., Miranda, R., & Kauranen, I. (2016). A global, comprehensive review of literature related to paper recycling: A pressing need for a uniform system of terms and definitions. *Waste Management, 48,* 64–71.

Francie, K. A., Pierre, K. J., Pierre, D., Victor, S., & Vladimir, P. (2015). Stochastic models and numerical solutions for manufacturing/remanufacturing systems with applications to the printer cartridge industry. *Journal of Manufacturing Systems, 37,* 662–671.

Geyer, R., & Van Wassenhove, L. N. (2005). The impact of constraints in closed-loop supply chains: The case of reusing components in product manufacturing. In B. Fleischmann, A. Klose (Eds.), *Distribution logistics. Lecture Notes in Economics and Mathematical Systems* (Vol. 544, pp. 203–219).

Govindan, K., Soleimani, H., & Kannan, D. (2015). Reverse logistics and closed-loop supply chain: A comprehensive review to explore the future. *European Journal of Operational Research, 240,* 603–626.

Jung, K. S., Dawande, M., Geismar, H. N., Guide, V. D. R., Jr., & Sriskandarajah, C. (2014). Supply planning models for a remanufacturer under just-in-time manufacturing environment with reverse logistics. *Annals of Operations Research, 232,* 1–49.

Kim, K., Song, I., Kim, J., & Jeong, B. (2006). Supply planning model for remanufacturing system in reverse logistics environment. *Computers & Industrial Engineering, 51,* 279–287.

Krikke, H., Hofenk, D., & Wang, Y. (2013). Revealing an invisible giant: A comprehensive survey into return practices within original (closed-loop) supply chains. *Resources, Conservation and Recycling, 73,* 239–250.

Li, X., Li, Y., & Govindan, K. (2014a). An incentive model for closed-loop supply chain under ERP law. *Journal of Operational Research Society, 65,* 88–96.

Li, X., Baki, F., Tian, P., & Chaouch, B. A. (2014b). A robust block-chain based tabu search algorithm for the dynamic lot sizing problem with product returns and remanufacturing. *Omega, 42,* 75–87.

Li, L., Li, C., Tang, Y., & Du, Y. (2017). An integrated approach of reverse engineering aided remanufacturing process for worn components. *Robotics and Computer-Integrated Manufacturing, 48,* 39–50.

Lund, R. (1998). Remanufacturing: United states experience and implications for developing nations. In World Bank Technical Paper Number 31. Retrieved May 08, 2018, from http://documents.worldbank.org/curated/en/792491468142480141/pdf/WTP310PUB0REPL00Box367874B00PUBLIC0.pdf.

Lund, R. T., Hauser, W. (2012). The database of remanufacturers. Retrieved May 08, 2018, from http://www.reman.org/Papers/Reman_Database_Lund.Pdf.

Maslennikova, I., & Foley, D. (2000). Xerox's approach to sustainability. *Interfaces, 30,* 226–233.

Mathirajan, M., Manoj, K., & Ramachandran, V. (2011). A design of distribution network and development of efficient distribution policy. *International Journal of Logistics Systems and Management, 9*(1), 108–137.

Mawandiya, B. K., Jha, J. K., & Thakkar, J. J. (2018). Optimal production-inventory policy for closed-loop supply chain with remanufacturing under random demand and return. *Operations Research International Journal, 18,* 1–42.

Mondal, S., & Mukherjee, K. (2012). Simulation of tyre retreading process—An indian case study. *International Journal of Logistics Systems and Management, 13*(4), 525–539.

Papachristos, G., & Adamides, E. D. (2014). Internal supply-chain competition in remanufacturing: Operations strategies, performance and environmental effects. *International Journal of Logistics Systems and Management, 19*(2), 187–211.

Parkinson, H. J., & Thompson, G. (2003). Analysis and taxonomy of remanufacturing industry practice. *Proceedings of the Institute of Mechanical Engineers, Part E: Journal of Process Mechanical Engineering, 217,* 243–256.

Rathore, P., Kota, S., & Chakrabarti, A. (2011). Sustainability through remanufacturing in India: A case study on mobile handsets. *Journal of Cleaner Production, 19,* 1709–1722.

Sarath, P., Bonda, S., Mohanty, S., & Nayak, S. K. (2015). Mobile phone waste management and recycling: Views and trends. *Waste Management, 46,* 536–545.

Shumon, M. R. H., Arif-Uz-Zaman, K., & Rahman, A. (2011). Prospects of remanufacturing: A Bangladesh perspective. *International Journal of Industrial Engineering, 18*(5), 254–259.

Souza, G. C. (2013). Closed-loop supply chains: A critical review, and future research. *Decision Sciences, 44*(1), 7–37.

Sun, X., Wu, M., & Hu, F. (2013). Two-period inventory control with manufacturing and remanufacturing under return compensation policy. *The Scientific World Journal, 2013,* 1–8.

Sustainability Accounting Standards Board. (2014). Auto parts research brief. Retrieved May 30, 2018, from http://www.sasb.org/wp-ontent/uploads/2014/09/TR0102_AutoParts_Industry_Brief.pdf.

Tang, O., & Teunter, R. (2006). Economic lot scheduling problem with returns. *Production and Operation Management, 15,* 488–497.

Torkaman, S., Ghomi, S. M. T. F., & Karimi, B. (2017). Multi-stage multi-product multi-period production planning with sequence-dependent setups in closed-loop supply chain. *Computers and Industrial Engineering, 113,* 602–613.

Torkaman, S., Ghomi, S. M. T. F., & Karimi, B. (2018). Hybrid simulated annealing and genetic approach for solving a multi-stage production planning with sequence-dependent setups in a closed-loop supply chain. *Applied Soft Computing, 71,* 1085–1104.

Teunter, R., Kaparis, K., & Tang, O. (2008). Multi-product economic lot scheduling problem with separate production lines for manufacturing and remanufacturing. *European Journal of Operational Research, 191,* 1241–1253.

Welle, F. (2011). Twenty years of PET bottle to bottle recycling—An overview. *Resources, Conservation and Recycling, 55,* 865–875.

Zahraei, S. M., & Teo, C. C. (2018). Optimizing a recover-and-assemble remanufacturing system with production smoothing. *International Journal of Production Economics, 197,* 330–341.

Zhang, J., Liu, X., & Tu, Y. L. (2011). A capacitated production planning problem for closed-loop supply chain with remanufacturing. *International Journal of Advanced Manufacturing Technology, 54,* 757–766.

Zhang, H., & Wen, Z.-G. (2014). The consumption and recycling collection system of PET bottles: A case study of Beijing, China. *Waste Management, 34,* 987–998.

Zheng, J., & Wu, J. (2016). Research on closed-loop supply chain with third party recycler based on consumer preference. *Journal of Service Science and Management, 9,* 171–181.

Chapter 4
Greening the Supply Chain Through CSR: A Model for Assessing Sustainability Using Fuzzy-Based PPP Approach

M. S. Sai Vinod, Udhaya S. Ravishankar, and N. Sivakumar

Abstract Climate change and global warming have hit the boardrooms of corporations more than ever before. Sustainability indices have become important criteria in today's parlance to assess the risk of investing or collaborating with any company as part of their supply chain strategy. Many international firms provide these indices used by corporations to assess sustainability and collaborate. In a competing scenario, there is a need for companies in developing economies like India to be on par with their counterparts across the world by obtaining good sustainability scores. Towards this, one of the measures the Government in India has taken is mandating transparency in Corporate Social Responsibility (CSR) reporting through The Companies Act 2013. Since then large amounts of money that have been spent in specific areas have been reported towards achieving the societal needs and as part of that towards greening the supply chain. Using data obtained from the annual reports of NIFTY indexed companies, this paper develops a fuzzy-logic-based model to obtain a sustainability score for companies. The inputs used in the model deal with the triple bottom line criteria—Profit, People and Planet (PPP). Using fuzzy logic, rules have been created that define sustainability. This chapter also discusses the patterns in the sustainability scores obtained with respect to the NIFTY indexed companies.

Keywords Sustainability measurement · Fuzzy logic · CSR · Green supply chain · PPP model

---

M. S. S. Vinod (✉) · U. S. Ravishankar · N. Sivakumar
Department of Management and Commerce, Sri Sathya Sai Institute of Higher Learning, Prasanthi Nilayam 515134, India
e-mail: mssaivinod@sssihl.edu.in; saivinod@gmail.com

U. Ramanathan and R. Ramanathan (eds.), *Sustainable Supply Chains: Strategies, Issues, and Models*, https://doi.org/10.1007/978-3-030-48876-5_4

1 Introduction

> Successful companies have a social responsibility to make the world a better place and not just take from it—National CSR portal, India.

The world has been grappling with the issues of climate change and global warming in the recent decades. As new environmental challenges are being thrown up every day, countries and companies are facing tremendous pressure to mitigate the same (Koshy 2019a). In a larger context and from a long-term perspective, a holistic environment sustainable approach is desirable—which in simple terms means avoidance of the depletion of natural resources in order to maintain an ecological balance. International bodies including the United Nations (UN) have been striving constantly to propagate and provide achievable targets for sustainability that balances environment protection, societal progress and economic benefit (UNEP Finance Initiative 2019). In the Conference of the Parties (COP21), Paris-2015 and earlier conferences consisting of group of leading industrial countries in the world, the Triple Bottom Line (TBL) approach has been emphasized, which consists of the following (UN 1987; UNEP 2015):

1. People, the social equity bottom line.
2. Planet, the environmental protection and enhancement bottom line and
3. Profit, the economic bottom line.

To have a balanced improvement in TBL, many countries have set specific mandate for corporations. In a developing nation like India, sustainability is far from achieved from various angles—policy-making, practice and performance monitoring. Though certain inroads have been made recently in the form of new environmental laws and The Companies Act, 2013, there is a need to monitor the practice and performance through a reliable index.

Across the globe, TBL approach is currently being monitored in its various dimensions through ratings like The Dow Jones Sustainability Indices, MSCI's Environment Social Governance (ESG) scores, Trucost's ESG analysis and Thomson Reuter's ESG scores (Bonini and Swartz 2014). In a globalized economy, there is a need for companies in developing economies like India to be on par with their counterparts all over the world. The implications of using such scores are in the lines of risk assessment of companies, reputation of the companies, product portfolio, resource allocation and energy mix management.

Sustainability has become a strategic necessity and is no longer just a strategic differentiator (Ioannou and Serafeim 2019). In this context, some businesses have made sustainability a part of their charter. For example, Lego, a toys manufacturer has stated its purpose as, "inspire and develop the builders of tomorrow" (Vermeulen 2019). Unilever has highlighted that 33% of the consumers prefer to buy from sustainable brands (Unilever 2017). In business-to-business (B2B) scenario, sustainability is becoming even more vital (Bonini and Swartz 2014). For example, Salesforce has championed the 1-1-1 philanthropic model. It means giving one percent of product, one percent of equity and one percent of employees' time to communities and the non-profit sector (Gavin 2019). The International Energy Agency has recognized the need for carbon capture, utilization and storage (CCUS) technology for a sustainable future (IEA 2019).

With the rapid enhancement of technology, analytical capabilities and emergence of concepts of machine learning and deep learning, sustainability can be achieved with more practical outlook and deeper insights. This work is a small step in this endeavour. The succeeding section identifies the gap and substantiates the need for the work by assessing literature in the field of study. The methodology section gives the sources of data collection, sample data and its computation in the form of fuzzy-based rating technique applied to the NIFTY indexed companies in India. It rates the companies in the dimension of profit, people and planet through relevant data obtained from annual reports. Rule-based fuzzy logic has been applied for each of these dimensions to obtain sustainability scores. The study has been taken up for the period 2014–15 to 2017–18. The results and discussion section provides a bird's-eye view of the obtained results. The limitations of the study and managerial implications of fuzzy-based methodology are also presented. An elaborated year-wise and company-wise fuzzy scores are provided as Annexures to the work.

The important contributions of this work include the following:

i. Incorporating CSR dimension in greening the supply chain.
ii. Selection of proxies to measure sustainability in the dimensions of profit, people and Planet, which is relevant to the current Indian context.
iii. Using rule-based fuzzy logic methodology to incorporate proxies for sustainability and score them.

2 Literature Review

The concepts of sustainability, its scoring, CSR, environment responsibility have been studied in many ways. This section provides key research that has shaped this work. Literature review in this section has been done on the concept of sustainability, measuring sustainability, sustainability in India, green supply chain, economic performance and CSR, and the role of fuzzy logic in measuring sustainability.

2.1 Concept of Sustainability

The Brundtland Commission in 1987 gave the widely accepted definition of sustainability, "meeting the needs of the present without compromising the ability of future generations to meet their own needs" (UN 1987). Sustainability in business generally addresses two main aspects, namely, the effect of business on environment and the effect of business on society (Spiliakos 2018).

Sustainability characteristics encompass aspects that go beyond accounting profits to capture social and environmental impacts of business actions. McKinsey reports that sustainability is a deepening concept amongst executives and is its use is constantly on the rise (McKinsey and Company 2017; Henderson 2018).

2.2 Measuring Sustainability

The measurement of sustainability is an attempt to burrow the paradox, as the definition is qualitative in nature construed by human reasoning. This requires computational models that embed qualitative information in their structures (Vinodh et al. 2011); (Öberseder et al. 2011). Measuring sustainability has gained traction in the recent years. Stakeholders like investors, regulators, customers and suppliers use sustainability indices to mitigate risk, plan strategy, foster innovation and manage ethical issues (S&P Global 2019).

Sustainability measurement involves multiple criteria, which are both quantitative and qualitative in nature. There has been an attempt to standardize the methodology and parameters to measure sustainability (Spiliakos 2018). One of the earliest models to measure sustainability was the five-capital model of sustainability proposed by Forum for the Future (KCX Consulting 1990). This is shown in Fig. 1.

In the recent times, several sustainability scoring frameworks have been developed. These include the Dow Jones Sustainability index, MSCI ESG index, Trucost ESG index and Thomson Reuters' ESG index.

The Dow Jones Sustainability index (DJSI) uses Corporate Sustainability assessment (CSA), which not only considers publicly available information but also demands companies to participate in assessment (Robecosam 2019). The aspects on tax compliance, corporate lobbying and other ethical parameters are extensively assessed through questionnaires. Top companies of the world like Nestle and Walmart

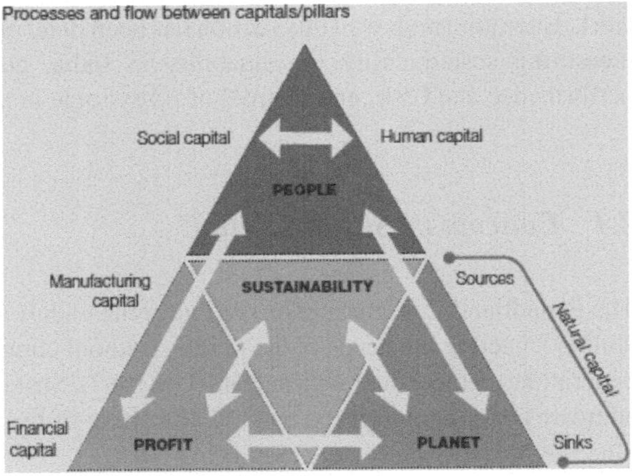

Fig. 1 Five capital model (The Forum for the Future 2019)

have volunteered themselves to be assessed on DJSI. MSCI ESG index uses the data from the public domain like Government and NGO datasets, company disclosure documents and media sources to measure sustainability (Morgan Stanley Capital International 2019). Trucost's computes ESG scores to assess sustainability using elaborate data on transparency of disclosure, innovation and risk, and energy mix used by companies (S&P Global 2019).

Thomson Reuters' uses ten comparable measures across companies, namely, resource use, emission, innovation, workforce, human rights, community, product responsibility, management, shareholders and CSR strategy. They also assess the controversies related to the companies in these measures. Their measure uses data from sources like annual reports, media, NGO webpages, company websites and CSR reports (Reinfinitiv 2019).

In last few decades, there has been an emphasis in many companies to incorporate environment management systems and to obtain certification for the same. For example, one of the worlds most recognized environmental management system, the ISO 14000 series, has the environment performance evaluation of ISO 14031 that deals with the following dimensions: environmental condition indicators, management performance indicators and operational performance indicators (ISO 2019). Table 1 highlights the parameters used to assess these indicators.

Other efforts to bring about sustainability in companies are through enhancing the reporting standards like The Global Reporting Initiative (GRI). GRI guidelines help to communicate the impact of business on issues like climate change, human rights and corruption. Adherence to GRI standards has been very positive in some pockets of the globe and has been making inroads in the Indian business context (GRI 2019). Research has found that, about 40% of companies in India publish their sustainability reports exclusively (Vinod and Sivakumar 2018a).

Table 1 Environment performance indicators in ISO 14031 (International Organization for Standardization 2019)

Indicators	Parameters used to assess the indicators
Environmental condition indicators	• Water quality of nearby lake • Regional air quality • Noise pollution level at peak periods • Atmosphere CO_2 emissions (ppm)
Management performance indicators	• Number and results of environmental audits conducted • Staff member training • Supplier assessments
Operational performance indicators	• Absolute energy consumption (KWh) • Waste per unit of output • Transportation volume • Volume of products shipped

2.3 *Sustainability Initiatives in India*

Indians, by virtue of their culture, have been good philanthropists and businessmen. The aspect of providing clean environment has been incorporated in the Indian Constitution. In fact, Article 51-A (g) under the directive principles of the Indian constitution states that

> It shall be duty of every citizen of India to protect and improve the natural environment including forests, lakes, rivers and wildlife and to have compassion for living creatures. (Government of India 2019a, b)

In the world scenario, India has been providing cooperation to international accords on climate change and sustainability. Starting from the Water Act of 1974 based on UN Conference on the Human Environment to COP-21 Paris accord-2015, a number of environment-related laws have been passed in India like

- The Air (Prevention and Control of Pollution) Act, 1981;
- The Environment Protection Act, 1986;
- The Environmental Impact Assessment Notification, 2006;
- The National Green Tribunal Act, 2010 and
- The Hazardous Waste Management Regulations, 2016.

The Companies Act 2013 provided teeth to the Government, mandating companies to spend two percent of their profits in CSR-related activities and to disclose the same in their annual reports. Yet another milestone in India was the launch of National CSR Portal in 2016 providing the details of the CSR activities undertaken by companies (Government of India 2019a, b). It ensured that Indian companies would meet the UN specified Sustainable Development Goals (SDGs) to a maximum extent.

The Government data for about twenty thousand Indian companies shows that around INR 100 billion was spent each year on CSR activities since 2014–15. A total of about INR 220 billion was spent by the NIFTY indexed companies alone, in India, from 2014–15 to 2017–18 on CSR activities. As seen in Fig. 2, the CSR spending year-on-year is on the rise (Vinod and Sivakumar 2017).

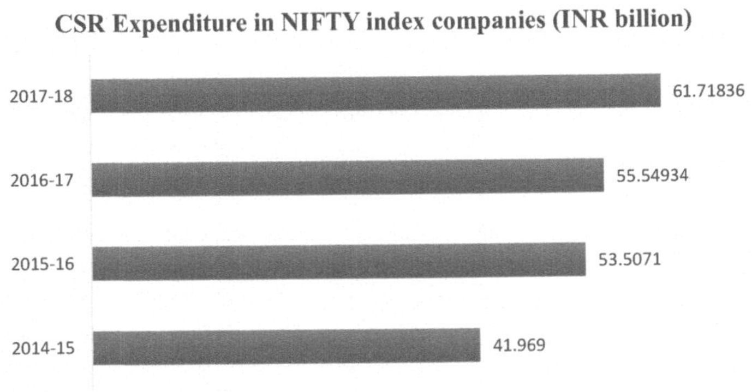

Fig. 2 CSR expenditure in NIFTY indexed companies

As per schedule 7 of the Companies Act 2013, CSR spending has to be in the following areas (Ministry of Corporate affairs 2013):

i. eradicating extreme hunger and poverty;
ii. promotion of education;
iii. promoting gender equality and empowering women;
iv. reducing child mortality and improving maternal health;
v. combating human immunodeficiency virus, acquired immune deficiency syndrome, malaria and other diseases;
vi. ensuring environmental sustainability;
vii. employment enhancing vocational skills;
viii. social business projects;
ix. contribution to the Prime Minister's National Relief Fund or any other fund set up by the Central Government or the State Governments for socio-economic development and relief and funds for the welfare of the Scheduled Castes, the Scheduled Tribes, other backward classes, minorities and women; and
x. such other matters as may be prescribed.

Amongst the above, ensuring environmental sustainability has been drawing about 10% of the CSR funds in the NIFTY indexed companies. This amounts to about INR 5 billion each year. Vinod and Sivakumar (2018b) state that environmental spending has been on the rise since 2014–15, which is encouraging. The amount has been spent in the areas like tree plantation, river rejuvenation, ensuring cleanliness in the locality, funding environment-related innovation, research on education and environmental challenges. The spending has been largely need based and local to the industry's location.

Though there has been a rise in the budget allocated for environmental ministry in India, it has not been utilized fully for the sanctioned projects (Press Trust of India 2019). For example, only 10 out of 100 projects have been completed under the Namami Gange mission (an initiative to clean up the Ganges River) with only INR 6.7 billion spent out of INR 28 billion sanctioned until May 2019 (Koshy 2019a, b, c). Research by key institutions in India like Indian Institute of Technology and Indian Institute of Science has shown that 12 Himalayan states in India are highly vulnerable to climate change (Koshy 2019a, b, c, d). These states are highly rich in natural resources and many businesses depend on them. Instead of helping in sustainable and holistic agriculture, multinational conglomerates have been dragging farmers in India to the court for rights infringement issues (Jebaraj 2019). Citizens are suffering with air-borne diseases, as India has become a host to 14 of the 20 most polluted cities in the world (Koshy 2019a, b, c, d). There is a lot of scope for corporates in India to contribute to the sustainability story of India.

2.4 Green Supply Chain, Economic Performance and CSR

Few studies have been made in the direction of greening the supply chain through CSR and vice versa, i.e. achieving the objectives of CSR through greening the supply chain.

Green supply chain includes set of activities undertaken by companies internally as well as externally to improve its environmental performance by including all stakeholders—suppliers, clients, partners and local communities, (Sarkis et al. 2010). In the same work, authors bring out the dimension of supply and demand sustainability in CSR networks in green supply chain. There is an emphasis to increase social embeddedness in green supply chain, which allows access to environmental management and increased partner cooperation.

The concept of compassionate operations was brought into the light with special focus of Millennial Development Goals (MDGs) by UN (Sarkis 2012). Here, one of the emphases was achieving environmental sustainability in developing nations along with other goals like poverty alleviation, universal primary education and gender equality.

There have been studies to know effects of green supply chain on profits and it was found that coordinating activities between manufacturers and suppliers in the environmental dimension will yield positive results (Zhu et al. 2012). In the same paper, the authors have adopted the role of social cost in measuring the effectiveness of GSC on company performance. Thus, societal impact of the operations plan of any organization, its implementation and results become vital to the study of sustainability. Companies need to adopt green practices at operational level as well as at strategic level to be effective and make an impact to the local community (Bhardwaj 2016).

CSR undertaken in the form of green supply chain may lead to legitimacy issues if not implemented in the proper way. A rigorous CSR policy and transparency will be reflected in the corporate financial performance (Wang and Sarkis 2017). Thus, implementation of CSR activities becomes a critical criterion to distinguish a legitimate company in contrast to a greenwashing company.

From the above-stated works, it can be concluded that green supply chain, economic performance and CSR must be integrated. It is also seen that there is a need to carry out such work in a developing nation perspective. The endeavour of this work is to fill the research gap by integrating the aspects of green supply chain, economic performance and implementation of SDGs as part of CSR in Indian context.

2.5 Fuzzy Logic

Multi-criteria decision-making methods are vital tools for any business manager (Mateusz et al. 2018). Methods like Analytical Hierarchical Process (AHP), VIKOR,

TOPSIS, Delphi, ANP and DEMATEL have been used in enhancing decisions relating to human resources management, risk and financial management, operation management, marketing, resource planning, quality management, health care, sustainability and environment (Mardani et al. 2016). Thus, from the literature, it is seen that operations research and allied methods have been used to measure sustainability especially in developing nations (White et al. 2010).

Fuzzy logic has found its relevance in the field of business, because of its inherent modelling structure which involves multi-criteria decision-making. It is a soft computing method that has the capability of scoring performances or describing any other entity by means of approximate reasoning. This method makes it particularly useful, especially when there lies utmost uncertainty in quantifying the entities, but there exists some qualitative understanding about them. In the areas like consumer behaviour, operations, retailing, behavioural finance and economics, fuzzy logic has been able to provide results with precision. Even in measuring sustainability some amount of inroads has been made using fuzzy-based methodology.

Muñoz et al. (2008) studied 52 Spanish companies based on TBL concept using return on shareholder funds, return on total assets, stakeholder orientation index and sustainability reporting index as proxies. The study used fuzzy logic to measure stakeholder orientation. Similarly, Escrig-Olmedo et al. (2015) developed a single-stage fuzzy logic system for rating the Corporate Environmental Performances (CEP) using criteria such as strategic intent, governance and management, engagement and operational performance based on data from Spanish companies as per GRI guidelines.

Sustainability Assessment by Fuzzy Evaluation (SAFE) model has been developed to measure the sustainability of nations and corporates, respectively (Andriantiatsaholiniaina et al. 2004). This model scores sustainability by considering both ecological and human factors, such as the environmental and social developments, fuzzified by approximate reasoning at each level. In the ecological factors, air, land, water and biodiversity are considered of fuzzification, whilst in the human factors political, health, wealth and knowledge are considered. The data required to enable such fuzzy sustainability scoring are emissions, wastes, crop usage, wildlife affected, compliance with international standards, revenue, fatality rates and R&D expenditure (Phillis and Davis 2009).

Grigoroudis et al. (2019) developed a fuzzy model for energy sustainability. This work based its sustainability scores on environment, energy, human systems, economy and security inputs. Under environment, for example, it considered air and land as inputs. With respect to air, Greenhouse Gas (GHG) emissions, NO_x emissions, SO_2 emissions and Particulate Matter ($PM_{2.5}$) were taken as inputs. Further, for land, forest change and energy from waste were considered.

A close look into the fuzzy models developed shows that

1. Many parameters and sub-parameters have been used to measure sustainability. To assign weights to each parameter in measuring sustainability may complicate the work further.

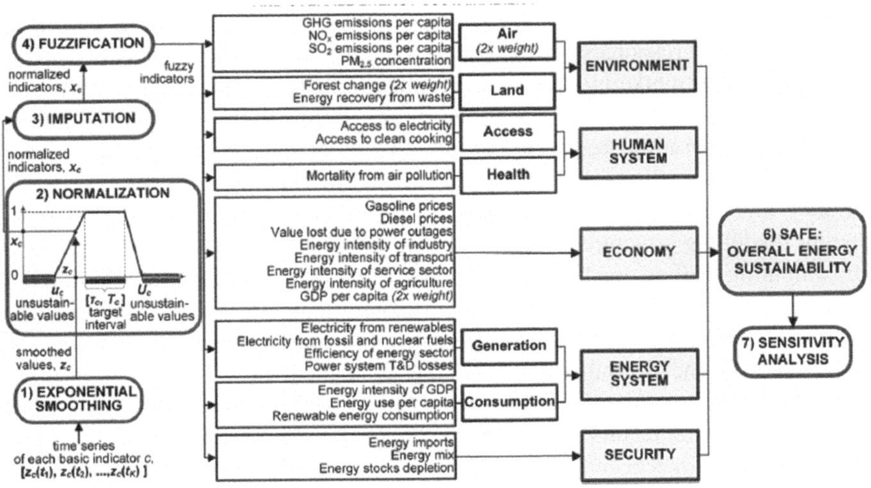

Fig. 3 Fuzzy model for energy sustainability by Grigoroudis et al. (2019)

2. Unavailability of data for all the parameters used by the works (cited above) in the Indian context. This is due to the fact that adherence to GRI standards of reporting in India is still underway (Kumar and Prakash 2019).

The data used for this work encompasses most of the parameters to score sustainability. This is done as part of CSR funding by companies.

The literature reviewed in this section point out to the need for developing a reliable sustainability measurement framework in the Indian context. The measurement parameters have to be available. It was also found that not many improvements have been made since Escrig-Olmedo et. al's 2015 paper, in the area of fuzzy scoring with respect to sustainability. Considering the Indian context, availability of data and methodology, this paper addresses this important research gap using a fuzzy logic approach.

3 Methodology of the Study

This section provides insights into the methodology used in calculating sustainability scores.

Sample: The NIFTY is one of the prominent stock market indices in India. It has a diverse group of 50 companies from sectors like banking, information technology, automobile, energy production, pharmaceutical, steel and cement. The companies listed in the NIFTY index have been used as a sample for this study.

Data collection: This work uses 4 years of data from 2014–15 to 2017–18 obtained from annual reports of the companies in the NIFTY index. The following data of the companies studied have been used:

Table 2 PPP proxies used to construct the fuzzy model

Orientation	Proxy used
Profit	Net profit
People	CSR spending on SDGs (SDG)
Planet	Amount spent on environment as part of CSR (GSC)

1. Profit of the companies,
2. Amount of profit spent in meeting Sustainable Development Goals (SDGs) through CSR and
3. Amount spent on environment as part of CSR, considered as amount spent on greening the supply chain (GSC).

Thus, the work gathered 600 (i.e. 50*4*3) Company Profit-SDG expenditure-GSC (Green Supply Chain) data points. From this data, percentage increase in profits, percentage increase in CSR spending and percentage increase in CSR amount spent on greening the supply chain were calculated year-on-year. This gave rise to 300 data points. This data was subjected to fuzzy logic scoring as explained in the next section.

Fuzzy-based sustainability scoring: The fuzzy logic scoring scheme developed in this work is similar to the works of Muñoz et al. (2008) and Phillis and Davis (2009). However, it takes parameters that are available in Indian context into consideration for the sustainability scoring. This work is made up of multi-stages as explained below.

Stage 1: Criteria selection

This work uses the PPP model similar to Munoz et al. (2008) in the Indian context with slightly different proxies as shown in Table 2.

Stage 2: Year-wise PPP orientation pattern mining

At the beginning, data is collected and presented as in Table 3. It gives a sample of one company's year-wise data collected from the annual report in each one of the parameters—profit, SDG and GSC.

Table 4 provides the percentage increase in each of the parameters used in the fuzzy model from Table 3.

PPP orientation and sustainability scores of companies are determined by means of fuzzy approximate reasoning. It is based on percentage increase in each of the parameters in Table 4, which in a way normalizes the scores (Marius et al. 2013) and makes the data ready for fuzzification.

The fuzzification engine is shown in Fig. 4. Specific rules that have been applied to identify the PPP orientation are

- Profit-orientation pattern: A company was identified as being profit oriented in a particular year if its profit increase was greater than the increase in SDG and GSC.

Table 3 Year-wise data collected from annual report

Amount in INR mn	2014–15			2015–16			2016–17			2017–18		
Company	Profit	SDG	GSC	Profit	SDG	GSC	Profit	SDG	GSC	Profit	SDG	GSC
X	43165	429.1	7	43230	867.2	145	47455	949.8	346.8	49915	1005.1	774.9

Table 4 Rule-based fuzzy scores based on increase in PPP

Company name	Increase in profit	Increase in SDG	Increase in GSC	Profit oriented	People oriented	Planet oriented	Sustainability
X	11.66	11.40	142.27	0.3813	0.3165	0.6179	0.4854

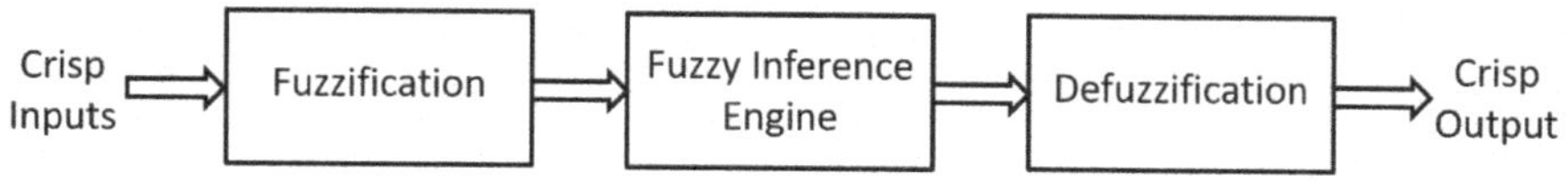

Fig. 4 Fuzzy sustainability scoring scheme

- People-orientation pattern: A company was identified as being people oriented in a particular year if its increase in SDG expenditure was greater than profits and GSC.
- Planet-orientation pattern: A company was identified as being planet oriented in a particular year if its increase in GSC expenditure was greater than profits and SDG.
- Sustainable orientation pattern: A company was identified as being sustainable in orientation in a particular year if its increase in all criteria was relatively equal (Fig. 4).

The rule sets used in this paper that define the scoring process are in fact much more intricate than just plain rule formulation as is typically done in most fuzzy-logic-related works. In the fuzzy scoring blocks shown in Fig. 3, there exist sub-stages within them that capture the aforementioned patterns, before finalizing the score.

Stage 3: Implementation

The percentage increases in profits, SDG and GSC computed from the data collected were taken as training set. This means that the fuzzy partitioning on each input criteria was derived by aggregating data from all companies and from all 3-year differences in this training set. Specifically, the Quartiles (Q) and Inter Quartile Ranges (IQR) to provide the necessary partitions were used. Table 3 summarizes this (Table 5).

The inputs are partitioned into five fuzzy sets, namely, VERY BAD, BAD, NEUTRAL, GOOD and VERY GOOD. Using the values in Table 3, the cross-over points for the sigmoid membership functions for the terms VERY BAD, BAD, GOOD and VERY GOOD were determined. For the NEUTRAL linguistic term, the corresponding values were taken for the centroid of the respective Gaussian membership function. Figure 5 shows this.

For the output fuzzy sets, triangular membership functions with equal widths across the range between 0 and 1 were used. Sustainability score was calculated based on year-on-year and company-wise comparison on the proxies selected.

Table 5 Input partition values based on quartiles from the data

	Very bad ($Q1 - 1.5 \times \boldsymbol{IQR}$)	Bad ($Q1$)	NEUTRAL ($Q2$)	Good ($Q3$)	Very good ($Q3 + 1.5 \times \boldsymbol{IQR}$)
% increase in profits	−42.2369	−0.25013	13.3056	27.7411	69.7278
% increase in SDG	−79.7106	1.4789	15.9603	55.6052	136.7948
% increase in GSC	−194.1993	−21.1961	5.2049	94.1394	267.1426

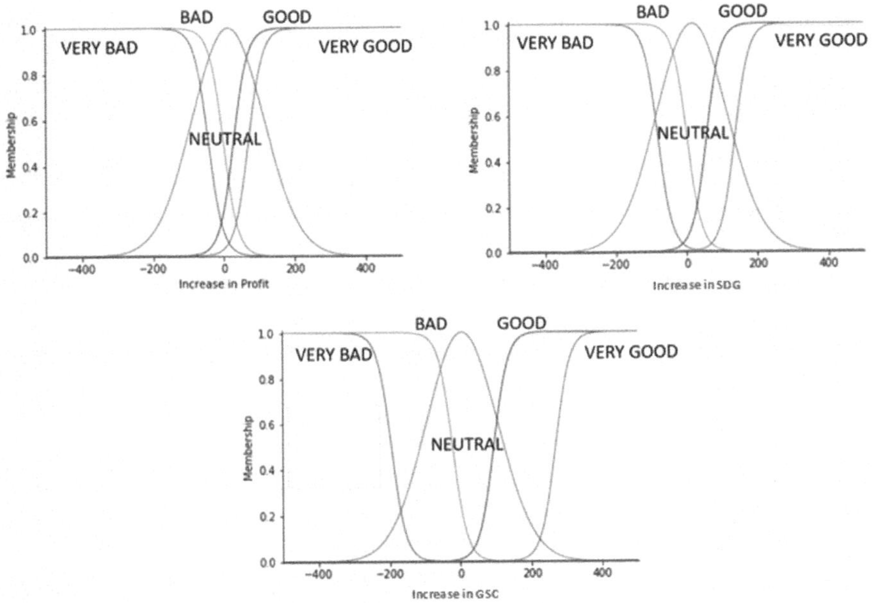

Fig. 5 Input fuzzy sets

4 Results and Discussion

This section provides important results obtained from the research. The average year-wise fuzzy scores for each of the orientation types are shown in Table 6.

An analysis of the scores of Table 6 shows the following:

- Sustainability score is highest in the years 2014–15 to 2015–16. This can be attributed to the consistency amongst the PPP scores.

Table 6 Average year-on-year PPP orientation of NIFTY indexed companies

Year	Profit	People	Planet	Sustainability
2014–15 to 2015–16	0.3908	0.4511	0.4352	0.4932
2015–16 to 2016–17	0.4094	0.3860	0.4544	0.4657
2016–17 to 2017–18	0.4049	0.3832	0.4730	0.4773
Average score	0.4017	0.4067	0.4542	0.4787

- Average profit- and people-orientation scores of NIFTY indexed companies are almost same. However, the year-on-year comparison reveals that there are variations in their values.
- Planet-orientation scores of the companies studied are higher compared to profit and people-orientation scores.
- Profit-orientation scores have been steady.
- People-orientation scores have been falling.
- Sustainability scores of the companies have been inconsistent in the period of study.

Annexures 1–3 give the year-wise PPP orientation and sustainability scores of NIFTY indexed companies.

Some of the inferences acquired from the results of the fuzzy-based sustainability scores include the following:

- The top sustainable companies come from both the service sector and manufacturing sectors. It is good to see this balance.
- Planet-orientation scores have been better than profit- and people-orientation scores. This implies that planet orientation via CSR has been on the rise and enforcement of company law has been good to make decent improvements with respect to environmental protection.

Some points of concern are

- Decrease in sustainability scores from 2014–15 to 2017–18.
- There is a lack of consistency in companies towards their orientation. A closer look of the scores reveals that none of the companies score consistently high in the profit, people and planet orientations.

4.1 Limitations of the Study and Scope for Further Work

There are some limitations to this work which are given as follows:

- This work has taken into account the data from 2014–15 onwards only, since the Companies Act 2013 was implemented.
- With only limited data, the inferences drawn cannot be generalized for the entire Indian industry.

There is a large scope for further work.

- The methodology proposed is very generic and can accommodate for the data in years to come.
- Dataset from specific companies can be used to improve the model and provide customized solutions.
- The sustainable fuzzy scores can be calculated based on magnitude of expenditure to understand the results better.
- Weights for each of the PPP criteria can be assigned to get different sustainability scores as part of further study.
- The proxies used for calculating the sustainability scores can be extended.
- More companies can be brought into the ambit of this study to get a better understanding.

5 Conclusion

Some of the suggestions and recommendations with managerial implications based on this study include the following:

1. Sustainability reporting practices of Indian companies under the Companies Act 2013 can be based on the fuzzy methodology proposed in this study.
2. Companies can choose their supply chain partners in India based on sustainability scores at domestic as well as international market places.
3. Legal compliance can be implemented using fuzzy-based sustainability scores.
4. Credit rating methodology can incorporate fuzzy-based sustainability scores as a part of their methodology.
5. Awards and encouragements can be instituted using fuzzy-based sustainability scores.
6. Last but not the least, fuzzy-based sustainability scores can be used to study overall Indian national economic sustainability.

This paper developed a fuzzy-based PPP orientation scoring framework. The sustainability scores of all the companies listed in the NIFTY index were calculated. The framework has the potential to improve sustainability measurement and scoring. If implemented in right earnest the framework can lead to a positive sustainable future.

Dedication This work is an offering of respect and love to the Revered Founder Chancellor of Sri Sathya Sai Institute of Higher Learning—Bhagawan Sri Sathya Sai Baba.

Annexure 1: PPP Orientation and Sustainability Scores of NIFTY Indexed Companies for 2014–15 to 2015–16

Company name	Profit oriented	People oriented	Planet oriented	Sustainability
Adani Ports and Special Economic Zone Ltd.	0.4049	0.4014	0.5333	0.4343
Ambuja Cements Ltd.	0.3509	0.5458	0.3890	0.4163
Asian Paints Ltd.	0.3967	0.5851	0.4114	0.5635
Aurobindo Pharma Ltd.	0.4673	0.5358	0.4167	0.6805
Axis Bank Ltd.	0.4278	0.3990	0.4870	0.4392
Bajaj Auto Ltd.	0.2572	0.3135	0.6875	0.5762
Bajaj Finance Ltd.	0.2808	0.7279	0.2500	0.5000
Bharat Petroleum Corporation Ltd.	0.4088	0.4453	0.5632	0.7058
Bharti Airtel Ltd.	0.5333	0.4271	0.3838	0.4808
Bharti Infratel Ltd.	0.2857	0.2509	0.7497	0.5000
Bosch Ltd.	0.3688	0.6185	0.2500	0.4403
Cipla Ltd.	0.3794	0.4931	0.3750	0.4181
Coal India Ltd.	0.2465	0.4546	0.5454	0.5000
Dr. Reddy's Laboratories Ltd.	0.4082	0.5118	0.3750	0.4730
Eicher Motors Ltd.	0.5000	0.5000	0.2500	0.5000
GAIL (India) Ltd.	0.2493	0.2575	0.7500	0.5000
HCL Technologies Ltd.	0.4470	0.5030	0.3701	0.5779
HDFC Bank Ltd.	0.4746	0.5154	0.2500	0.5000
Hero MotoCorp Ltd.	0.3282	0.6609	0.2500	0.5880
Hindalco Industries Ltd.	0.3776	0.3957	0.4119	0.4073
Hindustan Petroleum Corporation Ltd.	0.3333	0.3337	0.6663	0.5972
Hindustan Unilever Ltd.	0.4592	0.4026	0.3898	0.4205
Housing Development Finance Corporation Ltd.	0.2658	0.2602	0.7500	0.5000
I T C Ltd.	0.4203	0.4085	0.4720	0.4335
ICICI Bank Ltd.	0.5178	0.3960	0.3750	0.4169
Indiabulls Housing Finance Ltd.	0.3893	0.5941	0.2500	0.5871
Indian Oil Corporation Ltd.	0.4210	0.4521	0.4771	0.5248
IndusInd Bank Ltd.	0.4391	0.4758	0.4657	0.5840
Infosys Ltd.	0.2602	0.2499	0.7500	0.5000

(continued)

(continued)

Company name	Profit oriented	People oriented	Planet oriented	Sustainability
Kotak Mahindra Bank Ltd.	0.4209	0.4476	0.3750	0.4779
Larsen & Toubro Ltd.	0.3562	0.5837	0.3478	0.4078
Lupin Ltd.	0.4602	0.4791	0.3780	0.5762
Mahindra & Mahindra Ltd.	0.3924	0.3920	0.4531	0.4046
Maruti Suzuki India Ltd.	0.4747	0.5035	0.2808	0.5443
NTPC Ltd.	0.3465	0.6337	0.2500	0.5372
Oil & Natural Gas Corporation Ltd.	0.3797	0.3861	0.3612	0.3841
Power Grid Corporation of India Ltd.	0.2634	0.5852	0.3981	0.6148
Reliance Industries Ltd.	0.4299	0.3794	0.3887	0.3940
State Bank of India	0.2531	0.2510	0.7500	0.5000
Sun Pharmaceutical Industries Ltd.	0.2593	0.6475	0.2500	0.5000
Tata Consultancy Services Ltd.	0.4806	0.4409	0.3718	0.4860
Tata Motors Ltd.	0.3946	0.4035	0.3941	0.4105
Tata Steel Ltd.	0.3749	0.4488	0.3795	0.4060
Tech Mahindra Ltd.	0.5534	0.3531	0.3870	0.4499
UltraTech Cement Ltd.	0.3810	0.4077	0.4615	0.4269
UPL Ltd.	0.4432	0.3985	0.4403	0.4231
Vedanta Ltd.	0.4230	0.3741	0.3952	0.3978
Wipro Ltd.	0.4405	0.4165	0.4646	0.4416
Yes Bank Ltd.	0.4429	0.4662	0.5345	0.6185
Zee Entertainment Enterprises Ltd.	0.4729	0.4433	0.4083	0.4940

Annexure 2: PPP Orientation and Sustainability Scores of NIFTY Indexed Companies from 2015–16 to 2016–17

Company name	Profit oriented	People oriented	Planet oriented	Sustainability
Adani Ports and Special Economic Zone Ltd.	0.5015	0.4207	0.3750	0.4545
Ambuja Cements Ltd.	0.3880	0.3948	0.4168	0.3969

(continued)

(continued)

Company name	Profit oriented	People oriented	Planet oriented	Sustainability
Asian Paints Ltd.	0.4279	0.4846	0.3750	0.5198
Aurobindo Pharma Ltd.	0.3029	0.2516	0.7498	0.5000
Axis Bank Ltd.	0.5232	0.3826	0.4097	0.4466
Bajaj Auto Ltd.	0.3813	0.3165	0.6179	0.4855
Bajaj Finance Ltd.	0.3190	0.2507	0.6906	0.5765
Bharat Petroleum Corporation Ltd.	0.5907	0.2895	0.2500	0.4353
Bharti Airtel Ltd.	0.4759	0.2731	0.4258	0.3962
Bharti Infratel Ltd.	0.5094	0.3449	0.4369	0.4764
Bosch Ltd.	0.3929	0.4504	0.5588	0.6079
Cipla Ltd.	0.4772	0.3745	0.3750	0.3864
Coal India Ltd.	0.2400	0.5199	0.4698	0.5000
Dr. Reddy's Laboratories Ltd.	0.3927	0.4008	0.4330	0.4076
Eicher Motors Ltd.	0.5729	0.4194	0.2500	0.5102
GAIL (India) Ltd.	0.3746	0.3850	0.3615	0.3817
HCL Technologies Ltd.	0.2644	0.4892	0.5107	0.5000
HDFC Bank Ltd.	0.3291	0.3341	0.5016	0.3853
Hero MotoCorp Ltd.	0.4697	0.4481	0.4038	0.5040
Hindalco Industries Ltd.	0.3615	0.3811	0.4000	0.3863
Hindustan Petroleum Corporation Ltd.	0.5624	0.3998	0.2500	0.5995
Hindustan Unilever Ltd.	0.4542	0.4143	0.4416	0.4402
Housing Development Finance Corporation Ltd.	0.4151	0.5685	0.2577	0.5258
I T C Ltd.	0.4622	0.4117	0.4360	0.4394
ICICI Bank Ltd.	0.3892	0.4041	0.4330	0.4131
Indiabulls Housing Finance Ltd.	0.2731	0.2474	0.7500	0.4944
Indian Oil Corporation Ltd.	0.5230	0.4341	0.4304	0.5480
IndusInd Bank Ltd.	0.5432	0.4238	0.3745	0.4842
Infosys Ltd.	0.2700	0.2538	0.7500	0.5000
Kotak Mahindra Bank Ltd.	0.3709	0.4277	0.3746	0.3970
Larsen & Toubro Ltd.	0.2588	0.2500	0.6850	0.4880
Lupin Ltd.	0.5360	0.3756	0.4028	0.4462
Mahindra & Mahindra Ltd.	0.4079	0.3927	0.3963	0.4006

(continued)

(continued)

Company name	Profit oriented	People oriented	Planet oriented	Sustainability
Maruti Suzuki India Ltd.	0.4728	0.4158	0.4265	0.4478
NTPC Ltd.	0.3570	0.3729	0.4391	0.3771
Oil & Natural Gas Corporation Ltd.	0.2545	0.2515	0.7500	0.5000
Power Grid Corporation of India Ltd.	0.4522	0.4416	0.3750	0.4768
Reliance Industries Ltd.	0.4779	0.3943	0.3897	0.4101
State Bank of India	0.5998	0.2497	0.3996	0.5000
Sun Pharmaceutical Industries Ltd.	0.3814	0.2281	0.5637	0.4294
Tata Consultancy Services Ltd.	0.2831	0.2519	0.7450	0.5000
Tata Motors Ltd.	0.3922	0.4794	0.4559	0.4969
Tata Steel Ltd.	0.3694	0.3901	0.4147	0.3992
Tech Mahindra Ltd.	0.5773	0.3951	0.2925	0.5382
UltraTech Cement Ltd.	0.3860	0.4023	0.4503	0.4180
UPL Ltd.	0.2332	0.6041	0.3941	0.4457
Vedanta Ltd.	0.2593	0.6794	0.2500	0.5000
Wipro Ltd.	0.4557	0.4213	0.4596	0.4527
Yes Bank Ltd.	0.4783	0.4146	0.4959	0.6045
Zee Entertainment Enterprises Ltd.	0.2819	0.2932	0.4277	0.3555

Annexure 3: PPP Orientation and Sustainability Scores of NIFTY Indexed Companies from 2016–17 to 2017–18

Company name	Profit oriented	People oriented	Planet oriented	Sustainability
Adani Ports and Special Economic Zone Ltd.	0.4850	0.4287	0.4467	0.4768
Ambuja Cements Ltd.	0.3738	0.3268	0.3616	0.3619
Asian Paints Ltd.	0.4242	0.3277	0.5754	0.5372
Aurobindo Pharma Ltd.	0.4152	0.3976	0.5203	0.4357
Axis Bank Ltd.	0.3925	0.3958	0.4170	0.4048
Bajaj Auto Ltd.	0.3691	0.3175	0.6298	0.4634
Bajaj Finance Ltd.	0.4829	0.4821	0.2500	0.5000

(continued)

(continued)

Company name	Profit oriented	People oriented	Planet oriented	Sustainability
Bharat Petroleum Corporation Ltd.	0.4792	0.4908	0.2500	0.5000
Bharti Airtel Ltd.	0.2635	0.7498	0.2500	0.5000
Bharti Infratel Ltd.	0.5067	0.4385	0.3594	0.4971
Bosch Ltd.	0.2682	0.2507	0.7310	0.5000
Cipla Ltd.	0.3114	0.5093	0.4722	0.5556
Coal India Ltd.	0.3682	0.3341	0.3627	0.3589
Dr. Reddy's Laboratories Ltd.	0.3737	0.3767	0.4310	0.3853
Eicher Motors Ltd.	0.3861	0.2643	0.6139	0.6361
GAIL (India) Ltd.	0.3794	0.3966	0.4327	0.4090
HCL Technologies Ltd.	0.4283	0.3410	0.3644	0.3698
HDFC Bank Ltd.	0.4221	0.5779	0.2500	0.5000
Hero MotoCorp Ltd.	0.2763	0.2504	0.7500	0.5000
Hindalco Industries Ltd.	0.5044	0.3934	0.4515	0.4813
Hindustan Petroleum Corporation Ltd.	0.5780	0.4141	0.3883	0.5880
Hindustan Unilever Ltd.	0.4138	0.4088	0.5604	0.4684
Housing Development Finance Corporation Ltd.	0.3888	0.4284	0.5569	0.4839
I T C Ltd.	0.4355	0.4046	0.4433	0.4241
ICICI Bank Ltd.	0.2520	0.2502	0.7500	0.5000
Indiabulls Housing Finance Ltd.	0.2941	0.7500	0.2500	0.4963
Indian Oil Corporation Ltd.	0.5575	0.4337	0.3975	0.5998
IndusInd Bank Ltd.	0.5633	0.3286	0.3318	0.4092
Infosys Ltd.	0.2673	0.2507	0.7234	0.5000
Kotak Mahindra Bank Ltd.	0.3960	0.2566	0.5851	0.6266
Larsen & Toubro Ltd.	0.4088	0.3977	0.3745	0.4094
Lupin Ltd.	0.4815	0.4104	0.4262	0.4453
Mahindra & Mahindra Ltd.	0.3899	0.3879	0.5259	0.4173
Maruti Suzuki India Ltd.	0.5379	0.4280	0.3462	0.5666
NTPC Ltd.	0.3846	0.3748	0.5567	0.4298
Oil & Natural Gas Corporation Ltd.	0.3867	0.4035	0.3676	0.3994
Power Grid Corporation of India Ltd.	0.3375	0.2506	0.6620	0.5641

(continued)

(continued)

Company name	Profit oriented	People oriented	Planet oriented	Sustainability
Reliance Industries Ltd.	0.3869	0.3154	0.6127	0.5320
State Bank of India	0.4049	0.4015	0.3709	0.4129
Sun Pharmaceutical Industries Ltd.	0.3942	0.4406	0.3759	0.4195
Tata Consultancy Services Ltd.	0.2705	0.2506	0.7264	0.5000
Tata Motors Ltd.	0.3945	0.3790	0.4472	0.3993
Tata Steel Ltd.	0.3489	0.4293	0.5195	0.4356
Tech Mahindra Ltd.	0.4962	0.4187	0.4136	0.4619
UltraTech Cement Ltd.	0.2695	0.2509	0.7500	0.5000
UPL Ltd.	0.2819	0.2501	0.7500	0.5000
Vedanta Ltd.	0.5000	0.2502	0.5000	0.5000
Wipro Ltd.	0.4435	0.3968	0.3750	0.4101
Yes Bank Ltd.	0.5719	0.2507	0.3975	0.5938
Zee Entertainment Enterprises Ltd.	0.5000	0.5000	0.2500	0.5000

References

Andriantiatsaholiniaina, L. A., Kouikoglou, V. S., & Phillis, Y. A. (2004). Evaluating strategies for sustainable development: Fuzzy logic reasoning and sensitivity analysis. *Ecological Economics*, 149–172. https://doi.org/10.1016/j.ecolecon.2003.08.009.

Bhardwaj, B. R. (2016). Role of green policy on sustainable supply chain management: A model for implementing corporate social responsibility (CSR). *Benchmarking: An International Journal*. https://doi.org/10.1108/bij-08-2013-0077.

Bonini, S., & Swartz, S. (2014). *Profits with purpose: How organizing for sustainability can benefit the bottom line*. Retrieved June 1, 2019, from http://www.highmeadowsinstitute.org, http://www.highmeadowsinstitute.org/wp-content/uploads/2014/08/McKinsey_Profits-with-Purpose.pdf.

Escrig-Olmedo, E., Muñoz-Torres, M. J., Fernández-Izquierdo, M. Á., & Rivera-Lirio, J. M. (2015). Measuring corporate environmental performance: A methodology for sustainable development. *Business Strategy and the Environment*, 142–162. https://doi.org/10.1002/bse.1904.

Gavin, M. (2019, June 6). *3 examples of corporate social responsibility that were successful*. Retrieved June 11, 2019, from Harvard Business School Online: https://online.hbs.edu/blog/post/corporate-social-responsibility-examples.

Government of India. (2019a, March 1). *CSR in India*. Retrieved from https://csr.gov.in/CSR/.

Government of India. (2019b, June 1). *Environment protection under constitutional framework of India*. Retrieved from http://pib.nic.in, http://pib.nic.in/newsite/PrintRelease.aspx?relid=105411.

GRI. (2019, June 1). *G4 sustainability reporting guidelines*. Retrieved from https://www2.globalreporting.org, https://www2.globalreporting.org/standards/g4/Pages/default.aspx.

Grigoroudis, E., Kouikoglou, V. S., Phillis, Y. A., & Kanellos, F. D. (2019). Energy sustainability: A definition and assessment model. *Operational Research: An International Journal*. https://doi.org/10.1007/s12351-019-00492-2.

Henderson, R. M. (2018, February 12). *More and more CEOs are taking their social responsibility seriously*. Retrieved June 1, 2019, from Harvard Business Review: https://hbr.org/2018/02/more-and-more-ceos-are-taking-their-social-responsibility-seriously.

IEA. (2019, May 29). *Transforming industry through CCUS*. Retrieved June 16, 2019, from https://www.iea.org/, https://www.iea.org/publications/reports/TransformingIndustrythroughCCUS/?utm_content=bufferd635b&utm_medium=social&utm_source=twitter.com&utm_campaign=buffer.

International Organization for Standarsization. (2019, June 1). *Environmental management—Environmental performance evaluation—Guidelines*. Retrieved from https://www.iso.org, https://www.iso.org/standard/52297.html.

Ioannou, I., & Serafeim, G. (2019, February 11). *Yes, sustainability can be a strategy*. Retrieved June 1, 2019, from Havard Business Review: https://hbr.org/2019/02/yes-sustainability-can-be-a-strategy.

ISO. (2019, June 1). *ISO 14031: The business value of environmental performance evaluation*. Retrieved from https://www.slideshare.net, https://www.slideshare.net/PECBCERTIFICATION/iso-14031-the-business-value-of-environmental-performance-evaluation.

Jebaraj, P. (2019, May 3). *Pepsico agrees to withdraw cases against Gujarat farmers*. Retrieved May 3, 2019, from The HINDU: https://www.thehindu.com/business/Industry/pepsico-to-withdraw-case-against-gujarat-potato-farmers/article27013935.ece.

KCX Consulting. (1990). *Five capitals models*. Retrieved June 1, 2019, from http://www.kcxconsulting.com, http://www.kcxconsulting.com/index.php/about-kcx/capital-model.

Koshy, J. (2019a, May 18). Retrieved May 18, 2019, from The Hindu: https://www.thehindu.com/sci-tech/energy-and-environment/what-is-the-new-un-climate-risk-protocol-and-what-necessitated-it/article27172902.ece.

Koshy, J. (2019b, March 14). *A climate vulnerability index for India*. Retrieved June 1, 2019, from The HINDU: https://www.thehindu.com/sci-tech/energy-and-environment/a-climate-vulnerability-index-for-india/article26537153.ece.

Koshy, J. (2019c, February 5). *Emission levels rising faster in Indian cities than in China*. Retrieved May 3, 2019, from The HINDU: https://www.thehindu.com/sci-tech/energy-and-environment/emission-levels-rising-faster-in-indian-cities-than-in-china/article26178211.ece.

Koshy, J. (2019d, May 4). *Only 10 of 100 new sewage projects on Ganga have been completed*. Retrieved May 4, 2019, from The HINDU: https://www.thehindu.com/news/national/only-10-of-100-new-sewage-projects-on-ganga-have-been-completed/article27029780.ece.

Kumar, K., & Prakash, A. (2019). Examination of sustainability reporting in Indian banking sector. *Asian Journal of Sustainability and Social Responsibility*. https://doi.org/10.1186/s41180-018-0022-2.

Mardani, A., Zavadskas, E. K., Govindan, K., Senin, A. A., & Jusoh, A. (2016). VIKOR technique: A systematic review of the state of the art literature on methodologies and applications. *Sustainability*. https://doi.org/10.3390/su8010037.

Marius, P., Trandabat, D., & Trandabat, A. (2013). Assessment of corporate environmental performance based on fuzzy approach. *APCBEE Procedia*, 368–372. https://doi.org/10.1016/j.apcbee.2013.05.063.

Mateusz, P., Danutaa, M., Małgorzataa, Ł., Mariusz, B., & Kesraa, N. (2018). TOPSIS and VIKOR methods in study of sustainable development in the EU countries. *Procedia Computer Science*, 1683–1692. https://doi.org/10.1016/j.procs.2018.08.109.

McKinsey and Company. (2017, December). *Sustainability's deepening imprint*. Retrieved June 1, 2019, from Mckinsey: https://www.mckinsey.com/business-functions/sustainability/our-insights/sustainabilitys-deepening-imprint.

Ministry of Corporate affairs. (2013). *Schedule 7*. Retrieved June 1, 2019, from Ministry of Corporate Affairs: http://www.mca.gov.in/SearchableActs/Schedule7.htm.

Morgan Stanley Capital International. (2019, June 1). *ESG ratings*. Retrieved from https://www.msci.com, https://www.msci.com/esg-ratings.

Muñoz, M. J., Rivera, J. M., & Moneva, J. M. (2008). Evaluating sustainability in organisations with a fuzzy logic approach. *Industrial Management and Data Systems, 108*(6), 829–841. Retrieved from www.emeraldinsight.com/0263-5577.htm.

Öberseder, M., Schlegelmilch, B. B., & Gruber, V. (2011). "Why Don't Consumers Care About CSR?": A qualitative study exploring the role of CSR in consumption decisions. *Journal Business Ethics*, 449–460. https://doi.org/10.1007/s10551-011-0925-7.

Phillis, Y. A., & Davis, B. J. (2009). Assessment of corporate sustainability via fuzzy logic. *Journal of Intelligent and Robotic Systems*, 3–20. https://doi.org/10.1007/s10846-008-9289-3.

Press Trust of India. (2019, June 1). *Environment ministry budget increased By 20%, Gets Rs. 3,111 crore*. Retrieved from www.ndtv.com, https://www.ndtv.com/business/budget-2019-environment-ministry-budget-increased-by-20-percent-gets-rs-3-111-crore-1986962.

Reinfinitiv. (2019, June 1). *ESG score methodology*. Retrieved from https://www.refinitiv.com/, https://www.refinitiv.com/content/dam/marketing/en_us/documents/methodology/esg-scores-methodology.pdf.

Robecosam. (2019, June 1). *SP ESG index family*. Retrieved from https://www.robecosam.com, https://www.robecosam.com/csa/indices/sp-esg-index-family.html.

S&P Global. (2019, June 1). Retrieved from trucost ESG analysis: https://www.trucost.com/trucost-blog/trucost-and-sp-global/.

Sarkis, J. (2012). Models for compassionate operations. *International Journal of Production Economics*, 359–365. https://doi.org/10.1016/j.ijpe.2012.06.018.

Sarkis, J., Zhu, Q., & Lai, K.-H. (2010). An organizational theoretic review of green supply chain management literature. *International Journal of Production Economics*, 1–5. https://doi.org/10.1016/j.ijpe.2010.11.010.

Spiliakos, A. (2018, October 10). *What does "Sustainability" mean in business?* Retrieved June 1, 2019, from Harvard Business School Online: https://online.hbs.edu/blog/post/what-is-sustainability-in-business.

The Forum for the Future. (2019, June 1). *The five capitals*. Retrieved from https://www.forumforthefuture.org, https://www.forumforthefuture.org/the-five-capitals.

UN. (1987). *1987: Brundtland report, our common future*. Retrieved June 1, 2019, from UN—Milestones in sustainable development: https://www.are.admin.ch/are/en/home/sustainable-development/international-cooperation/2030agenda/un-_-milestones-in-sustainable-development/1987–brundtland-report.html.

UNEP. (2015, November). *UNEP climate action*. Retrieved June 1, 2019, from Sustainable Innovation Forum 2015: http://www.cop21paris.org/.

UNEP Finance Initiative. (2019, May). *Changing course*. Retrieved May 18, 2019, from A comprehensive investor guide to scenario-based methods for climate risk assessment, in response to the TCFD: https://www.unepfi.org/wordpress/wp-content/uploads/2019/05/TCFD-Changing-Course.pdf.

Unilever. (2017, May 1). *Report shows a third of consumers prefer sustainable brands*. Retrieved June 1, 2019, from https://www.unilever.com, https://www.unilever.com/news/press-releases/2017/report-shows-a-third-of-consumers-prefer-sustainable-brands.html.

Vermeulen, F. (2019, May 8). *Companies don't always need a purpose beyond profit*. Retrieved June 1, 2019, from Harvard Business Review: https://hbr.org/2019/05/companies-dont-always-need-a-purpose-beyond-profit.

Vinod, S., & Sivakumar, N. (2017). Categorizing motivations of corporate environmental responsibility as part of CSR in Indian firms. In *3rd International Conference on Economic Growth & Sustainable Development: Emerging Trends*. Mysuru: SDMIMD. Retrieved from https://www.researchgate.net/publication/321071116.

Vinod, S., & Sivakumar, N. (2018a). Corporate environmental responsibility of Indian companies comprising the NIFTY index: Analysis and categorization. In *Biznes I Zarzadzanie a Bezpieczenstwo W Polsce I Na Świecie* (pp. 275–289). Politechnika Opolska.

Vinod, S., & Sivakumar, N. (2018b). Predictive analysis of corporate environmental responsibility as part of CSR in NIFTY indexed companies using Markov chain analysis. In *Sixth International Conference on Business Analytics and Intelligence*. Bangalore.
Vinodh, S., Eazhil Selvan, K., & Hari Prakash, N. (2011). Evaluation of sustainability using fuzzy association rules mining. *Clean Technologies and Environmental Policy*, 809–819. https://doi.org/10.1007/s10098-011-0356-2.
Wang, Z., & Sarkis, J. (2017). Corporate social responsibility governance, outcomes, and financial performance. *Journal of Cleaner Production*. https://doi.org/10.1016/j.jclepro.2017.06.142.
White, L., Smith, H., & Currie, C. (2010). OR in developing countries: A review. *European Journal of Operational Research*, 1–11. https://doi.org/10.1016/j.ejor.2010.02.015.
Zhu, Q., Sarkis, J., & Lai, K.-H. (2012). Examining the effects of green supply chain management practices and their mediations on performance improvements. *International Journal of Production Research*, 1377–1394. https://doi.org/10.1080/00207543.2011.571937.

Chapter 5
Six Sigma Marketing: An Innovative Approach to Achieve Strategic Goals of Sustainable Supply Chain

K. Muralidharan and Neha Raval

Abstract With the changing market scenario, contemporary organizations are in dire need of managing their different organizational functions most efficiently and cost-effective way. One such intense customer-focused function is marketing. The intense competition between global and local players has brought many new responsibilities to the marketing function, which includes addressing environmental and climate change issues through marketing functions. The extent of spread of Information Technology has also contributed positive changes to this business scenario as never before. With improved communications, addressing the need of heterogeneous customer base, getting customer insights, etc. became easy for the business establishments. This is where the measurement-based marketing approaches like Six Sigma Marketing (SSM) came into being. SSM is a fact-based data-driven disciplined approach to grow market share by providing targeted product/markets with superior value. The benefit of integrating Six Sigma into marketing process includes better information (management by fact) to make better decisions. It is known that Six Sigma as a process improvement approach has the legacy to bring significant improvement in different business functions. With computer-mediated digital environment, digital marketing is proposing fertile land for Six Sigma implementation to realize its optimum potential and vice versa. In line with the above arguments and constructs, we propose a case study based on SSM applied on a small traveling firm in India to improve its marketing efforts. Following DMAIC (define-measure-analyze-improve-control) approach, we propose how major phases of marketing activity like awareness, consideration, retention, and conversion can be improved through Six Sigma approach. Considering different metrices of digital marketing, this SSM approach explains how sigma level of the marketing efforts can be used to influence customer perception and provide pleasurable experiences. We also provide an illustration to show the usefulness of SSM in sustainable supply chain.

K. Muralidharan (✉) · N. Raval
Department of Statistics, Faculty of Science, The Maharajah Sayajirao University of Baroda, Vadodara 390 002, India
e-mail: muralidharan.k-stat@msubaroda.ac.in

N. Raval
e-mail: njr.sky@gmail.com

U. Ramanathan and R. Ramanathan (eds.), *Sustainable Supply Chains: Strategies, Issues, and Models*, https://doi.org/10.1007/978-3-030-48876-5_5

Keywords Six Sigma · Supply chain management · Performance matrix · Sigma level · Sustainable products · Critical to quality · Voice of customer · Key process inputs

1 Introduction

To cope up with ever-changing market scenario, organizations required to manage each of their business functions effectively and efficiently. Marketing segment has the unique position of repute among all other organizational functions with capacity to bring customers to the door step of the organization. Effective management of marketing function can put organization at competitive advantage in the market place. Still marketing is facing many challenges as a business function. As mentioned by Aaker (2011), marketing is still required to improve based on following challenges:

- In the situation of market sluggishness, marketing has great responsibility to be innovative enough to create such an enhancement of offering that cannot be evaded by customers.
- Marketing has to be strategic instead of tactic through customer insights, value proposition, and brand strategy.
- Getting rid of product, country, and functional silos, marketing need to foster communication and cooperation instead of competition and isolation.
- Marketing need to inject energy and involvement into their brand to achieve brand equity.
- Marketing need to use different available tools such as digital platform and social media tactically.

This list of required changes offers great room for marketing function to improve its functionality. Improving effectiveness and efficiency of marketing function can help organizations to promote their offerings in much convincible way.

Six Sigma as a customer-focused improvement approach addresses two important dimensions: (1) being *effective* through unique value proposition in the market and (2) being *efficient* by reducing variability of Critical-to-Quality (CTQ) characteristics. These characteristics of Six Sigma are providing opportunity for its expansion over other organizational functions except manufacturing. Different organizational functions are also marching toward applying this approach to improve upon their effectiveness and efficiency. As mentioned by Pestorius (2007), sales and marketing are the latest functions to join Six Sigma bandwagon.

Marketing as an organizational function has always been debated for its productivity and end results (Rust et al. 2004). Six Sigma with reputation of profit generations in manufacturing setup comes handy to marketing and sales professional to address its productivity issues. Six Sigma as a data-driven process improvement approach can help marketing to leverage its full potential to manage firm-offering-customer link profitably and balance the benefits of resources and capabilities of marketing process in a better way (Muralidharan and Raval 2017).

Though Six Sigma is carrying reputation of sophisticated quality improvement program, its integration with marketing has remained an under explored area relatively. For many years, Six Sigma and Marketing professionals are working in isolation, with no motivation of integration between two fields. This is largely because of marketing manager's conventional understanding about Six Sigma as "*it is for production.*" On the other hand, by realizing the potential of Six Sigma concepts, experts are ready to explore new horizon of Six Sigma implementation looking beyond a particular segment of organization. The global level competition has further boosted the marketing process to a new level of thinking, leaving marketing professionals to become proactive to incorporate instant changes to their day-to-day activities. These fanatical efforts to integrate marketing segment to quality improvement programs like Six Sigma have given rise to a vibrant field called Six Sigma Marketing (SSM).

2 Six Sigma Marketing: The Literature Perspective

There are very few literatures available on this concept so far; however, there are plenty of concepts/books/literatures on Six Sigma and Marketing subjects available in isolation. Therefore, we understand the necessity of providing some of the available literatures and provide some critical account of both the fields, so that the professionals can understand the tenets of integration of Six Sigma and Marketing. Some books limited to Six Sigma Marketing are as follows:

- *Sales and Marketing: The Six Sigma way* by Webb and Gorman (2006): In this book, author proposed process-based Six Sigma approach with its utility in marketing, where the authors demonstrate Six Sigma application in marketing based on improving performance of a company's website named Technical Resume Group (TRG). Considering origin of Six Sigma in production, authors provided unique dimension to the sales and marketing process. Author proposed very interesting analogy between production process and sales and marketing. According to the authors, people in the market place is analogous to raw material in production. Finding, winning, and keeping customers in marketing are analogous to procurement, assembly, and shipping goods. As production adds value to raw material similarly marketing adds value by solving problems of people.
- *Six Sigma Marketing: From cutting costs to growing market share* by Reidenbach (2009): In this book, author discussed about value-based Six Sigma Marketing approach. As satisfied customers are not always the loyal customer in this cutting-edge competition era, the marketing professionals should be creative in their delivery process. The DMAIC approach with reference to value-based criteria is discussed in this book. Author discusses different Six Sigma tools and their usage with reference to customer value management.
- *Applying the science of Six Sigma to the art of Sales and Marketing* by Pestorius (2007): In this book, author discuss data-based Six Sigma approach in different

marketing activities like recruitment of sales person, new product sales, sales representative competencies, field-visit process, etc. Author discussed each of this activity with reference to quality management of Six Sigma approach.

- *Six Sigma for marketing process: An overview for Marketing Executives, Leaders and Managers* by Creveling et al. (2007): In this book, authors discuss Six Sigma application at three levels of marketing: Strategic, tactical, and operational. For all these levels, their gate and termination point with intermediate steps are discussed. Tools used in all process steps are discussed in detail. According to the authors, marketing value is found to be a function of the degree to which it develops knowledge and skills in connecting the customers to the product and to financial accountability. Hence, using Six Sigma for any one of the marketing phase is not sufficient. Applying Six Sigma at all three levels of marketing, strategic, tactical, and operational levels, is important to realize significant result.

Thus, from a critical evaluation of the abovementioned books and some other literatures, we perceive Six Sigma Marketing (SSM) as "*a fact based data driven disciplined approach to growing market share by providing targeted product/markets with superior value*" (see also Harry and Schroeder 2005; Muralidharan 2015). Due to exploratory nature of this topic, very few academic articles have been published on this topic so far. Few remarkable studies in the form of article are by Salzarulo and Krehbiel (2012), Niemes (1999), Das et al. (2006), Muralidharan and Raval (2015) and the references contained therein. Let us now discuss the importance of Six Sigma philosophy and understand why it is so important in quality improvement in marketing process.

3 The Six Sigma Philosophy: Marketing Perspective

There are various perceptions about the Six Sigma concepts. Six Sigma is a structured problem-solving approach to reduce variations and thereby improve quality. The metric-based definition of Six Sigma is the most widely recognized in comparison to other definitions. The metric definition says that, out of one million opportunities there arise 3.4 defects, which implies high level of quality in the process. This definition is aimed at keeping tight control over variation of parameters which are the Critical to Quality (CTQ) of the final product or service. In order to keep defect rate as low as possible, it is important to keep variation of CTQs as low as possible. This concepts was first introduced by Motorola, and according to them, a Six Sigma process leads to 99.9999998% area of the normal curve (a quality distribution) within the specification limits. Only 0.0000002% area outside specification limits results into only 2 DPMO. This further increase to 3.4 DPMO due to process mean shift (see Table 1 for the DPMO correspond to various sigma levels).

The working philosophy of Six Sigma follows a structured DMAIC (*define-measure-analyze-improve-control*) approach to streamline quality improvement

Table 1 DPMO based on different sigma levels

Specification limit	DPMO	
	Targeted distribution	Shifted distribution
$\pm 1\sigma$	317310	691462
$\pm 2\sigma$	45500	308537
$\pm 3\sigma$	2699	66807
$\pm 4\sigma$	63.34	6210
$\pm 5\sigma$	0.57	233
$\pm 6\sigma$	0.002	3.4

efforts (Pandey et al. 2000; Hoerl 2004; Muralidharan 2015). This is actually a generalization of PDCA (*plan-do-check-act*) cycle to manage business processes scientifically (Deming 2008). The DMAIC methodology provides framework to understand how to integrate various statistical tools for process improvement (Montgomery and Woodall 2008). After each phase of DMAIC methodology, there is a review process to examine performance of each phase. This review process is important to understand achievement of each phase and to maintain time schedule of project. One of the reasons that DMAIC is so successful is that it focuses on the effective use of statistical tools combined with management principles (George 2002; Mast and Bisgaard 2007; Snee 2010; Muralidharan and Raval 2017).

Process-based approach to the organizational activities is one of the key edges of Six Sigma implementation. Six Sigma provides plenty of scope for improvement by promoting process-based organizational activities like marketing. Process-based approach is capable of linking customer value to core organizational processes and hence provides competitive advantage to the organization, and hence there is lot of scope for improving marketing process activity. As mentioned by McCormack and Johnson (2001) "*managing a business means managing its process*," and therefore at most importance should be given to their organized processes. Process-based approach was originally preferred over functional approach in order to overcome departmental silos of functional approach. Process mapping in relation to other organization activities is the beginning of establishing relationship between all activities, measurement, and objects involved in the realization of any output. Process mapping helps to provide visible workflow; improve communication; provide common frame of reference to all involved in the process; and provides pathways to customer satisfaction by indicating functions, input, process, and output that organization uses to provide value to the customer (Biazzo 2002; Damelio 2011; Hooper 2001).

For a marketing process, the customer satisfaction score, customer loyalty measurement, repeat purchase intentions, etc. are examined for process effectiveness and assessments. Here the efficiency of the process is based on relativity of resources used versus results achieved. Measurements here required examining reliability of process. Reliability can be best addressed through variation and hence measurement of variation is important here. As Snee (1990) mentioned, employment of statistical thinking and measurements is based on the following three principles: (1) all process

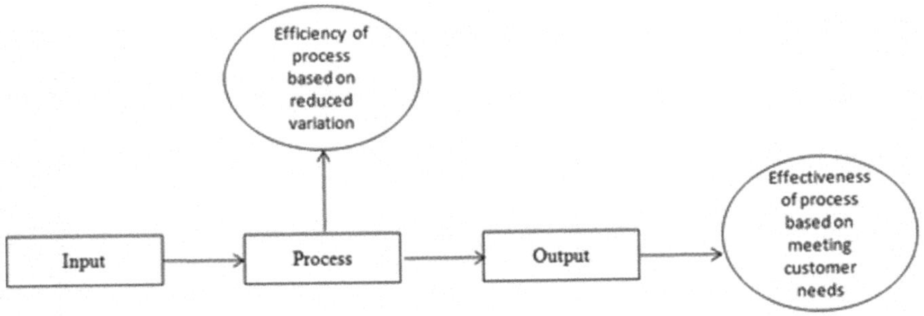

Fig. 1 Effectiveness and efficiency measurements of marketing process

occurs in a system of interconnected stages, (2) variation exists in all processes, and (3) understanding and analyzing variation are keys to success. Figure 1 shows such a process which results in identifying the variation inherent in the process and the necessary inputs needed to meet the customer requirements. The overall assessment about the process performance becomes a by-product of this situation. Therefore, to enhance the efficiency of the marketing process, the Six Sigma quality professionals and marketing experts should mutually identify the core areas of expertise, which works best for the organizational growth and value creation.

Last 30 years have shown a rapid growth of the Six Sigma philosophy and its applications in production, service, automobile, and healthcare services. The concept is also widely used as sustainable business practice tool to realize the market consolidation and overall growth of the organization. From structured manufacturing environment to service environment (George 2000), from production to transactional business functions (Pestorius 2007), from shallow defect reduction objective to holistic value creation objective (Montgomery and Woodall 2008)—Six Sigma has undergone massive transformation. However, Six Sigma as a quality improvement approach for marketing has not progressed well. Such a thing is possible through the adoption of process-based approach and measurement-based decision-making as discussed above. Further, Six Sigma tools can facilitate to identify the problems associated with the marketing process; plan the necessary changes necessary in the process; and implement the unique solution to manage all marketing activities with improved quality as perceived by the producer and consumer. Hence, there is a strong requirement of integrating quality programs like Six Sigma in marketing process for quality assessment and improvement of services and products. The remaining part of this chapter articulates on the integration of Six Sigma and marketing concepts for an improved service and product. Toward this, first we need to understand the current marketing scenario as prevailing now.

4 Current Marketing Scenario

The goal of marketing function is to reach potential customer to satisfy their needs in the best profitable way. To meet this goal and to adjust with external factors, an organization needs to explore internal resources in optimum way (Ifezue 2005). Amalgamation of these internal and external factors is leading to drastic changes in organizational structure and so in marketing. The growth and change in marketing can be further examined through external and internal factors, where

- *Internal factors* are those factors organizations have direct control over. These are in-house factors that are affecting marketing function like management style, organization's financial capability, technological advancement, research, and development.
- *External factors* are those factors organizations does not have direct control over. These factors include customer requirements; competitors; and economic, social, cultural, and legal changes.

Figure 2 shows the changing scenario of current marketing process, where economic conditions, organizational structure, customer requirements, and spread of information technology are the catalysts responsible for creating a vibrant marketing scenario. SSM is definitely a new step in this direction, as it can facilitate information- or measurement-based decision-making in every activity. These components are further detailed in the subsequent section.

4.1 Changing Economic Condition

The changing economic condition in last few decades can be best explained through one phenomenon called globalization. Globalization has changed market environment as never before. Impact of globalization is so fierce that it is affecting every business in each corner of the world. Organizations from all around the globe are

Fig. 2 Contributing factors of current marketing scenario

participating in international trade, as oppose to limited players dominated in international trade. This scenario resulted into steep market competition and hence changing organizational structure completely. We strongly feel that Six Sigma concept integrated with marketing process has lot of scope for improving the organizational goal in this current environment to deliver quality services and products. This can facilitate a strong communication with customers to understand their need closely and develop efficient supply chain to propose competitive prices against local players and products.

4.2 Increased Level of Customer Sophistication

Marketing being responsible for revenue growth, required to be attentive toward changing need of customers. As mentioned by Muralidharan (2015), customer focus should be considered as major *leading indicator* for marketing to achieve successes, where leading indicator is defined as "*the indicator that precedes the occurrence of something.*" Since customer focus is the prerequisite to achieve customer loyalty which paves way to increased revenue, lower acquisition cost, lower retention cost, and increased profitability which are considered as major leading indicators for marketing. Customer focus has also changed immensely in last few decades. Compared to traditional promotion-based profit-making marketing to recent relationship-based customer-focused marketing, marketing has advanced more than anything with reference to customer focus. The changing customer focus with reference to different organizational dimensions is mentioned in Table 2.

As discussed in the previous section, the advanced physical features of offering focussed on customer value and emotional bonding between customer and offering results into delighted total customer experience. According to Mascarenhas et al.

Table 2 Different dimensions of traditional and contemporary marketing scenario

Dimension	Traditional	Contemporary
Business goal	Profit is prime business focus	Profit is the outcome of customer delight
Customer orientation	Customer satisfaction is the locus of control of marketing efforts	Total customer experience is the locus of control of marketing efforts
Marketing approach	Focus on promoting offering's features	Focus on understanding importance of offering in customer's life cycle
Marketing tactics	Based on physical aspects of offerings like quality, quantity, functionality, availability, price, information, delivery	Based on educating and understanding customers and thereby creating customer value

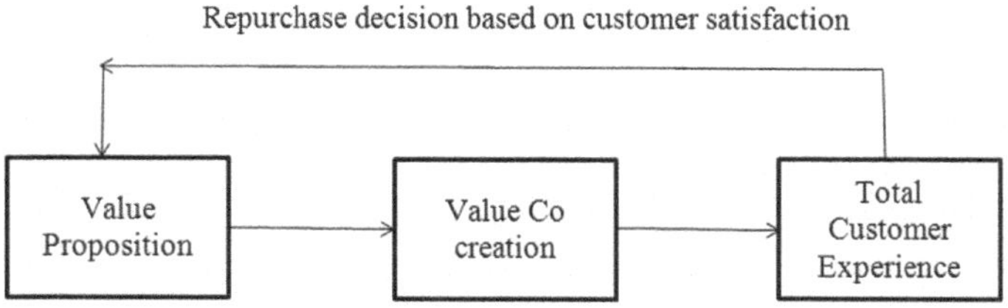

Fig. 3 Customer touch points based on "value" aspect

(2006), "Total Customer Experience" is the way to achieve lasting customer loyalty and hence assuring market position and consolidation (see Fig. 3).

4.3 Changing Organizational Structure

It is important to understand the changes happening in organizational structure of modern era since it determines how organization operates and performs, and consequently how it affects marketing practices. Rising competition and need of winning customer loyalty necessitate close relationship with customers. Hence, prime need of organizational setup is to be more responsive and flexible to customer requirements. The traditional organizational structure based on specialization, functional separation, standardization, and hierarchical control makes it inflexible and unresponsive to the frequent changes in the field of marketing. Whereas the modern organizational structure is based on flexibility, innovation, and responsiveness as their core criteria as mandated by changing phases of business functions. Existence of such modern organization is based on their core product or service offering and hence called *cellular* organization. In this new form of organizational setup, (i) functional separation is replaced by blur functional boundaries, (ii) work specialization is replaced by work integration, and (iii) standardization is replaced by variety.

4.4 Increased Reliance on Information Technology

The far-reaching spread of Information Technology is changing business scenario. According to ITU (2016) report, around 47% of the world population is using Internet by the end of 2016. The major users of Internet from developed parts of the world are from Europe, America, and CIS with contribution of 81% to the whole Internet user base. Another cluster joining this IT bandwagon is developing countries from Asia and Pacific with contribution of 40% to the Internet user base.

As discussed in previous sections, survival of the firm in contemporary market is determined by being competitive, the extent of IT-enabled services and products, focus on Total Customer Experience (TCE), and supporting flexible organizational structure. All of the above components required to establish strong channel of communication between different stakeholders. However, companies are looking for unique environments in which

- They got opportunity to expand their business in different parts of the world and small players can also venture in the competitive market.
- They can keep eyes on their customers and can establish close and constant communication with customers.
- They can have free access to information between different business units and they can empower themselves with latest market information.

The above needs of new generation business environment are well addressed through World Wide Web. World Wide Web is providing unique platform for business activities. Following characteristics of web-based environment propose enriching environment for business activities to nurture:

- Unlike traditional one-to-many communication model, this new computer-mediated environment facilitates many-to-many communication model and hence create network of communication.
- Traditional campaigns are designed based on the assumption of homogeneous customer requirements. In this mode of communication, customers merely act as passive listener. They do not get any opportunity to interact with firm. This new web-based environment allows customers to be proactive on digital platform. Customers can interact with firms, with other customers, and can give their feedback on digital platform. Hence, this new platform addresses heterogeneous users who are in search of varied experience. Engaging customers in whole channel of marketing instead of merely understanding their need is the major objective fulfilled by digital platform.
- Traditional campaign starts with media commercials addressing mass of customers, whereas digital campaign starts with customer insights and tries to create a digital customer experience that they admire. Campaigns generally create brand through in-store experience, which is now created on digital platform. Hence, organizations need to understand how they can bring in-store experience to digital platform.
- Instead of bombarding customers with information, this platform is based on motivation of customers to find particular site or piece of information. Here marketing professionals should play a vital role in providing relevant and quality information as per the taste and preference of the customers. Many Six Sigma tools can be used for finding preference and classifying the customers according to their product/service choice and importance.

4.5 Components Contributing to Current Marketing Scenario

Based on the above discussion we can identify components of four major factors contributing to current marketing environment. The economic changes in market resulted in competitive market, changing customer requirements resulted in the need of delighted TCE, changing organizational structure resulted in network-based business model, and development in Information Technology leads to integrated communication. All these components together define current marketing scenario as shown in Fig. 4. It is observed that the structural relationship contributes a high amount of transparency and quality in the marketing process.

5 Integration Between Six Sigma and Marketing

It is now appropriate to integrate Six Sigma and Marketing based on the above components. This integration between Six Sigma and Marketing paves ways for establishing a good customer confidence with sustainable business practices provided that the top-level commitment is in accordance with the organizational objective. One can observe that, Six Sigma as a process improvement approach is expanding its horizon from data-rich manufacturing environment to service environment, whereas marketing is carving ways from superficial customer perception environment to data-rich digital environment. Hence, Six Sigma converges toward service environment which is the fundamental nature of marketing, and marketing converges toward data-rich environment which is the fundamental nature of Six Sigma. This complementary convergence of two fields provides great opportunity for possible integration between Six Sigma and Marketing. Pictorially, the convergence of these two fields toward a new discipline called Six Sigma Marketing is shown in Fig. 5.

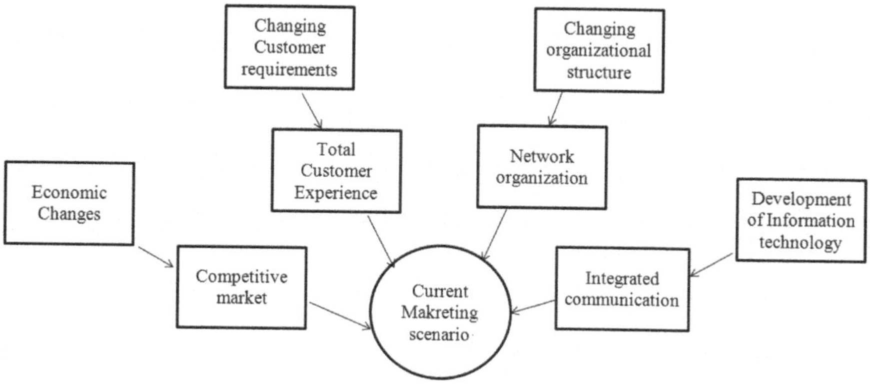

Fig. 4 Components contributing to current marketing scenario

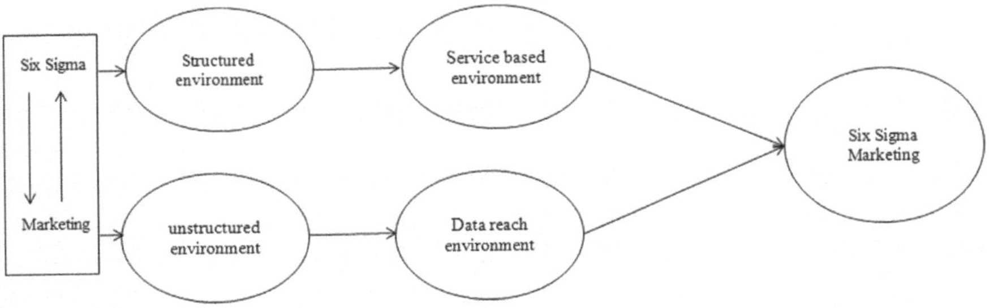

Fig. 5 Integration basis for Six Sigma and marketing

From the figure, it is observed that the measurement-based information and quality delivery is what makes the bonding between Six Sigma and Marketing a successful philosophy. And the philosophy is more of applied and experimental rather than mere perception. As per the new regulations and compliances of company's act, all marketing functions are now responsible to inculcate feeling of trust among prospective customers for their offerings and rely on quality delivery of services and processes

Thus, SSM is a fact-based data-driven disciplined approach to grow market share by providing targeted product/markets with superior value and quality conscious services. The benefit of integrating Six Sigma into marketing process includes better information to make better decisions. In a way, SSM replaces management by objectives with management by facts through a rigourous structural problem-solving approach. However, integration between the science of Six Sigma to the art of marketing needs proper measurement-based assessment and evaluations. Based on progress of both the fields over a period of time and contemporary practices, Muralidharan and Raval (2013, 2016) studied various SSM models that can be used to improve upon the issues faced by marketing as a business function.

Hence, a successful SSM can provide a powerful tool to leverage functionality of each of the organizational components and make them more effective for building long-lasting customer confidence. The following characteristics of SSM make it unique compared to other existing approaches:

- It focuses on fact-based decision-making at each phase of marketing process.
- It proposes well-defined structure under which fact-based approach can flourish.
- It promotes root cause analysis model to address marketing problems.
- It provides measurement-based evaluation of each phase of marketing process.
- It leverages sigma-level-based objective improvement criteria.

5.1 *Importance of Data in SSM*

As discussed in the previous section, components contributing to current marketing environment are competitive market condition, focus on total customer experience, network-based organization, and integrated communication. Achieving competitive advantage based on each of this component requires sound decision-making based on real-time data. See Muralidharan and Raval (2017b) for a case-based study on improving the productivity and marketing quality of a travel-related website based in India. Using various Six Sigma tools, authors have provided some best possible solutions to improve the customer conversion and provided turnkey solutions for travel-related problems.

Marketing has no credential if it fails to understand customers. Understanding customers require collecting information from all different sources. Marketers nowadays have great access to proactive and reactive data which can be used to build effective marketing strategy.

- *Proactive data*: proactive data are collected to understand customers' anticipated need. In marketing, surveys, interviews, group discussions, and direct customer observation are great sources of proactive data.
- *Reactive data*: reactive data are generated based on the result of some activity. Sources of reactive information are customer complaint, website traffic, warranty claims, safety of sites, etc.

Advancement in technology is proposing great ease to collect proactive or reactive information. Because of this technological advancement it is now possible to approach individual from any part of the world through online survey or scheduling interview to collect proactive data. Recent marketing environment has changed the difference between different marketing activities like promotion, selling, order taking, and collecting customer feedback. Organizations are deploying different ways to connect and communicate to customer base like websites, social media, digital ads, email marketing, and so on. Customers are becoming active part of each and every activity. Traditionally, effectiveness of marketing communication was measured through sales or direct customer response only. Today, the scenario has changed completely. Effectiveness of marketing communication can very well be tracked through customer's response on online campaign or advertising. Digital platform is the great source of passive information. Organization using online platform can get information about the demographic characteristics of their customers, how they react to their campaign in the form of number of likes and shares, how much time they are spending on digital platform, how they are interacting on digital platform, etc. Each conversation between customer and firm is recorded on digital platform. Digital platform is generating so much of information and using this information to design effective marketing strategy is the major challenge to marketer today. In one of the case-study-based papers, the authors of this chapter have already recommended the digital platform for increasing customer loyalty and trust for improving the overall business and growth of the organization (Muralidharan and Raval 2018).

5.2 Six Sigma Marketing: The Implementation

Since interactive marketing environment is non-instructive in nature, people make choices based on communication. People themselves are searching for information, and hence the only thing remaining in the hand of marketer is providing them with relevant information. Based on this unique conceptual framework of digital marketing, one can influence the customer very easily and get their endorsement for the product or service. This framework is discussed in great detail by Muralidharan and Raval Neha (2016).

Figure 6 depicts the customer involvement of interactive marketing at various stages and their interrelationships. With this great leverage in digital mode of marketing, it is important to provide customer value to attract, engage, and win customers. Understanding what customer value and designing marketing strategy based on their interest is the unbeatable way to get success in this digital era. Not wasting money and resources in creating digital features of business that is not valued by your customer is important part of contemporary marketing strategy (Verma et al. 2004).

As it is known that, Six Sigma as a process improvement approach has the legacy to bring significant improvement in quality of various business functions. With computer-mediated digital environment on warpath, digital marketing is proposing fertile land for Six Sigma implementation to realize its optimum potential and vice versa. From marketing point of view, customers are passing through three major phases of decision cycle. They are

- *Awareness*: In this phase, customers are merely aware about offerings, and can change with respect to new offerings. They are called rational customers.
- *Consideration*: In this phase, customer realizes value in offerings and intends to purchase it. They are called prospective customers.
- *Conversion*: In this phase, customer has already purchase offerings. They are called existing customers.

Customers who are aware of quality programs like Six Sigma, Lean, Kaizen, TQM, etc. in the organization will definitely consider the product with complete trust and satisfaction. Even companies who specify their sigma value (a parameter of quality level) along with their product and services also enjoy appreciation and regards. Such aimed marketing efforts inspire customers to move into a loyal and permanent customer (see Fig. 7). This is called the cycle of lead generation to purchase. Companies are adopting digital face to interact with their customers and to create strong channel of communication. Customers are relying upon firm's digital platform to take decision about possible purchase. Hence, it is important for the firm to understand what customer value in their digital face.

One of the major parts of any company's digital marketing effort is their online presence through website. Quality of website plays an important role to drive business and hence creating revenue. As explained by Bai et al. (2008), website quality

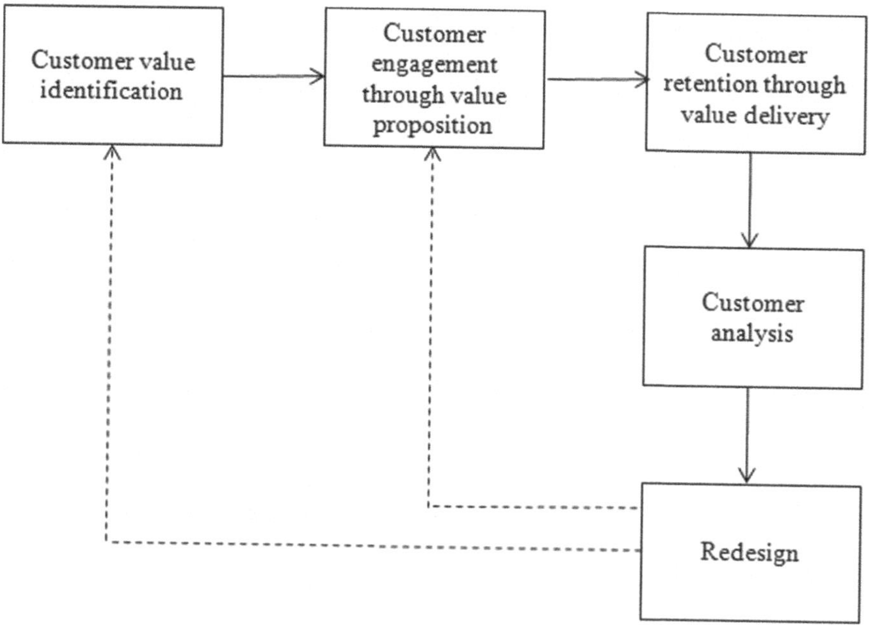

Fig. 6 Conceptual framework of interactive SSM

is related with customer satisfaction and hence ultimately customer purchase intentions. Many authors have described different dimensions of website quality based on its functionality and effectiveness. Perdue (2001) considers site navigation, attractiveness, and information content as important characteristic of website. Szymanski and Hise (2000) focus upon convenience, site design, and financial security as important part of web-based experience. Ranganathan and Ganapathy (2002) consider ease of navigation on website, time taken for navigation and page download, and use of multimedia on website for exciting customers, and possibly they will finally convert into a loyal customer for you.

Figure 8 shows the connecting threads between customer decision stages and corresponding relative marketing efforts as perceived by the marketers. The customer decision stages run through many decision stages, which ultimately convert to the final decision regarding the purchase of the product or service. This was facilitated by various stages of the tactical and strategic marketing efforts coupled with integration of various marketing characteristics like customer engagement, customer realization, customer retention, value proposition, value co-creation, etc. In this process, the marketing activity translates into a psychological process yielding total customer experience as a pleasurable one.

The trust and confidence generated through various marketing efforts then translates into complete adoption of the services offered by the organization. So the power of quality efforts shadows over the marketing process for realizing complete transparency in the organization. This is the uniqueness of Six Sigma in marketing.

Fig. 7 Customer purchase funnel

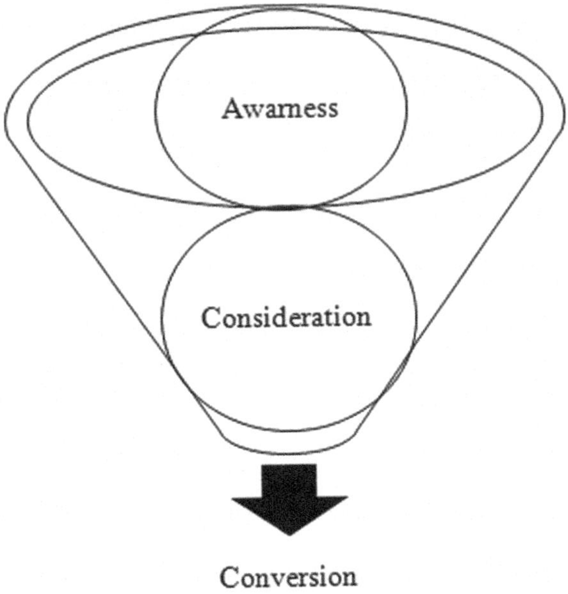

5.3 SSM DMAIC Approach

The DMAIC philosophy of Six Sigma applies to both process improvement and product improvement and even design redesign efforts. The key elements in a DMAIC project are team discipline, structured use of metrics and tools, and execution of a well-designed project plan that has clear goals and objectives. This applies to marketing component as well. Below we discuss the DMAIC approach of SSM for enabling quality and efficiency in the process.

5.3.1 Define

This phase involves the defining and clarifying customer requirements (both internal and external). Those requirements are then get translated and parsed into supplier (or technical) specifications. Here each specification becomes deliverable and gets assigned to the person accountable for producing the deliverable. The necessary activities to complete that deliverable get defined and the appropriate support tools are identified. Finally, all these steps are summarized in a project plan according to the agreed-on methodology, and it moves to marketing professionals for their comments and approvals. This phase contributes to SSM based on

- Identification of the market opportunity and clear positioning of company.
- Identification project that is in best interest of all stakeholders.

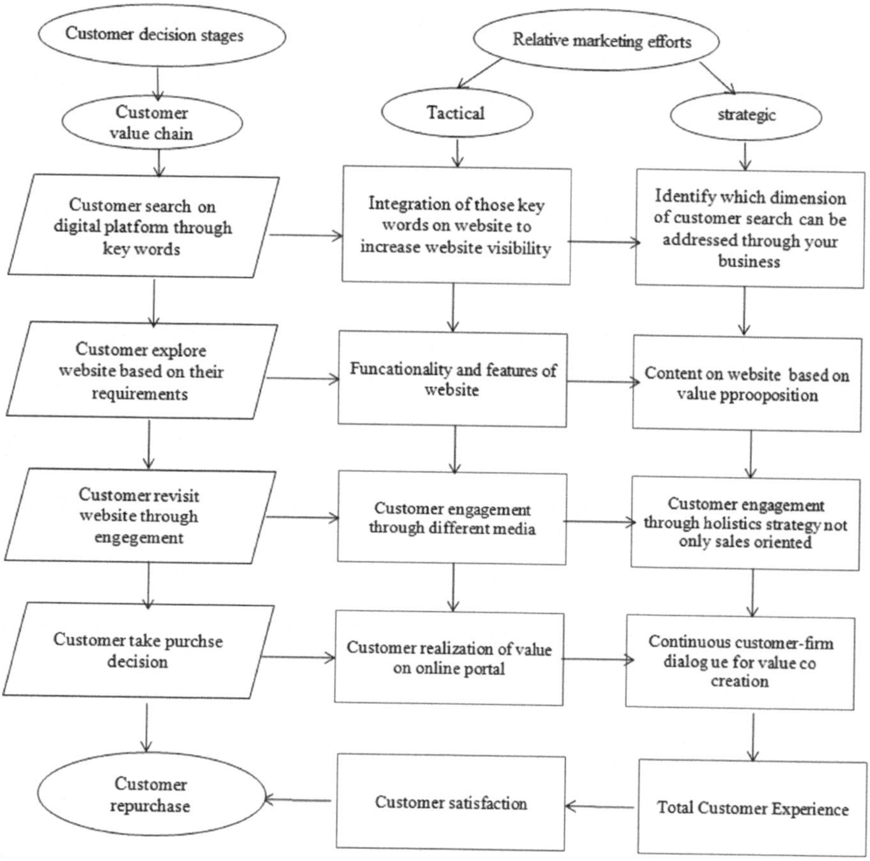

Fig. 8 Customer decision stages in relative to marketing efforts

- Clearly defined project objectives with possible advantage attached with attaining it.
- Examination of processes that can best address project objectives.
- Identification of appropriate KPI that affects process performance significantly.

5.3.2 Measure

A reliable data is essential for SSM. Through which performance can be measured and improvements can be made. The first step in measuring and improving performance is to determine what data exists, where that data is, what data is needed, and how to obtain the data. Customer purchase activity, marketing program results and conversion rates, actual costs for programs and people, lead quality data and lead cost, win/loss ratio defections that occur in the buying process, etc. are the core information that help professional to make decisions. The more it is quantified, the better the decision be. *Measure* phase facilitates.

- Operational definition of KPI identified in *Define* phase.
- Criteria about how, where, and when to collect data.
- Optimum utilization of proactive and reactive data.
- Examine baseline performance of process based on measurements.

5.3.3 Analyze

The SSM professionals must frequently use seven quality tools (histogram, Pareto chart, Gantt chart, cause-and-effect diagram, failure mode effect analysis, affinity diagram, and control chart) to have an up-to-date assessment of their marketing processes. Even a normality plot or Box-plot technique for quantitative measurement is a powerful tool for assessing whether the process is under statistical control or not. Performance improvement results from deriving insight through the analysis of the data helps the professionals to validate gaps in requirements versus current metrics with vital causes. By analyzing the data and understanding what it means, facilitates the marketing people to determine the degree of impact it is having on the organization, and redesign processes that will improve performance. Further, *Analyze* phase helps to

- Prioritize root causes and identify the most contributing one.
- Establish a proper process model in terms of deterministic or probability model.
- Suggest possible relationships and associations between control variables.
- Identify the most contributing input/output variables through confirmatory analysis.
- Test the significance of various input/output variables.
- Conduct a revised root cause analysis to identify the vital causes.
- Judge the current performance with the customer requirements.
- Compute initial sigma level and map with the ongoing process.
- Suggest necessary adjustments and allow the process to improve.

5.3.4 Improve

The main purpose of applying Six Sigma to marketing component is to determine how to improve productivity, performance, and processes. Data analysis should result in valuable insights that generate possibilities for improvement. These possibilities for improvement can include enhancements in tools, systems, processes, and skills. Even though change is disruptive, developing new ways to approach the market enables the marketing organization to play a more strategic role. Hence, improve phase involves developing solutions targeted at confirmed causes. The objective also extends to verify that the confirmed causes are statistically and practically significant and to optimize the process or product/service with the improvements.

At this stage from a microlevel, the SSM professionals are capable for suggesting an improvement implementation plan, develop potential improvements or solutions for root causes, develop evaluation criteria, and prioritize solution options for each root cause identified through the improve phase. Further, it facilitates to

- Examine solutions with a short-term and long-term approach.
- Weigh the costs and benefits of "quick-hit" versus more difficult solution options.
- Select and implement the improved process and metric.
- Measure results and conduct a designed experiment as per the marketing requirements.
- Validate solutions for improvement using statistical analysis.
- Evaluate whether improvements meet targets.
- Evaluate for marketing risk.

5.3.5 Control

It is believed that marketing as an organizational activity prides itself on its creativity, viability, and feasibility. But the time has come for marketing to document its processes and best practices and to apply these consistently in order to optimize marketing opportunities. The phases discussed so far, enabled us to identify a viable model and help us to quantify the nature of the relationship between the important variables of the process output. The statistical process control technique can now be employed with considerable effectiveness for monitoring and surveillance of the process. Thus, applying Six Sigma to marketing will increase marketing's ability to deliver on

- Market requirements as per customer's perception,
- Improve the efficiency and effectiveness of the marketing planning process,
- Successfully manage marketing operations,
- Provide transparency into marketing processes, and
- Improve the collaboration between marketing and other groups within the business.

If improvement is not natural, then one can resort to other variants of Six Sigma improvement like Designed for Six Sigma (DFSS) and Quality Function Deployment approach, and so on. DFSS ensures that the new product or service meets stakeholder's needs, provides a high level of performance, and is robust to process variations. While DMAIC is a reactive approach fixing past problems and monitoring the ongoing process, DFSS is used in a proactive way to avoid problems in the first place. DFSS approach enables launching new products and services on time and on budget and target gaining incremental revenues sooner, achieving greater market share, and ensuring that the company generates differentiated products and services that target customer and stakeholder needs. This methodology works with the following five steps:

- *Define*—Identify the customer and project

- *Identify*—Define what the customers want, or what they do not want
- *Design*—Design a process that will meet customer's needs.
- *Optimize*—Determine process capability and optimize design.
- *Verify*—Test, verify, and validate design.

6 Sustainability Practices Through Lean in Six Sigma Marketing

Improving business processes through Lean or any other approach is not the only goal of modern organizations. Changing environmental conditions of our planet has made organizations to think about how they can contribute to pollution prevention and environmental opportunities (Muralidharan 2015; Chugani et al. 2017).

According to Muralidharan (2015), *Green Six Sigma* (GSS) can be defined as "*the qualitative and quantitative assessment of the direct and eventual environmental effects of all processes and products of an organization. The activities involve the systematic usage of infrastructure and manpower; optimum use of technology and accountability of sustainable business practices.*"

Some of the specific qualitative assessments that need special attention are

- Source of information regarding raw materials and machinery.
- Strength of the company in terms of its size, manpower, machinery, and methods.
- Major breakthrough achievements in the past and present.
- Availability of the complete information on innovations carried out by the company.
- Availability of environment-friendly and sustainable developments carried out by the company.
- Waste disposal schemes employed in the organization.
- Sources of energy conservation and fuel consumption.
- Methods of reducing noise pollution.
- Environmental performance measurement carried out by the organization.
- Process control and quality control measures practiced in the company.
- Tax paid and company liabilities specific to environmental issues.
- Reactive rather and proactive maintenance activities.

Apart from the above qualitative information, there are many expected and unexpected environmental effects taking place in an organization, which can be measured in some quantitative scale. Some of them are

- Number of collaborations between companies and governments,
- Number of major accidents, catastrophes that have taken place,
- Number of projects conducted on healthy business practices and sustainable business practices,
- Number of bugs reported in computer programs and software,

- Number of training programs held for educating the employees,
- Amount and of level of green house gas emissions,
- Amount of carbon released by the company, and
- Amount of waste produced and disposed in a day in the organization.

The above organization specific assessments will have high impact on its marketing and sales process as well. They can be further evaluated through the Lean principles. Lean was not initially aimed to address environmental opportunities, but waste eliminating nature of Lean makes it suitable to contribute to it. Persistent attempt of Lean to reduce waste makes it potential approach to drive sustainability efforts. Lean approach is highly focused on reducing seven types of wastes as discussed above. These wastes can be seen through environment perspective to join the bandwagon of Green movement. As explained in (EPA 2009):

- *Overproduction*: Overproduction of units lead to high consummation of raw material and energy. Also hazardous material results in extra emissions, disposal, and extended exposure of hazardous material to workers.
- *Waiting time*: Wastage of energy during waiting period like power failure, voltage fluctuation, lightning, cooling, or heating. Potential damage of units or raw material during such a long waiting time causes wastages of energy, cost, and time.
- *Unnecessary movement/transport*: Energy requirements during unnecessary movement of goods and manpower are many often unnoticed and can be a cause for delays and depression. Special packaging is required to make goods transferable in such situations. The unnecessary movement and delay of transportation can also cause Greenhouse Gas (GHG) emission and carbon management a serious issue for pollution increase and can result in permanent shutdown of the company.
- *Overprocessing or incorrect processing*: More consumption of raw material and parts is as good as repeating the process again and again, and can lead to accumulation of wastes in terms of knowledge and machinery causing a serious threat for environment.
- *Excess inventory*: Excess inventory required additional place for storage and hence extra energy consumption for storage place. Excess inventory also results in damaged parts due to long-term storage. This results in more material and energy to replace damaged parts. Close monitoring of abundance and scarcity is the best way of removing these wastes.
- *Defects*: They are wastage of raw material and energy to produce defective part. Space and energy requirements to store defective units and to recycle or dispose defectives are a major concern these days. By promoting green methods of evaluations like reuse, reduce, and recycle policy, we can manage defects for a better environment.

Hence, Lean approach which focuses on waste reduction can refocus its efforts based on sustainability issues incorporated in the best possible way. Such approach is very well known as "Green Lean" or "Clean Lean" in literature. As seen before, Six Sigma as a process improvement approach focuses on defect reduction and hence facilitates optimum use of resources. This leads to lower production cost and high profitability. Revere and Black (2003) define Six Sigma DMAIC plan considering energy conservation at its core. According to them, the following aspect of sustainability can be integrated in the DMAIC approach:

- *Define*: Energy management team should be formed to keep sustainability project align over different phases like measure, analyze, improve, and control. Proper budget should be allotted to run energy management program successfully.
- *Measure*: To examine the energy consumption of organization and to explore improvement opportunities, energy audit should be done.
- *Analyze*: Data collected through audit can be analyzed to understand current energy consumption status, to understand budget allocation and to examine energy-saving opportunities.
- *Improve*: Based on information analyzed, energy-saving goals can be decided to improve upon current energy consumption.
- *Control*: Once improvement is realized—sustainability of project implementation is achieved through control phase.

7 Case Study

This case study is about supply chain management of healthcare products of a company (name suppressed) established in Ahmadabad, India. The company began producing pharmaceutical products for mother and child healthcare and subsequently introduced products in all dosage forms like tablets, capsules, liquids, and injectable substances. The company has presence in more than 20 countries directly and in more than 15 countries indirectly, covering Asia, Africa, and Far East countries. The current focus of the company is to expand their business to sustainable products through some efficient supply chain method. An SSM project is carried out to assess the customer satisfaction and the product acceptance from healthcare service personals. The product selected is Ampoules and Vials manufacturing, and the manufacturing process includes washing, dry heating sterilization (DHS), filling, visual inspection, and packaging. The maximum number of defects is identified at washing and DHS process stages. The rejection hugely affects the supply chain of the company, as customer (health service personal here) complains and material cost (as both the materials are perishable items) is very high.

As a part of increasing customer sophistication by creating value of their products and services, it was necessary to make an assessment about the percentage rejection of Ampoules and Vials. The company was following the traditional method of quality check for quite some time including acceptance sampling plans (a statistical tool for control and monitor defects) for defect identification. This was not yielding the expected results. Therefore, the process owner (name not mentioned because of confidentiality) thought of initiating a DMAIC project to identify the causes of defects and critical-to-quality parameters affecting the rejection. In consultation with the statistician and marketing people, a project charter is prepared to initiate the project. Based on the problem statement and goal, the process capability and initial sigma level based on current state of process is calculated, and is around 2.6σ for all process combined. As per the inventory status, this amounts to roughly 270 rejections out of 10000 of the Ampoules, according to the company.

During the measure phase, it was decided to check the washing process in every hour interval. The ampoules are checked medically for the defects like black particle, white particle, glass particle, fiber, volume, sealing, and breakage. The cause-and-effect (C&E) diagram (not reported here) is carried out to know the sources of variation and its impact. It has been found that a majority of the causes are due to setting the ampoules wrongly in the washing needles and mishandling of the washed ampoules while placing on the tray. Some causes identified during the packaging process are (i) required GSM of label is 70–75 but it comes in range of 90–95 from the labeling company, (ii) due to printing of label on the wrong side of bind of the paper, the label gets less removed, and (iii) the solution is still present on the outside of ampoules even after the process of cleaning them. The ampoules still gets broken at this stage due to wear and tear of all the processes. By and large the variations were all due to human and supply chains. Therefore, the primary concern has now shifted to (i) reduce number of defects in ampoules, (ii) improve sigma level of the process, and (iii) decrease transit time of the verification and control check.

In the analysis phase, a Failure Mode Effect Analysis (FMEA) was carried to understand the occurrence and severity of the failures (defects here). Based on the Risk Priority Number (=Severity * Detection * Occurrence), the causes were further prioritized for next course of action. This analysis indicated that out of two machines doing the job, the second machine is producing defects in huge numbers. If the first machine takes 5 min for assembling then the second machine takes about 11 min. This was very crucial. To test the significance of washing for less than 5 min and defectives it produces, we have carried out a chi-square analysis, and found that the chances of defectives will be more, if the washing is done beyond 5 min (p-value = 0.02). It was also observed that the material supply for the ampoule also needs to relook, as there are chances of defects occurring due to a particular vendor supplying the same. Some recommendations were made to improve the quality with immediate effect. They are shown in Table 3.

Table 3 FMEA recommendations

Potential failure mode	Recommendations	Reason
Wrong GSM	Company should think of labeling	Contract label company is not providing as per specifications
Wrong side printing	Contract label company may be assisted to correct it	Contract label company is not providing as per specifications
Solution on Amp	More effective cleaning required (use of lukewarm water)	Cost-effective solution need
Breakage of Amp	Strict surveillance required	The current surveillance not satisfactory

Table 4 Final sigma level

Process	Sigma level	
	Before	After
Washing (M_1)	3.22	3.39
Washing (M_2)	2.61	2.85
Cleaning	2.25	2.49
Packaging	2.79	4.00

In Improve phase, the visual inspection was made strict and acceptance sampling procedure was monitored for every process. The handling of ampoules during washing process, labeling process, and packaging process were put under strict surveillance and the defects were noted. After a series of confirmatory analysis (testing of hypothesis) and monitoring activities (control chart), the sigma level of each process was computed. It is seen that the improvement is satisfactory, but need to sustain. See Table 4 for details.

It is very clear that there are many issues with the product quality of the company, which are partially due to the product quality and partially due to supply chain of the materials provided by the vendors. Although substantial improvements can be made possible, the company needs to work on a holistic approach, as the product is used for direct human application.

8 Conclusions

Data forms the essential part of fact-based decisions, whether it is marketing or sales. The relevance of data characteristics, data quality, data reliability, and data relevancy are important features of marketing and sales. To make a marketing process meaningful, one must make common sense a common practice. There is nothing strange in it. Webb and Gorman (2006) consider marketing and sales as a production process. According to them, quality and process improvement methods have roots in manufacturing where they add value in very obvious and tangible ways. In contrast, sales add value by providing information and resolving issues, both of which are intangible.

Hence, Six Sigma tool as a process improvement approach can contribute significantly to energy conservation and safe environment. This facet of Six Sigma is reported as "Green Sigma" or "Green Six Sigma" in literature (see also Muralidharan and Ramanathan 2013). Thus, the use of Lean and Six Sigma approaches gives decision-makers better information and helps drive uncertainty out of marketing. Companies can better align product ideas with solid market opportunities and better balance and manage their offerings portfolio by using Lean and Six Sigma to significantly increase the probability of marketplace success. These approaches offer a common structure and language that will facilitate communication throughout the tactical product development and commercialization process between marketing and engineering.

References

Aaker, D. (2011). Five challenges facing marketing. *Havard Business Review.*

Bai, B., Law, R., & Wen, I. (2008). The impact of website quality on customer satisfaction and purchase intentions: Evidence from Chinese online visitors. *International Journal of Hospitality Management, 27.*

Biazzo, S. (2002). Process mapping techniques and organisational analysis: Lessons from sociotechnical system theory. *Business Process Management Journal, 8*(1), 42–52.

Chakrabarty, A., & Chuan Tan, K. (2007). The current state of Six Sigma application in services. *Managing Service Quality: An International Journal, 17*(2), 194–208. https://doi.org/10.1108/09604520710735191.

Chugani, N., Kumar, V., Garza-Reyes, A. J., Rocha-Lona, L., & Upadhyay, A. (2017). Investigating the green impact of lean, Six Sigma, and lean Six Sigma: A systematic literature review. *International Journal of Lean Six Sigma, 8*(1).

Creveling, C. M., Hambleton, L., & McCarthy, B. (2007). *Six Sigma for marketing processes an overview for marketing executives.* Leaders and Managers: Prentics Hall.

Damelio, R. (2011). *The basics of process mapping* (Second edn.), Productivity Press.

Das, N., Gauri, S. K. R., & Das, P. (2006). Six Sigma principles in marketing: An application. *International Journal of Six Sigma and Competitive Advantage, 2*(3), 243–262.

Deming, W. E. (2008). *Out of the crisis*, Affiliated East-West Press Pvt. Ltd.

EPA. (2009). *The Environmental Professional's Guide to Lean & Six Sigma: United States Environmental Protection Agency.*

Field, A. (2009). *Discovering statistics using SPSS*, Sage.

George, M. L. (2000). *Lean Six Sigma for services*, McGraw-Hill.

George, M. L. (2002). *Lean Six Sigma: Combining Six Sigma quality with lean production speed*, McGraw-Hill Education.

Gummesson, E. (1994a). *Relationship marketing—From 4Ps to 30 Rs.* Stockholm.

Harry, M. J., & Schroeder, R. (2005). *Six Sigma the breakthrough: Management strategy revolutionizing the world's top corporations.* Currency.

Hoerl, R. (2004). One perspective on the future of six-sigma. *International Journal of Six Sigma and Competitive Advantage, 1*(1).

Hooper, J. H. (2001). The process approach to QMS In ISO 9001 and ISO 9004. *Quality Progress.*

Ifezue, A. N. (2005). The role of marketing in economic development of developing countries. *Innovative Marketing, 1*(1), 15–20.

(ITU) International Telecommunication Union. (2016). *ICT Facts and Figures 2016: The International Telecommunication Union.*

Mascarenhas, O. A., Kesavan, R., & Bernacchi, M. (2006). Lasting customer loyalty: A total customer experience approach. *Journal of Consumer Marketing, 23*(7), 397–405.

Mast, D., & Bisgaard, S. (2007). The science in Six Sigma. *Quality Progress, 41*(1), 25–29.

McCormack, K. P., & Johnson, W. C. (2001). *Business process orientation: Gaining the E-business competitive advantage*, CRC Press.

Montgomery, D. C., & Woodall, W. H. (2008). An overview of Six Sigma. *International Statistical Review, 76*(3), 329–346. https://doi.org/10.1111/j.1751-5823.2008.00061.x.

Munro, R. A., Maio, M. J., Nawaz, M. B., Ramu, G., & Zrymiak, D. J. (2009). *The certified Six Sigma green belt*, Pearson Power.

Muralidharan, K., & Ramanathan, R. (2013). Green statistics: A management perspectives. *Journal of Industrial Engineering, 6*(9), 15–20.

Muralidharan, K., & Raval, N. (2013). Six Sigma: Some marketing essentials. *Internatioal Jouranl of Marketing and Human Resource Managment, 4*(2), 1–12.
Muralidharan, K., & Raval, N. (2015). Assessment of value added activities: Lean versus Six Sigma. In *Paper presented at the 3rd Annual convention and Competition on Lean Six Sigma.*
Muralidharan, K., & Raval, N. (2016). *The science of Six Sigma and art of marketing* (ICQRIT paper), Communicated for publication.
Muralidharan, K., & Raval, N. (2017a). Six Sigma marketing and productivity improvement. *A Quarterly Journal of the National Productivity Council, 58*(1), 107–114.
Muralidharan, K., & Raval, N. (2017b). A case study to improve website quality through Six Sigma Marketing, Communicated for publication.
Muralidharan, K., & Raval, N. (2018). Scientific outlook of lean in Six Sigma Marketing. In *Proceeding of Seventh International Conference on Lean Six Sigma. Paper presented at the Seventh International Conference on Lean Six Sigma.* Dubai.
Niemes, J. (1999). Taking sales success to new heights with Six Sigma. *National Productivity Review, 18*(3), 37–41.
Pande, P. S., Neuman, R. P., & Cavanagh, R. R. (2000). *The Six Sigma way: How GE, motorola, and other top companies are honing their performance*, McGraw-Hill Education.
Perdue, R. R. (2001). Intent site evaluations: The influence of behavioral experience, existing images, and selected website characteristics. *Journal of Travel and Tourism Marketing, 11*(2–3), 21–38.
Pestorius, M. J. (2007). *Applying the science of Six Sigma to the art of sales and marketing*, ASQ.
Ranganathan, C., & Ganapathy, S. (2002). Key dimensions of business-to-customer website. *Information & Management, 39,* 457–465.
Reidenbach, E. (2009). *Six Sigma marketing: From cutting costs to growing market share*, Amer Society for Quality.
Revere, L., & Black, K. (2003). Integrating Six Sigma with total quality management: A case example of measuring medication errors. *Journal of Healthcare Management, 48*(6), 377–391.
Rust, R. T., Ambler, T., Carpenter, G. S., Kumar, V., & Srivastava, R. K. (2004). Measuring marketing productivity: Current knowledge and future directions. *Journal of Marketing, 68,* 76–89.
Salzarulo, P. A., & Krehbiel, T. C. (2012). Six Sigma sales and marketing: Application to NCAA basketball. *American Journal of Business, 27*(2), 113–132.
Sheth, B. (2016). *Fundamentals of marketing*. iVIPANAN.
Snee, R. D. (1990). Statistical thinking and its contribution to total quality. *The American Statistician, 44*(2), 116–121.
Snee, R. D. (2010). Lean Six Sigma—Getting better all the time. *International Journal of Lean Six Sigma, 1*(1), 9–29.
Szymanski, D. M., & Hise, R. T. (2000). e-Satisfaction: An initial examination. *Journal of Retailing, 76*(3), 309–322.
Verma, R., Iqbal, Z., & Plaschka, G. (2004). Understanding customer choices in E-financial services. *California Management Review, 4*(46), 43–67.
Webb, M. J., & Gorman, T. (2006). *Sales and marketing: The Six Sigma way*. Kaplan.

Chapter 6
Evaluating Long-Term Sustainability of Supply Chains Using An Evolutionary Game Theory Framework

Sujatha Babu and Usha Mohan

Abstract Sustainability in supply chains is typically studied across one or more dimensions such as environmental, social, economic, culture and governance. Traditionally sustainability in supply chains has focused on environmental dimensions, while a few have attempted to focus on social and economic dimensions without really integrating them. There has been only a small effort to define sustainability by integrating all relevant dimensions (a holistic approach). Our model proposes to fill this gap. We identify sustainability of a supply chain with the equilibrium of the system over a long (but finite) period of time after integrating the various dimensions. Thus, it necessitates looking at factors that can cause a shift in the equilibrium. Towards this, we propose to build a strong theoretical framework to integrate, explain and predict sustainability for supply chains using cross-disciplinary effort. In our theoretical framework, evolutionary game theory serves as the pure conceptual theory-building tool, the metrics are qualitative in nature and the indicators are quantitative statistical measures. The use of evolutionary game theory concepts allows us to understand how sometimes trivial actions by members of the supply chain can trigger cascading effects that can move the system away from equilibrium. One of the salient aspects of our model is its complete scalability in terms of changes to the dimensions and metrics. As an example, we explain and predict social and economic sustainability (in tandem) for a public health insurance supply chain using evolutionary game theory.

Sujatha Babu—Work done during the scholar's stay in IIT Madras.

S. Babu
Chennai Mathematical Insitute, Chennai, India
e-mail: sujathab@gmail.com

U. Mohan (✉)
Indian Institute of Technology Madras, Chennai 600036, India
e-mail: ushamohan@iitm.ac.in

U. Ramanathan and R. Ramanathan (eds.), *Sustainable Supply Chains: Strategies, Issues, and Models*, https://doi.org/10.1007/978-3-030-48876-5_6

Keywords Supply chain management · Sustainability · Evolutionary game theory · Theoretical framework · Public health insurance

1 Game Theory in Supply Chains

As the stakeholders in the supply chain have conflicting goals and vested interests, game theory serves as a viable tool to provide solutions to the supply chain management aspects. Gérard et al. (1999) discuss various applications of game theory in supply chain modelling. They use a game-theoretic framework and demonstrate how a Nash equilibrium differs from the optimal solution of a supply chain. One of the most interesting problems arising in supply chains is the problem of coordination among stakeholders in a supply chain. Cachon and Netessine (2006) discuss how game theory is a critical tool for analysing supply chains, again emphasizing the importance of finding a unique Nash equilibrium in designing coordination mechanisms. We refer to Chen (2000, 2005), Aydin and Porteus (2009) and Cho et al. (2009) for notable research on supply chain contracts within a game-theoretic framework. We also refer to Mahesh Nagarajan and Greys Sošić (2008) for a survey on cooperative bargaining games with the objective of allocating profit among the stakeholders in a supply chain. A comprehensive survey of more than 130 papers of supply chain-related game-theoretical models, by classifying based on solution concepts and different areas of supply chain, can be found in Mingming Leng and Mahmut Parlar (2005).

Based on literature surveys by Carter and Rogers (2008), Stefan Seuring (2013) and Marcus Brandenburg et al. (2014), amongst others, it has been observed that a majority of the literature on sustainability in supply chains has been usually defined along environmental dimensions in terms of "green" concepts. Green supply chain management has emerged as a way to tackle the environment and economic dimensions in a supply chain. A tool to analyse long-term behaviour is evolutionary game theory (EGT). Qing-hua Zhu and Yi-jie Dou (2007), Sikhar Barari et al. (2012) and Seyed Gholamreza et al. (2011) have analysed interactions between supplier and retailer in green supply chains in terms of using the EGT framework. They, however, look at the sustainability of the supply chain in terms of economic dimension alone. Though they study short-term and long-term effects using evolutionary game theory, they do not study the effects of policy changes on the supply chain. The other critical social dimension has not been integrated either. A closely related concept in green supply chain management is that of closed-loop supply chains (CLSC), where the focus is on creating a reverse flow and reverse logistics. One of the key issues in such CSLC is to understand the long-term behaviour of companies adopting green manufacturing practices. Jian Li et al. (2013) looks at a closed-loop supply chain of manufacturers and retailers where the retailers are engaged in recycling waste that is then repurchased by the manufacturers. However, the scope of their study is limited as they look only at economic factors to study the sustainability of the CLSC using evolutionary game theory.

Carter and Rogers (2008) specify that sustainability lies in the intersection of the performance across all three dimensions. There is very little literature dealing with integrated sustainability across all dimensions. The meagre amount of literature that looks at supply chain sustainability does not seem to effectively address the question of determining sustainability integrated across all stakeholders and across all three (or four) dimensions. They also do not allow an analysis of the impact of new/emerging factors on the sustainability of the supply chain. The few attempts have fallen short of clearly defining an analytical framework to model the supply chain towards studying sustainability. The above papers and surveys restrict themselves in one or more of the following aspects: (1) They restrict themselves to two-player games while there can be any number of players in reality; (2) They only address matrix games restricted to a simple payoff based on cost and reward, as opposed to payoff functions that consider various metrics; (3) They do not study the effects of mutations and learning; (4) They do not study dynamics that can exist within a population.

In this chapter, in the following section, the supply chain is first modelled as an evolutionary game as shown in Babu and Mohan (2017). The stability concept in evolutionary game theory is equated to the sustainability of the supply chain. Both linear (matrix form) and non-linear payoff functions for the supply chain are considered. The conditions under which sustainability is reached are also identified. By modelling the supply chain as an evolutionary game, the impact that the stakeholders have on the trajectory of the supply chain in order to lead it towards sustainability is demonstrated. This fragment is further extended to include the stochastic aspect of the evolutionary games. The stability concept in evolutionary stochastic games is again equated to the sustainability of the supply chain.

2 Some Relevant Topics in Game Theory

We start with a brief introduction to game theory. Consider a simple game between two *players* (hereafter referred to as player-1 and player-2). Without loss of generality, assume that both players want to maximize their *payoffs*. Each player has a finite set of *actions* available to them, which may or may not be the same for both players. Each player independently and simultaneously chooses an action from their respective set of actions. The payoff that each player receives is dependent on the actions selected by each player. Thus the payoffs for the entire game can be represented as a *matrix* where cell (i, j) in the matrix represents the payoffs when player-1 and player-2 play action i and j, respectively. Player-1 is referred to as the *row chooser* and player-2 as the *column chooser*. The game is played once. That is, players choose their strategies, obtain their respective expected payoffs and the game ends. Such games are called *two-player one-shot bimatrix games*.

For example, consider the following two-person game. Let the actions available to player-1 be Top (T) and Bottom (B), and the actions available to player-2 be Left (L) and Right (R). Let the payoff matrix for the game be as follows.

$$\begin{array}{c} \\ T \\ B \end{array} \begin{array}{c} \begin{array}{cc} L & R \end{array} \\ \begin{pmatrix} 3,1 & 2,3 \\ 1,0 & 3,5 \end{pmatrix} \end{array}$$

The first and second entries in each cell refer to player-1 and player-2's payoff for that specific set of actions. For instance, when player-1 and player-2 select actions T and R, respectively, they receive a payoff of 2 and 3, respectively.

If player-1 chooses action T, the maximum payoff that he can receive is 3. However, player-2 will choose action R in response to player-1's action T in order to maximize his payoffs. This results in player-1 receiving a lower payoff of 2 if he selects action T. If instead player-1 chooses action B, player-2 will choose action R to maximize his profits. Thus the two players receive a payoff of 3 and 5, respectively, in this case. Thus it is not to player-1's benefit to play action T. This leads to the concept of *Nash equilibrium*. A strategy (that is, an action pair) is a Nash equilibrium if it is suboptimal for the players to deviate from it. As it is suboptimal for player-1 and player-2 to deviate from actions B and R, respectively, (B, R) is the Nash equilibrium for the game, where they both receive the maximum payoff of 3 and 5, respectively.

There are further extensions to the above simple scenario. For instance, *n-person games* involves a game with more than two players. If the game is played for more than one time period and includes a sequence of moves, it is called a *repeated game*. If the game is played over a period of time, the future payoffs may be discounted by a discount factor. In this case, the game is referred to as a *discounted* game, else it is referred to as an *undiscounted game*. Another class of games is stochastic games and evolutionary games, which are described in the following sections.

2.1 Evolutionary Game

Evolutionary game theory had its beginnings in the seminal paper by two theoretical biologists, Maynard Smith and Price (1973) who introduced a new solution concept, evolutionarily stable strategy, to define stability and equilibrium in a population over a period of time when the population is invaded by a small number of mutants. Evolutionary game theory looks at interactions between one or more population of players who all display some behaviour. Players are selected randomly from these populations to play a single-shot classical game. This is repeated over a longer time period. The players may choose the same action or choose different actions each time. As the play progresses, new actions may become available to one or more population due to externalities that may arise (such as new policies, competitive products, emerging technologies, etc). Strategies with higher payoffs are adapted by the rest of the players and spread through the population. The aim is to determine if and when the game reaches a stable equilibrium (known as evolutionarily stable strategy) under these conditions.

In such a setting, it is important to understand how the adoption of strategies by the players in the population change over time based on the payoffs. There are many approaches to studying the dynamics involved in the evolution of the populations. We focus on one such dynamics—the *replicator dynamics* proposed by Taylor and Jonker (1978) who gave a dynamic approach (replicator dynamics) to the static evolutionary stability concept of Maynard Smith and Price (1973). In brief, the replicator dynamics looks at the per capita rate of change in the percentage of the population using various strategies. The per capita rate of change for a specific strategy is proportional to the difference between the expected payoff for the players who choose that strategy and the average payoff of the population. Thus more players in a population will use a strategy that results in a positive per capita rate of change.

The following definitions from Hofbauer and Sigmund (2003) are provided for a single homogeneous population of players with payoff matrix A, where the players have n actions available to them. The definitions can be extended to multiple populations.

Definition 2.1 State of the population: Let x_i indicate the proportion of players playing action i at time t. Then the state of the population indicates the proportion of players playing each of the n actions and is represented as $x = (x_1, \ldots, x_n)$.

Definition 2.2 Replicator equation: The replicator equations indicate the per capita rate of change in the population based on the difference between the expected payoff to players selecting a specific action i and the average payoff for the state. That is, the replicator equation is given by

$$\dot{x}_i/x_i = (Ax)_i - x^t Ax \tag{1}$$

Definition 2.3 Evolutionarily stable strategy (ESS): A strategy r is an ESS if no other strategy p introduced in a small quantity ϵ $(0 < \epsilon < 1)$ can invade the population. That is,

$$p^t A(\epsilon p + (1-\epsilon)r) < r^t A(\epsilon p + (1-\epsilon)r) \tag{2}$$

where $\epsilon > 0$ is termed as the invasion barrier. Rearranging the terms, Eq. 2 can be stated equivalently as the following two conditions:

(1) Equilibrium condition: $r^t Ar \geq p^t Ar$ for all p.
(2) Stability condition: If $r^t Ar = p^t Ar$, then $r^t Ap > p^t Ap$.

Remark:

(1) The equilibrium condition states that the strategy r must do at least as well as strategy p when played against strategy r. The stability condition states that if both strategy r and strategy p do equally well against strategy r, then strategy r must do better than strategy p against strategy p itself.
(2) While the above has been defined for invasion by a single mutant, multiple mutations have been discussed in Anirban Ghatak et al. (2012).

2.2 Stochastic Game

Shapley (1953), in his seminal paper, first defined a finite zero-sum two-player stochastic game with a finite state space S, and finite action space A_1 and A_2 for player-1 and player-2, respectively. Based on the current state $s \in S$ of the game, the two players select an action simultaneously (say, $a \in A_1$ and $b \in A_2$, respectively). The immediate reward $r(s, a, b)$ is a function of the current state of the system and the two actions of the players, where $r(.)$ is a continuous and bounded function on $S \times A_1 \times A_2$. Based on the Markovian transition probability $q(s'|s, a, b)$, the game now moves to state $s' \in S$. The game is played at discrete points of time (say, once a day) over the infinite future in this manner. As it may happen that player-1's accumulated income $\sum_{n=0}^{\infty} r(s_n, a_n, b_n)$ may not converge, the nth day income is discounted by β^{n-1} where the discount factor β is a preassigned number in $[0, 1)$. This automatically makes the ∞-series converge. As in the matrix game, player-1 wants to maximize his/her discounted total (expected) income, while player-2 wants to minimize the same. Note that this total income will depend on the initial state $s \in S$.

Stochastic games are typically represented as $(S, A_1, A_2, r, q, \beta)$. We provide a limited set of definitions for the purpose of this chapter.

Definition 2.4 Stationary strategy: Let S be the state space and P_{A_1} be the set of probability distributions on player-1's action set A_1. A stationary strategy for player-1 is a Borel measurable mapping $f : S \to P_{A_1}$ that is independent of the history that led to the state $s \in S$. Similarly, we define a stationary strategy for player-2 as a Borel measurable mapping $g : S \to P_{A_2}$ that is independent of the history that led to the state $s \in S$.

Definition 2.5 Behavioral strategy: Consider a general policy π given by $\pi = (\pi_1, \pi_2, \ldots \pi_n, \ldots)$ where π_n is the strategy set for day n and $\pi_n(\ldots|h_{n-1})$ is the conditional distribution dependent on the history h_{n-1}. Such a strategy is a behavioral strategy.

For discounted finite stochastic games, it is sufficient to look at stationary strategies.

Definition 2.6 Absorbing state: A state s in a stochastic game is an absorbing state if the game remains in state s forever once it reaches s.

Definition 2.7 Discounted payoff: Consider the non-zero-sum discounted stochastic game $\Gamma = (S, A_1, A_2, r_1, r_2, q, \beta)$. Let $r_1^{(n)}(s_0, f, g)$ and $r_2^{(n)}(s_0, f, g)$ be the expected immediate reward at the n-th stage to player-1 and player-2, respectively, when the game starts at state s_0 and they use stationary strategy f and g, respectively.

Then, the discounted payoffs (also known as the β-discounted payoff) for player-1 and player-2 are as follows.

$$I_\beta^{(1)}(f, g)(s_0) = \sum_{n=0}^{\infty} \beta^n r_1^{(n)}(s_0, f, g)$$

$$I_\beta^{(2)}(f, g)(s_0) = \sum_{n=0}^{\infty} \beta^n r_2^{(n)}(s_0, f, g)$$

For the zero-sum stochastic game (that is, when $r_1 = -r_2 = r$, say), the β-discounted payoff for player-1 is given by

$$I_\beta(f, g)(s_0) = \sum_{n=0}^{\infty} \beta^n r^{(n)}(s_0, f, g)$$

Definition 2.8 Optimal strategies and value: A pair of stationary strategies (f^o, g^o) is optimal in the zero-sum discounted stochastic game if for all $s \in S$

$$I_\beta(f, g^o)(s) \le I_\beta(f^o, g^o)(s) \le I_\beta(f^o, g)(s) \tag{3}$$

for all $f \in P_{A_1}^S$ and for all $g \in P_{A_2}^S$. Here P_{A_1} and P_{A_2} are the set of probability distributions on player-1 and player-2's action sets A_1 and A_2, respectively.

In other words,

$$\begin{aligned} I_\beta(f^o, g^o)(s) &= \sup_f \inf_g I_\beta(f, g)(s) = \inf_g I_\beta(f^o, g)(s) \\ &= \sup_f I_\beta(f, g^o)(s) = \inf_g \sup_f I_\beta(f, g)(s), \text{ for all } s \in S. \end{aligned} \tag{4}$$

Under assumptions of a finite state and action space, Shapley (1953) proved the existence and uniqueness of the value v_β and a pair of optimal strategies (f^o, g^o). The value for all $s \in S$ is given by

$$\begin{aligned} v_\beta(s) &= \sup_f \inf_g I_\beta(f, g)(s) = \inf_g \sup_f I_\beta(f, g)(s) \\ &= \inf_g I_\beta(f^o, g)(s) = \sup_f I_\beta(f, g^o)(s) \\ &= I_\beta(f^o, g^o)(s) \end{aligned} \tag{5}$$

While the value is unique, the optimal strategies may not be unique.

For a non-zero-sum two-player finite bimatrix game, (x^*, y^*) is a Nash equilibrium pair if

$$x^{*T} A y^* \geq x^T A y^* \text{ for all strategies } x \text{ of player - 1, and}$$
$$x^{*T} B y^* \geq x^T B y \text{ for all strategies } y \text{ of player - 2}$$

For the non-zero-sum discounted stochastic game with two players, a pair of stationary strategies (f^o, g^o) is an equilibrium pair for the minimization problem if

$$I_\beta^{(1)}(f^o, g^o)(s) \leq I_\beta^{(1)}(f, g^o)(s)$$
$$I_\beta^{(2)}(f^o, g^o)(s) \leq I_\beta^{(2)}(f^o, g)(s) \tag{6}$$

for all f, for all g, and for all s.

2.3 Evolutionary Stochastic Game

Evolutionary stochastic games (ESG) incorporate aspects of both evolutionary and stochastic games. As per Flesch et al. (2013), the game involves one or more populations, where the individuals of the population(s) interact in finitely many environments (equivalent to the notion of states in stochastic games). Further, the impact of each of these states on the fitness of the individuals in the population is determined by the transition probabilities between these states. In one sense, evolutionary games can be thought of as single-state evolutionary stochastic games. Hence, similar to evolutionary games with a single state, the proportion of the population types with higher fitness will increase in the environment. Flesch et al. (2013) further require the payoff and transition matrices to be symmetric as the players are facing the same strategic situation in all states.

Let $\{e_1, e_2, \ldots, e_{|X_P|}\}$ indicate the pure stationary strategies for population X_P. Then the type of an individual species identifies a specific action that the individual may take for each of the states. Let $x_d(s, i)$ and $e_k(s, i)$ indicate the probability on action i in state s for the stationary strategies x_d and e_k, respectively. Then,

$$x_d(s, i) = \sum_{k=1}^{|X_P|} d_k.e_k(s, i)$$

We say that $d = \{d_1, \ldots, d_{|X_P|}\}$ is a population distribution over the set of pure stationary strategies that uniquely defines a stationary strategy x_d. The fitness of the individual can be defined as the average of all the payoffs accumulated during his lifetime. In an evolutionary game, the payoff matrix is used to determine whether a strategy is ESS. In ESG, they consider the β-discounted reward or the limiting average reward (as defined for stochastic games) instead. They also extend the approach to replicator dynamics along the same lines. We provide a few relevant definitions from Flesch et al. (2013) under restrictions of symmetric payoff and transition matrices.

Definition 2.9 Replicator equation: The replicator equation for the evolutionary stochastic game is given by

$$\dot{d}_{l_i}/d_l = \gamma(e_l, e_d) - \gamma(x_d, x_d), \text{ for } l = 1, 2, \ldots, |X_P| \tag{7}$$

where d indicates the distribution over the set of pure stationary strategies e_l, and x_d is the stationary strategy induced by d.

Definition 2.10 Evolutionary stable strategy (ESS): A stationary strategy $\bar{x}$ is an ESS if for all stationary strategies $y \neq x$, we have

$$\gamma(y, \bar{x}) \leq \gamma(\bar{x}, \bar{x}) \tag{8}$$

$$\gamma(y, \bar{x}) = \gamma(\bar{x}, \bar{x}) \Rightarrow \gamma(\bar{x}, y) > \gamma(y, y) \tag{9}$$

Remark 1 If we consider a specific population distribution d, then the proportion $x_d(s, i)$ indicates the fraction of the population that choose action i in state s.

Remark 2 Flesch et al. (2013) show that every symmetric irreducible stochastic game admits a symmetric stationary equilibrium (x', x').

3 The Analytic Model

This section briefly describes how to model the supply chain as an evolutionary game. The stakeholders of the supply chain form the populations in the evolutionary game. The interactions between the players can be of two types: inter-population interaction (between players who belong to different populations) and intra-population interaction (between players who belong to the same population). For example, if you consider the two populations in a supply chain to be retailers and manufacturers, the interaction between a retailer and a manufacturer is considered inter-population interaction. If the supply chain has multiple retailers, then the interactions between two retailers would constitute an intra-population interaction. The inter-population and intra-population interactions determine the finite set of actions available to each player in the population. The type of supply chain determines the dimensions to be considered and the metrics for each dimension. It is important to note that the dimensions and the metrics for each stakeholder may vary. The model allows each stakeholder population to define their own set of relevant dimensions and metrics, and then study the impact of all these as a whole. The payoff function for each population is then created as a function of the metrics relevant to the population. In the case of a centralized supply chain, the individual payoff functions can then be integrated into a single payoff function for the entire supply chain. For a decentralized supply chain, it can be treated as a multi-population game. For instance, a two-tier decentralized supply chain (consisting of manufacturers and retailers) may be viewed as a two population evolutionary game where the payoff function for the supply chain could be indicated by the manufacturer and retailer's profits.

The trajectory of the payoff function for the supply chain on the solution space (that is, the standard simplex space for the game) indicates the presence or absence of a sustainable point. This trajectory graphs the rate of change in the supply chain's payoff function as the population composition changes. The evolutionarily stable point is in agreement with our notion of sustainability since a stable strategy cannot be invaded by mutants in small proportions. It is interesting to note that an evolutionary game may have none, one or more ESS.

The following subsection depicts how to model the supply chain as an evolutionary game.

3.1 Building the Multi-population Game

Consider a supply chain with N finite size populations. Let n_p to be the number of dimensions for population p, $(p = 1, 2, \ldots, N)$. Let $n_{p,i}$ be the total number of metrics considered for dimension i of population p. The metrics for dimension i of population p is given by $m_1^{(p,i)}, m_2^{(p,i)}, \ldots, m_{n_{p,i}}^{(p,i)}$.

Denote the real-valued C^1 payoff function for population p as

$$f_p : \mathbb{R}^{(n_{p1}+n_{p2}+\ldots+n_{p,n_p})+1} \rightarrow \mathbb{R}$$

Then f_p is given as follows:

$$f_p(m_1^{(p,1)}, \ldots, m_{n_{p1}}^{(p,1)}, m_1^{(p,2)}, \ldots, m_{n_{p2}}^{(p,2)}, \ldots, m_1^{(p,n_p)}, \ldots, m_{n_{p,n_p}}^{(p,n_p)}, t) \tag{10}$$

Let us represent the proportion of players playing various actions at time t by the state of the evolutionary game s. Let the game start in an initial state s_0. The payoff function for each population can be determined based on the historic data for the metrics across each dimension, after ignoring outliers. If the game starts in an initial state s_0, the payoff for the supply chain is then constructed based on the population payoffs as $f(s_0, f_1(\ldots), \ldots, f_N(\ldots), t)$. Replicator dynamics can then be applied to arrive at the stable points for the game. Using replicator dynamics requires that each of these payoff functions be a continuous function with first-order partial derivatives (i.e. a C^1 function).

The following subsection details the analytic framework for the same as detailed in Babu and Mohan (2017).

3.2 The Analytic Framework

Without loss of generality, consider evolutionary games with two populations only. The result can be easily extended to N-population evolutionary games. Interactions that arise in the two populations can be inter-population interactions and intra-population interactions. The analytic framework is restricted to inter-population interactions only. As interactions amongst players from the same population can be thought of as a special case of implementation of the framework, the framework can be extended to intra-population interactions too. The analysis is based on the most commonly used dynamics, namely, replicator dynamics. The analysis is restricted to a finite size population, finite state and finite action space for the two populations. We describe the framework and analysis using the following example of the two populations in a two-tier supply chain representing the manufacturer and the retailer (say).

Let us look at sustainability across economic, environmental and social dimension. Let the actions for the manufacturer indicate whether the manufacturer produces low (L) or high (H) quality product. Economic dimension could include metrics such as cost of production and profitability. So a high-quality product could lead to a lower profit margin than a low-quality product. Environmental dimension could include whether the manufacturer follows green processes for production. When the manufacturer follows green processes, the product turns out to be a high-quality product. The social dimension could include metrics relevant to inclusion in the workforce. Along similar lines, let the actions available for the retailer be S and N, indicating whether they stock or do not stock the product. The metrics for economic, environmental and social dimensions may be determined for the retailer too. Further, let s_0 be the starting state for the supply chain, where s_0 indicates the proportion of low-quality product versus high-quality product produced by the manufacturer at the start time, and the percentage of products that are stocked. Let f_1 and f_2 indicate the payoff function for the manufacturer and the retailer, respectively, where the payoff function indicates the scale of profit in the supply chain based on the metrics across all three dimensions as indicated by Eq. 10. The composite payoff function for the supply chain is given by $f(s_0, f_1, f_2, t)$. The payoff functions may have a linear (that is, matrix) or a non-linear representation as shown in the following subsections. The conditions under which the supply chain reaches sustainability for both these representations are given below.

3.2.1 The Matrix Form

A linear profit function may arise when the demand (Q) is deterministic. For instance, let P be the selling price per unit and C be the cost per unit for the retailer. Then the profit for the retailer is a linear function given by $Q \times P - C \times Q$. A similar linear function can be arrived at for the manufacturer. As these linear functions can be represented in a matrix form, we look at a matrix representation as follows.

In the bimatrix representation, let the matrices A and B represent the payoff for the manufacturer and retailer as shown below:

$$A = \begin{matrix} & S & N \\ L & a & b \\ H & c & d \end{matrix}, \quad B = \begin{matrix} & L & H \\ S & e & f \\ N & g & h \end{matrix}$$

Without loss of generality, assume that one action for each player leads the supply chain to sustainability. For such a bimatrix representation, the following proposition shows that the supply chain reaches sustainability only when all the players are completely committed to the single action (that is, pure strategy) that leads to sustainability. If the number of actions available to the player is greater than 2 and the player has a choice of more than one action that can lead to sustainability, then the player can select one amongst such progressive actions with a non-zero probability. In other words, the players will never play actions that do not lead to sustainability in the long run.

Proposition 3.1 *Consider a supply chain with asymmetrical payoffs, where the payoff matrices for the manufacturer and retailer are as given above. Such a supply chain never reaches sustainability in the interior of its strategy space. The sustainable points for the supply chain are always present in the hull of the strategy space.*

Proof Let the proportion of low-quality products and high-quality products produced by the manufacturer be indicated by x and $1 - x$, respectively, and the proportion of products stocked by the retailer be indicated by y. Thus, the state of the population game is $(X, Y) = ((x, 1 - x), (y, 1 - y))$. The replicator dynamics for the two population game is given by the functions $f_1(x, y)$ and $f_2(x, y)$ for each of the population as follows.

$$f_1(x, y) = \dot{x} = x(1 - x)[(AY)_1 - X^t AY] = x(1 - x)[y(a - b - c + d) + b - d] \tag{11}$$

$$f_2(x, y) = \dot{y} = y(1 - y)[(BX)_1 - Y^t BX] = y(1 - y)[x(e - f - g + h) + f - h] \tag{12}$$

The rest points are obtained by solving for x and y when $\dot{x} = 0$ and $\dot{y} = 0$. Thus, the corner rest points are $x = 0, x = 1, y = 0, y = 1$ indicating the pure strategies of the population. Consider the scenario when $0 < x < 1$. If $y = 0$ or $y = 1$, then the rest point occurs when the retailer plays only pure actions. Similarly, consider the scenario when $0 < y < 1$. If $x = 0$ or $x = 1$, then the rest point occurs when the manufacturer plays only pure actions. The interior rest point is given by

$$F = \left(\frac{h - f}{e - g + h - f}, \frac{d - b}{a - c + d - b} \right)$$

The Jacobian at the interior rest point is given by

$$J_F = \begin{pmatrix} \frac{\delta f_1}{\delta x} & \frac{\delta f_1}{\delta y} \\ \frac{\delta f_2}{\delta x} & \frac{\delta f_2}{\delta y} \end{pmatrix}_{|F}$$
$$= \begin{pmatrix} 0 & \frac{(a-c+d-b)(h-f)(e-g)}{(e-g+h-f)^2} \\ \frac{(e-g+h-f)(d-b)(a-c)}{(a-c+d-b)^2} & 0 \end{pmatrix}$$

The stability of the rest point is based on whether the eigenvalues (λ) of the Jacobian matrix are real or complex numbers, and distinct or repeated. Solving for the eigenvalues of the above Jacobian matrix yields the following:

$$\lambda^2 = \frac{(a - c)(d - b)(e - g)(h - f)}{(a - c + d - b)(e - g + h - f)} \tag{13}$$

The relation between the variables in the cells of the payoff matrix determines the sign of λ^2. For example, $a < c, d < b, g < e, f < h$ indicates that it is more profitable for the manufacturer to produce high-quality goods when the retailer stocks more, while it is more profitable for the retailer to stock more when the manufacturer produces low-quality goods. In this case, the eigenvalues are complex and the trajectory of the game is along periodic orbits around the rest point. So though the manufacturer can produce low-quality goods with a positive probability, there is a lower incentive for him to do so and the supply chain will never reach sustainability. Along similar lines, when $a < c, d < b, g < e, f > h$, the eigenvalues are real and the phase portrait indicates that the rest point is a saddle point. Also the eigenvalues are never repeated. Hence, the completely mixed strategy F is not evolutionarily stable, and there are no mixed ESS for the evolutionary game. The only point where sustainability is reached is at the corners of the simplex $S_2 \times S_2$ (that is, at $x = 0, x = 1, y = 0, y = 1$). That is, when $h = f, e = g, d = b$, and $a = c$, respectively. ■

The above result also corresponds with the numerical example provided in Sect. 5 where the ESS is reached only at the corners of the simplex. Proposition 3.1 is true in general for a supply chain with symmetric payoffs. The following corollary (stated without proof) looks at a special form of symmetric games with a stable interior rest point. Consider a special case of symmetric games where $A = cB^t$, where c

is a constant. The value $c = -1$ indicates a zero-sum game and $c = 1$ indicates a partnership game. As per Karl Schlag (1994), if $c < 0$ and if there exists an interior rest point for the evolutionary game where each player has more than two strategies, then there exists a Nash–Pareto pair and is the only condition under which the interior rest point is locally stable.

Corollary: 1 If the symmetric game is of the form $A = cB^t$ where $c < 0$ and has an interior rest point, then the rest point is a stable rest point. ■

3.2.2 Non-linear Payoff Function

In reality, not all payoffs neatly fit into a matrix format as they may be non-linear. For the purpose of this demonstration, consider a specific format of non-linear payoff functions for the manufacturer and retailer that arises as follows. Consider an inverse demand function ($Q(P)$) where P is the price and Q is the quantity that is a function of price and a cost function ($C(Q)$) for the manufacturer. Then the payoff to the manufacturer is given by $Q(P) \times P - C(Q)$, and hence leads to a quadratic function for the manufacturer. A similar quadratic function can be arrived at for the retailer. Proposition 3.2 considers a generic form of the quadratic payoff function for the manufacturer and retailer. The results can, however, be extended to any non-linear function.

Proposition 3.2 *Let the non-linear payoff functions for the manufacturer and retailer be $f_1(x) = px^2 + qx + r$ and $f_2(y) = uy^2 + vy + w$, respectively. Then the only sustainable rest points are at $x = 0, x = 1, y = 0, y = 1$. Further, there are no interior rest points. In fact, the sign of $(q-1)^2 - 4pr$ and $(v-1)^2 - 4uw$ indicates the path of the supply chain.*

Proof Since the game has rest points x^* when $f(x^*) = x^*$ or $f(x^*) = 0$, without loss of generality consider the case when $f(x^*) = x^*$. The corner rest points are $x = 0, x = 1, y = 0, y = 1$. The interior rest point is given by

$$F = \left(\frac{-(q-1) \pm \sqrt{(q-1)^2 - 4pr}}{2p}, \frac{-(v-1) \pm \sqrt{(v-1)^2 - 4uw}}{2u} \right)$$

The eigenvalues of the Jacobian at the interior rest point are given by $1 \pm \sqrt{(q-1)^2 - 4pr}$ and $1 \pm \sqrt{(v-1)^2 - 4uw}$. First consider the eigenvalues $1 \pm \sqrt{(q-1)^2 - 4pr}$. As the eigenvalues are not pure imaginary, the interior rest point is not a sustainable point. In particular, when $(q-1)^2 - 4pr < 0$, or $0 < (q-1)^2 - 4pr < 1$, the rest point is a repeller. When $(q-1)^2 - 4pr > 1$, the rest point turns out to be a saddle point. The same analysis can be extended to the second set of eigenvalues. ■

In particular, when $p = u = 0$, the game translates to a matrix game. Also, the analysis can be extended to any non-linear function format along the same lines.

Section 4 briefly describes a public health insurance supply chain. This sets the foundation for the example in Sect. 5 relating to National Health Services (NHS), UK to demonstrate this model.

4 Application of the Framework: A Public Health Insurance Service Supply Chain

The World Health Organization (WHO) defines the goal of universal health coverage as creating an environment where people can obtain needed health services without any out-of-pocket expenses. Out-of-pocket expenses include any expense paid by a person due to treatment-related procedures that are not covered by any insurance. Many public health insurance (PbHI) schemes are designed to achieve universal health coverage. This thus necessitates the study of sustainability of PbHI schemes holistically across all dimensions.

Let us start with a typical PbHI supply chain. The government offers healthcare coverage to eligible individuals (hereafter called consumers). The government contracts with one or more insurance companies and with healthcare service providers (public and/or private). Thus, government or regulatory bodies, insurance companies, care providers, and consumers constitute the stakeholders. This is shown in Figs. 1 and 2, which are partial adaptations of the two most commonly cited value chains, namely, Burns (2002) and Chakravarty (2014).

Of course, this is not an exhaustive list of stakeholders. If the insurance covers prescription drugs, then pharmaceutical companies also become stakeholders in the supply chain. For purposes of demonstrating this model, restrict the stakeholders to the consumer, insurance company and care provider with a small subset of metrics relating to social and economic dimension for each population. Figure 3 lists some commonly used metrics for social and economic dimension for a healthcare supply chain.

Section 5 demonstrates the model for National Health Services (NHS), UK . In this example, we infer the starting state based on the time series data. We use this to analyse the prediction error by comparing it with the actual state.

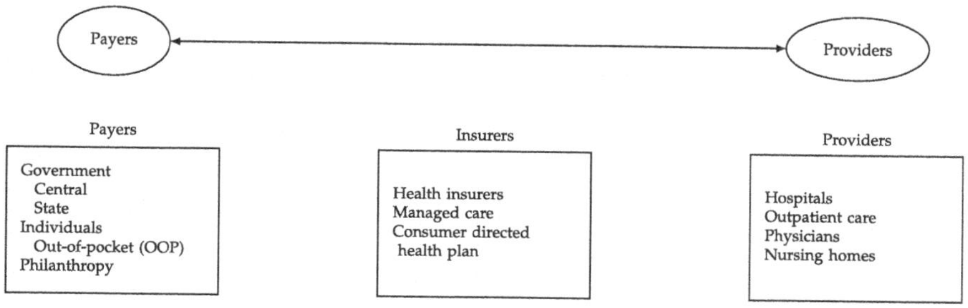

Fig. 1 Modification to healthcare value chain (Burns 2002)

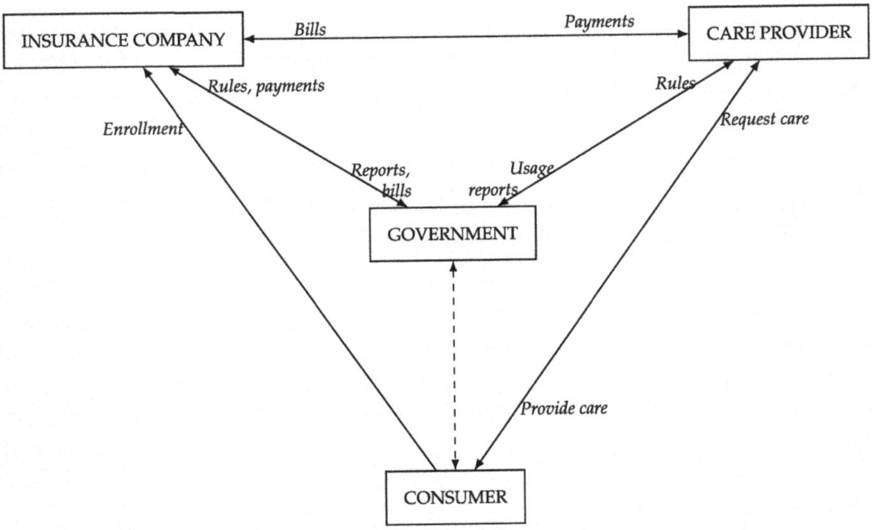

Fig. 2 Modification to a generic healthcare network (Chakravarty 2014)

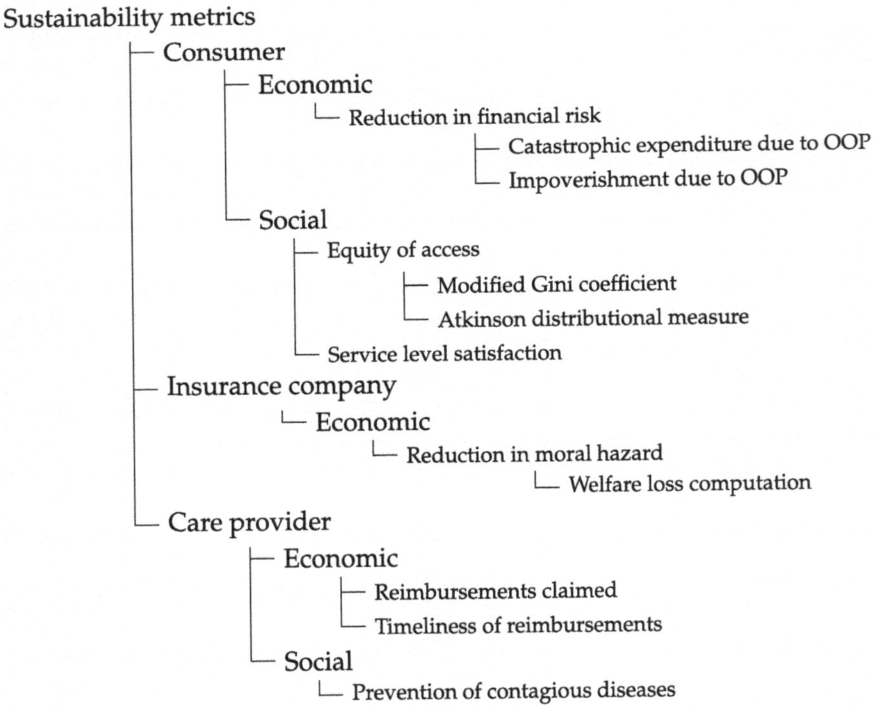

Fig. 3 Sustainability metrics for a generic PbHI scheme

The framework is used to answer the following two questions.

(1) In terms of the prediction error, does the predicted model become sustainable at a vastly different time from the actual system?
(2) For a given a set of starting payoffs and population state, is the system sustainable by itself; and if yes, when? What is the effect of implementing a new policy (equivalent to introducing a new payoff) that will cause the system to reach sustainability faster?

5 A Numerical Example: NHS, UK

The National Health Services (NHS) in United Kingdom is a PbHI scheme that has been in existence for a length of time. To determine the sustainability factors for NHS, we consider a subset of publicly available data about NHS restricted to two dimensions (economic and social) for two stakeholders (consumers and providers) as shown in Fig. 4.

The following sections detail the data used for each of these metrics.

5.1 Defining the Evolutionary Game Parameters

The state space for the provider population captures the proportion who have a high rating due to high productivity efficiency and high patient safety. Similarly, state space for the consumer population captures the proportion of consumers who find the NHS satisfactory and hence use it. These states in turn define the strategies available to the players in each population.

The payoff function for each population is built, and then the composite payoff function for the game is built. The payoff function for care providers is built based on the individual payoff matrix and the state function. The state function for care providers $f_H : \mathbb{R}^2 \to \mathbb{R}$ is a C^1 function across two dimensions with the following

Fig. 4 Sustainability metrics sample for NHS, UK

Sustainability metrics
- Provider
  - Economic
    - Efficiency in productivity
  - Social
    - Patient safety
- Consumer
  - Social
    - Median inpatient waiting time
    - Public satisfaction with NHS

Table 1 Provider-related NHS data

Year	Productivity efficiency[a]	Patient safety[b]
2004	96	42
2005	96	403
2006	98	659
2007	97	834
2008	97	909

[a]Office For National Statistics (2015)
[b]In'000 NHS (2011)

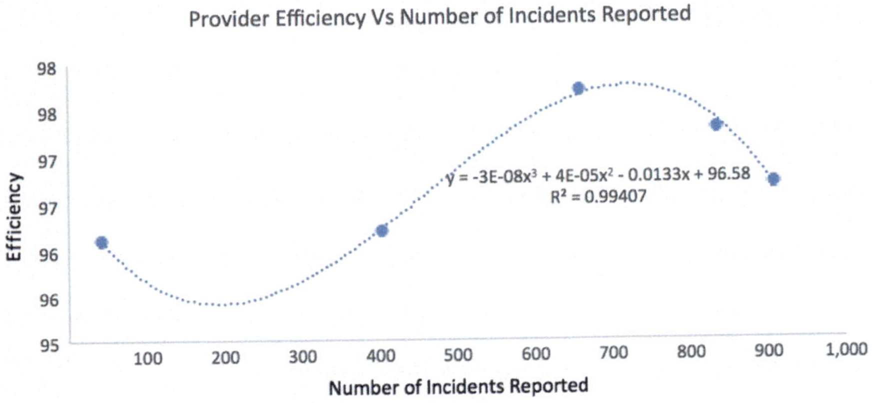

Fig. 5 C^1 function for provider population (NHS)

parameters:x_H-Patient safety (social metric); t-Time, along with efficiency in productivity (economic metric). Based on the data available from NHS (Table 1), the metrics are integrated across two dimensions resulting in the following function for providers (Fig. 5):

$$f_H(x_H, t) = -3 \times 10^{-08} x_H(t)^3 + 4 \times 10^{-05} x_H(t)^2 - 0.0133 x_H(t) + 96.58 \tag{14}$$

Let $f_C: \mathbb{R}^2 \to \mathbb{R}$ be a C^1 function for the consumers with the following parameters: x_C-Median inpatient waiting time (social metric); t-Time. The payoff function determines the proportion of the population that is satisfied with NHS (social metric). Based on the data available from NHS (Table 2), the linear function for consumers shown in Fig. 6 is the following:

$$f_C(x_C, t) = -1.3229 * x_C(t) + 57.795 \tag{15}$$

The starting state for the two populations at any time t can be found from Eqs. 14 to 15. The strategies for the provider are as follows: (1) provide low-quality service leading to low rating, and (2) provide high-quality service leading to high rating.

Table 2 Consumer-related NHS data

Year	Inpatient waiting time (in weeks)[a]	Public satisfaction survey (%)[b]	Year	Inpatient waiting time (in weeks)[a]	Public satisfaction survey (%)[b]
1994	14	44	2001	13	39
1995	12	37	2002	13	40
1996	12	36	2003	12	44
1997	13	35	2004	10	43
1998	15	41	2005	8	48
1999	13	47	2006	7	49
2000	13	42	2007	6	51

[a]In weeks (Department of Health 2013)
[b]% of population surveyed (Appleby and Phillips 2009)

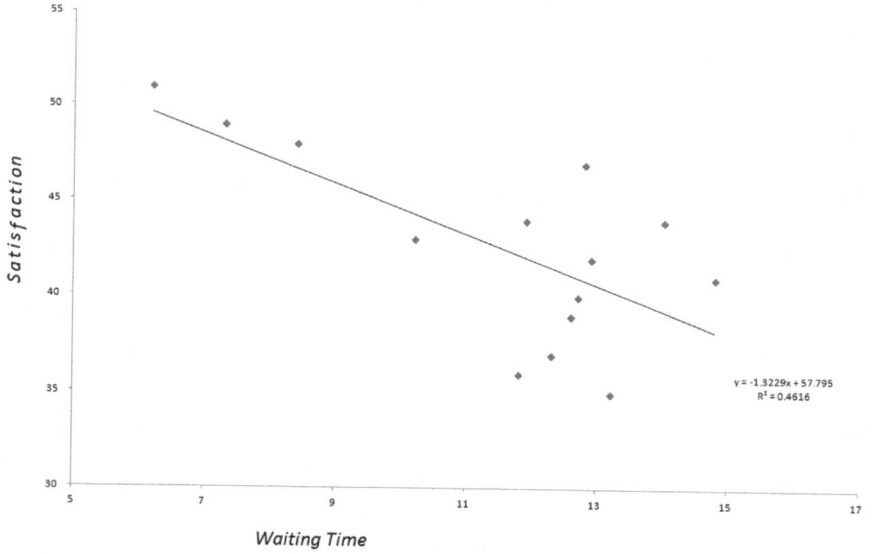

Fig. 6 C^1 function for consumer population (NHS)

The two strategies available to the consumer are as follows: (1) use the services of NHS, and (2) do not use the services of NHS. Note that this in turn is indicated by the proportion of satisfied consumers. The composite C^1 function $f : \mathbb{R}^4 \rightarrow \mathbb{R}$ is given by $f(s_0, f_H, f_C, t)$ where s_0 denotes the initial state of the evolutionary game. Tracing the rate of change in the trajectory of this composite average payoff function for the population identifies stability.

The following section details the experimental setup.

5.2 Experimental Setup

The questions answered through the experimental setup has been listed in Sect. 4. Towards this, the experiment is set up as follows. Given a starting state and payoff matrix for the provider and consumer population, the evolution of the population is traced for 15,000 time periods. To study the effect of introduction of policy changes (reflected as new payoff matrix for either or both populations), a decision is taken to introduce a new payoff matrix at this point to study the evolution of the population for another 15,000 time periods. Thus the evolution of the population is tracked for a total of 30,000 time periods.

Table 3 lists the input payoffs in the experimental setup. Three types of provider payoffs are considered: Undesirable (U), Good (G), Best (B). Here G and B are payoffs that give the provider an incentive to offer better quality service for consumers who come through the NHS, while U does not. Similarly, three types of consumer payoffs are considered: Undesirable (U), Good (G), Best (B). As before G and B are payoffs that provide the consumer with an incentive to avail of NHS while U does not. For instance, when the consumer payoff is U, the consumer is not motivated to use the provider in the NHS network. When the consumer payoff is G, he is indifferent to availing of medical services within or outside the NHS network. When the consumer payoff is B, he has an incentive to use the provider in the NHS network. The same holds for the provider payoff in terms of the quality of service provided.

The entries in Table 3 are read as follows. For the provider payoff matrix $B = [1\ 2;\ 3\ 1]$: the provider receives 1 unit when he operates under low efficiency and the consumer uses the network; 2 units when he operates under low efficiency and the consumer does not use the network; 3 units when he operates under high efficiency and the consumer uses the network; and 1 unit when he operates under high efficiency and the consumer does not use the network. The other payoff matrices for the provider are read similarly. In a similar vein, the consumer payoff matrix $B = [3\ 3;\ 1\ 0]$ looks at the payoff to the consumer using the network or not using the network when the provider provides high quality of service and low quality of service. The payoff for the consumer when they use the NHS network is higher than when they avail of services outside the network when the provider provides high-quality service and low-quality service. The other payoff matrices for the consumer are read similarly.

Tables 4 and 5 list the input states in the experimental setup. Five initial population states indicated by $S1$, $S2$, $S3$, F and A are considered based on the proportion of population engaging in a specific action. To understand the notation in this table,

Table 3 Sample input payoffs (NHS)

Payoff types	Providers	Consumers
Undesirable (U)	[2 0; 1 3]	[1 3; 2 0]
Good (G)	[2 0; 1 1]	[2 3; 1 0]
Best (B)	[1 2; 3 1]	[3 3; 1 0]

Table 4 Input states for computing prediction error (NHS)

States	Providers	Consumers
Forecasted (F)	0.95, 0.05	0.4986, 0.5014
Actual (A)	0.97, 0.03	0.51, 0.49

Table 5 Input states for checking sustainability (NHS)

States	Providers	Consumers
S1	0.9, 0.1	0.2, 0.8
S2	0.5, 0.5	0.5, 0.5
S3	0.2, 0.8	0.9, 0.1

consider the example of the provider state $S1 = [0.9, 0.1]$. This indicates that 90% of the provider population provide low-quality service, while the remaining 10% provide high-quality service. The consumer state $S1 = [0.2, 0.8]$ indicates that 20% of the consumers use the NHS network, while the remaining 80% do not. The other states are read similarly. In Table 4, the two states-forecasted (F) and actual (A)-will be used to compute the prediction error. In Table 5, $S1$ and $S3$ represent the different ends of the spectrum and $S2$ lies in between. These states are used to answer our questions on sustainability of the system.

Since the provider and consumer payoff can be changed at a selected point in the run, this results in iterating through 3^2 combinations of provider payoff, 3^2 combinations of consumer payoff, and 5 different starting population states, leading to a total of $3^2 \times 3^2 \times 5$ combinations. As some of these payoff changes may not be feasible (for example, if the payoff for the provider in the first half of the run is the best payoff, there will be no reason to introduce an undesirable payoff in the second half of the run), these infeasible runs have not been considered and are indicated as "Infeasible" in Table 7.

The code was implemented using Matlab (version R2013a). The first part of the numerical experiment looks at prediction errors based on the actual starting state (A) versus the forecasted starting state (F). At time $t = 1$, different feasible combinations of provider and consumer payoffs along with the starting states F and A are considered ($3^2 \times 3^2 \times 2$ combinations, of which some combinations may be infeasible as mentioned earlier). At time $t = 15001$, a decision is taken to introduce a new provider and/or consumer payoff to study the effect of a policy change on the system.

The second part of the experiment studies whether the system reaches sustainability; and if yes, when. At time $t = 1$, different feasible combinations of provider payoffs, consumer payoffs and starting states ($S1$, $S2$ and $S3$) are considered ($3^2 \times 3^2 \times 3$ combinations, of which some combinations may be infeasible as mentioned earlier). The run proceeds as earlier.

5.3 *Results and Analysis Using Replicator Dynamics*

Replicator dynamics is used to track the evolution of the population to answer the questions relating to prediction error and sustainability.

5.3.1 Prediction Error

The actual states for the two populations are given in Tables 1 and 2. The forecasted starting state is derived from Eqs. 14 to 15. For example, in 2007, 97% of the provider population showed high productivity efficiency (as per Table 1). However, the fitted equation for providers (Eq. 14) indicates that only 95% of the provider population showed high productivity efficiency that year. Hence, the actual state of the provider population ($[0.97, 0.3]$) is different from the forecasted starting state ($[0.95, 0.05]$). This implies that the time taken by the system to reach ESS will be different for the actual starting state and the forecasted starting state, and hence the average payoff to the population will differ too. Prediction error techniques help identify the magnitude of this difference across various provider-consumer feasible payoff combinations. Table 6 provides the Mean Squared Error (MSE) for the feasible combinations of input parameters. It can be seen that the value of MSE is less than 0.05 across all the feasible combinations of payoffs.

5.3.2 Sustainability and Evolutionarily Stable Strategies

As mentioned earlier, Tables 3 and 5 list the different types of payoffs and states being considered to study the sustainability of the system. The output of the experimental run is listed in Table 7 for selected time periods. This table helps study the sustainability for various feasible combinations of provider and consumer payoff and starting state. The simplex and evolution graph of two runs are provided as a sample, namely, (1) Figure 7: when the provider and consumer start in state S_1 and have payoff U for the entire run. In this case, the system is not sustainable as there is no ESS. (2) Figure 8: when the provider and consumer start in state S_2 with payoff U, and move to payoff G and B, respectively, at time 15001. In this case, the system is sustainable and does reach an ESS.

As shown in Fig. 7, though the system starts with 97% of the consumers using NHS, the percentage dwindles and then rises again. This cycle continues for the entire run and indicates the effects of the consumer payoff, the provider payoff, and the provider's own starting state. Thus, it is clear that the system is not sustainable. In Fig. 8, the system does cycle without moving towards sustainability for the first 15,000 time units. Then due to the new payoff that is implemented, the system moves towards the entire consumer population using NHS (i.e. sustainability). This is indicated by the small arc reaching the point $(1, 1)$ in the simplex graph.

Table 6 Prediction error—MSE (NHS)

Payoff at time t = 1		Payoff at time t = 15001		MSE	
Provider	Consumer	Provider	Consumer	For provider	For consumer
U	U	U	U	0.03154	0.02851
U	U	U	G	0.00959	0.00821
U	U	U	B	0.00948	0.0082
U	U	G	U	0.01619	0.03056
U	U	G	G	0.00901	0.00821
U	U	G	B	0.00896	0.0082
U	U	B	U	0.00667	0.00821
U	U	B	G	0.00667	0.00821
U	U	B	B	0.00667	0.0082
U	G	U	G	0.00006	0
U	G	U	B	0.00006	0
U	G	G	G	0.00006	0
U	G	G	B	0.00006	0
U	G	B	G	0.00074	0
U	G	B	B	0.00074	0
U	B	U	B	0.00003	0
U	B	G	B	0.00003	0
U	B	B	B	0.00095	0
G	U	G	U	0.02312	0.04475
G	U	G	G	0.00453	0.00839
G	U	G	B	0.00372	0.00855
G	U	B	U	0.00282	0.00818
G	U	B	G	0.00438	0.00856
G	U	B	B	0.00437	0.00862
G	G	G	G	0.00002	0
G	G	G	B	0.00002	0
G	G	B	G	0.00081	0
G	G	B	B	0.00081	0
G	B	G	B	0.00001	0
G	B	B	B	0.00084	0
B	U	B	U	0.00005	0.00002
B	U	B	G	0.00005	0.00028
B	U	B	B	0.00005	0.0002
B	G	B	G	0.00092	0.00001
B	G	B	B	0.00092	0.00001
B	B	B	B	0.00076	0

Table 7 Sustainability of the system (NHS)

Payoff at time = 1		Initial State	Payoff at time = 15001		Average population payoff at time (in 1000s)							Earliest time	Remarks
Provider	Consumer		Provider	Consumer	1	5	10	15	20	25	30	To reach ESS	
U	U	S1	U	U	0.2, 2.5	0.67896, 0.35395	0.01178, 0.35716	1.35644, 2.74605	3.86232, 1.75944	2.99456, 1.99029	2.88546, 0.39663	Infinity	No ESS
U	U	S2	U	U	1, 2	0.16201, 0.31981	0.98582, 2.01321	0.16479, 0.31716	1.97399, 1.0285	3.68522, 1.83234	1.96258, 1.04197	Infinity	No ESS
U	U	S3	U	U	1.6, 1.1	0.07008, 0.0816	0.03669, 1.9799	1.11174, 0.58557	3.83267, 1.97821	1.29547, 1.53579	3.73753, 1.50211	Infinity	No ESS
U	U	S1	U	G	0.2, 2.5	0.67896, 0.35395	0.01178, 0.35716	1.35644, 2.74605	4.83212, 0.91624	4.99877, 0.99939	4.99999, 1	29912	
U	U	S2	U	G	1, 2	0.16201, 0.31981	0.98582, 2.01321	0.16479, 0.31716	4.99693, 0.99918	4.99998, 0.99999	5, 1	25157	
U	U	S3	U	G	1.6, 1.1	0.07008, 0.0816	0.03669, 1.9799	1.11174, 0.58557	4.98288, 0.99146	4.99988, 0.99994	5, 1	27545	
U	U	S1	U	B	0.2, 2.5	0.67896, 0.35395	0.01178, 0.35716	1.35644, 2.74605	5.78343, 0.92782	5.99841, 0.99947	5.99999, 1	29745	
U	U	S2	U	B	1, 2	0.16201, 0.31981	0.98582, 2.01321	0.16479, 0.31716	5.99784, 0.99929	5.99998, 0.99999	6, 1	25032	
U	U	S3	U	B	1.6, 1.1	0.07008, 0.0816	0.03669, 1.9799	1.11174, 0.58557	5.97451, 0.9915	5.99982, 0.99994	6, 1	27520	
U	U	S1	G	U	0.2, 0.9	0.67896, 0.3443	0.01178, 0.12298	1.35644, 1.3675	3.64392, 1.8623	1.26927, 1.88132	2.13818, 0.61679	Infinity	No ESS
U	U	S2	G	U	1, 1	0.16201, 0.16061	0.98582, 0.99968	0.16479, 0.16065	1.35238, 1.90799	1.66052, 0.59589	3.60132, 1.8938	Infinity	No ESS
U	U	S3	G	U	1.6, 0.9	0.07008, 0.05056	0.03669, 0.6722	1.11174, 0.56577	3.82768, 1.98083	1.33031, 1.99014	0.65251, 1.27388	Infinity	No ESS
U	U	S1	G	G	0.2, 0.9	0.67896, 0.3443	0.01178, 0.12298	1.35644, 1.3675	4.92334, 0.96234	4.99945, 0.99973	5, 1	29053	
U	U	S2	G	G	1, 1	0.16201, 0.16061	0.98582, 0.99968	0.16479, 0.16065	4.99713, 0.99929	4.99998, 0.99999	5, 1	25015	
U	U	S3	G	G	1.6, 0.9	0.07008, 0.05056	0.03669, 0.6722	1.11174, 0.56577	4.98306, 0.99155	4.99988, 0.99994	5, 1	27534	
U	U	S1	G	B	0.2, 0.9	0.67896, 0.3443	0.01178, 0.12298	1.35644, 1.3675	5.90058, 0.96688	5.99931, 0.99977	5.99999, 1	28920	

Table 7 (continued)

Payoff at time = 1		Initial	Payoff at time = 15001		Average population payoff at time (in 1000s)							Earliest time	Remarks
Provider	Consumer	State	Provider	Consumer	1	5	10	15	20	25	30	To reach ESS	
U	U	S2	G	B	1, 1	0.16201, 0.16061	0.98582, 0.99968	0.16479, 0.16065	5.998, 0.99934	5.99998, 1	6, 1	24957	
U	U	S3	G	B	1.6, 0.9	0.07008, 0.05056	0.03669, 0.6722	1.11174, 0.56577	5.97471, 0.99157	5.99982, 0.99994	6, 1	27512	
U	U	S1	B	U	1.7, 1.1	0.34913, 1.02327	0.24007, 0.13477	2.05677, 2.72394	3.00011, 0.00027	3.00005, 0	2.99998, 0	22134	
U	U	S2	B	U	1.5, 2	0.24021, 0.32261	1.50645, 1.9855	0.2389, 0.32544	3.00084, 0.00171	2.99992, 0	2.99987, 0	22594	
U	U	S3	B	U	1, 2.5	0.06608, 0.12064	1.32605, 0.70889	0.57567, 1.67751	3.00003, 0.00007	2.99987, 0	2.9999, 0	21015	
U	U	S1	B	G	1.7, 1.1	0.34913, 1.02327	0.24007, 0.13477	2.05677, 2.72394	3.00017, 0.0001	3.00005, 0	2.99998, 0	22123	
U	U	S2	B	G	1.5, 2	0.24021, 0.32261	1.50645, 1.9855	0.2389, 0.32544	3.00119, 0.0006	2.99999, 0	2.99991, 0	22452	
U	U	S3	B	G	1, 2.5	0.06608, 0.12064	1.32605, 0.70889	0.57567, 1.67751	3.00008, 0.00004	3.00019, 0	2.99997, 0	20631	
U	U	S1	B	B	1.7, 1.1	0.34913, 1.02327	0.24007, 0.13477	2.05677, 2.72394	3.00023, 0.00008	3, 0	2.99998, 0	21602	
U	U	S2	B	B	1.5, 2	0.24021, 0.32261	1.50645, 1.9855	0.2389, 0.32544	3.00171, 0.00057	3.00001, 0	2.99996, 0	22351	
U	U	S3	B	B	1, 2.5	0.06608, 0.12064	1.32605, 0.70889	0.57567, 1.67751	3.00011, 0.00004	2.99999, 0	2.9998, 0	21158	
U	G	*	U	U								*	Infeasible
U	G	S1	U	G	0.2, 2.5	0.08897, 0.05167	0.00063, 0.00036	0, 0	5, 1	5, 1	5, 1	14213	
U	G	S2	U	G	1, 2	0.05113, 0.02687	0.00036, 0.00019	0, 0	5, 1	5, 1	5, 1	13642	
U	G	S3	U	G	1.6, 1.1	0.06201, 0.03108	0.00043, 0.00022	0, 0	5, 1	5, 1	5, 1	13846	
U	G	S1	U	B	0.2, 2.5	0.08897, 0.05167	0.00063, 0.00036	0, 0	6, 1	6, 1	6, 1	14213	

Table 7 (continued)

Payoff at time = 1		Initial	Payoff at time = 15001		Average population payoff at time (in 1000s)							Earliest time	Remarks
Provider	Consumer	State	Provider	Consumer	1	5	10	15	20	25	30	To reach ESS	
U	G	S2	U	B	1, 2	0.05113, 0.02687	0.00036, 0.00019	0, 0	6, 1	6, 1	6, 1	13642	
U	G	S3	U	B	1.6, 1.1	0.06201, 0.03108	0.00043, 0.00022	0, 0	6, 1	6, 1	6, 1	13846	
U	G	*	G	U								*	Infeasible
U	G	S1	G	G	0.2, 0.9	0.08897, 0.04688	0.00063, 0.00033	0, 0	5, 1	5, 1	5, 1	14213	
U	G	S2	G	G	1, 1	0.05113, 0.026	0.00036, 0.00019	0, 0	5, 1	5, 1	5, 1	13642	
U	G	S3	G	G	1.6, 0.9	0.06201, 0.03103	0.00043, 0.00022	0, 0	5, 1	5, 1	5, 1	13846	
U	G	S1	G	B	0.2, 0.9	0.08897, 0.04688	0.00063, 0.00033	0, 0	6, 1	6, 1	6, 1	14213	
U	G	S2	G	B	1, 1	0.05113, 0.026	0.00036, 0.00019	0, 0	6, 1	6, 1	6, 1	13642	
U	G	S3	G	B	1.6, 0.9	0.06201, 0.03103	0.00043, 0.00022	0, 0	6, 1	6, 1	6, 1	13846	
U	G	*	B	U								*	Infeasible
U	G	S1	B	G	1.7, 1.1	0.04927, 0.13586	0.00034, 0.00096	0, 0.00001	4.90506, 0.95253	3.0018, 0.0009	3, 0	14213	
U	G	S2	B	G	1.5, 2	0.02644, 0.07713	0.00019, 0.00055	0, 0	4.94692, 0.97346	3.00329, 0.00164	3.00001, 0.00001	13642	
U	G	S3	B	G	1, 2.5	0.03105, 0.09303	0.00022, 0.00065	0, 0	4.93404, 0.96702	3.00266, 0.00133	3, 0	13846	
U	G	S1	B	B	1.7, 1.1	0.04927, 0.13586	0.00034, 0.00096	0, 0.00001	5.8576, 0.95253	3.00271, 0.0009	3, 0	14213	
U	G	S2	B	B	1.5, 2	0.02644, 0.07713	0.00019, 0.00055	0, 0	5.92038, 0.97346	3.00493, 0.00164	3.00002, 0.00001	13642	
U	G	S3	B	B	1, 2.5	0.03105, 0.09303	0.00022, 0.00065	0, 0	5.90106, 0.96702	3.00399, 0.00133	3, 0	13846	
U	B	*	U	U								*	Infeasible
U	B	*	U	G								*	Infeasible

Table 7 (continued)

Payoff at time = 1		Initial	Payoff at time = 15001		Average population payoff at time (in 1000s)							Earliest time	Remarks
Provider	Consumer	State	Provider	Consumer	1	5	10	15	20	25	30	To reach ESS	
U	B	S1	U	B	0.2, 2.5	0.02737, 0.01398	0.00019, 0.0001	0, 0	6, 1	6, 1	6, 1	13023	
U	B	S2	U	B	1, 2	0.03974, 0.01992	0.00028, 0.00014	0, 0	6, 1	6, 1	6, 1	13391	
U	B	S3	U	B	1.6, 1.1	0.06098, 0.0305	0.00043, 0.00021	0, 0	6, 1	6, 1	6, 1	13841	
U	B	*	G	U								*	Infeasible
U	B	*	G	G								*	Infeasible
U	B	S1	G	B	0.2, 0.9	0.02737, 0.01378	0.00019, 0.0001	0, 0	6, 1	6, 1	6, 1	13023	
U	B	S2	G	B	1, 1	0.03974, 0.01988	0.00028, 0.00014	0, 0	6, 1	6, 1	6, 1	13391	
U	B	S3	G	B	1.6, 0.9	0.06098, 0.03049	0.00043, 0.00021	0, 0	6, 1	6, 1	6, 1	13841	
U	B	*	B	U								*	Infeasible
U	B	*	B	G								*	Infeasible
U	B	S1	B	B	1.7, 1.1	0.01388, 0.04116	0.0001, 0.00029	0, 0	5.9571, 0.9857	3.00926, 0.00309	3.00001, 0	13023	
U	B	S2	B	B	1.5, 2	0.0199, 0.05962	0.00014, 0.00043	0, 0	5.93817, 0.97939	3.0064, 0.00213	3.00001, 0	13391	
U	B	S3	B	B	1, 2.5	0.0305, 0.09148	0.00021, 0.00064	0, 0	5.90457, 0.96819	3.00414, 0.00138	3, 0	13841	
G	U	*	U	B								*	Infeasible
G	U	S1	G	U	0.2, 0.9	1.02397, 0.77934	0.11149, 0.51332	0.90562, 1.25785	3.37975, 1.74382	1.29088, 1.69672	3.09848, 1.18411	Infinity	No ESS
G	U	S2	G	U	1, 1	0.17418, 0.46816	0.77815, 1.16094	0.29909, 0.36652	1.35026, 1.58538	3.13772, 1.39774	1.76567, 1.78766	Infinity	No ESS
G	U	S3	G	U	1.6, 0.9	0.06158, 0.05175	0.00597, 0.73538	0.37953, 1.18398	3.01692, 0.45233	3.89942, 1.94822	1.69032, 1.99336	Infinity	No ESS
G	U	S1	G	G	0.2, 0.9	1.02397, 0.77934	0.11149, 0.51332	0.90562, 1.25785	4.93602, 0.97074	4.99955, 0.99979	5, 1	28782	

Table 7 (continued)

Payoff at time = 1		Initial	Payoff at time = 15001		Average population payoff at time (in 1000s)							Earliest time	Remarks
Provider	Consumer	State	Provider	Consumer	1	5	10	15	20	25	30	To reach ESS	
G	U	S2	G	G	1, 1	0.17418, 0.46816	0.77815, 1.16094	0.29909, 0.36652	4.99269, 0.99822	4.99995, 0.99999	5, 1	25946	
G	U	S3	G	G	1.6, 0.9	0.06158, 0.05175	0.00597, 0.73538	0.37953, 1.18398	4.44625, 0.7389	4.99506, 0.99759	4.99996, 0.99998	31982	
G	U	S1	G	B	0.2, 0.9	1.02397, 0.77934	0.11149, 0.51332	0.90562, 1.25785	5.93813, 0.97943	5.99957, 0.99986	6, 1	28411	
G	U	S2	G	B	1, 1	0.17418, 0.46816	0.77815, 1.16094	0.29909, 0.36652	5.9955, 0.99851	5.99996, 0.99999	6, 1	25764	
G	U	S3	G	B	1.6, 0.9	0.06158, 0.05175	0.00597, 0.73538	0.37953, 1.18398	5.67783, 0.89395	5.99758, 0.99919	5.99998, 0.99999	30798	
G	U	S1	B	U	1.7, 1.1	1.0467, 1.80331	0.97089, 0.62481	2.06289, 2.16346	3.00839, 0.03391	3.00013, −0.00001	2.99994, 0	25268	
G	U	S2	B	U	1.5, 2	0.84924, 0.64234	1.9328, 1.93909	0.5835, 0.66561	3.00081, 0.00171	2.99992, 0.00001	2.99996, 0	22806	
G	U	S3	B	U	1, 2.5	0.07271, 0.11333	1.46777, 0.74135	2.1782, 1.56351	0.99865, 1.99675	0.99999, 1.99998	1, 2	25872	
G	U	S1	B	G	1.7, 1.1	1.0467, 1.80331	0.97089, 0.62481	2.06289, 2.16346	3.00127, 0.00065	3.00005, 0	2.99983, 0	22079	
G	U	S2	B	G	1.5, 2	0.84924, 0.64234	1.9328, 1.93909	0.5835, 0.66561	3.00082, 0.00041	3.00005, 0	2.99999, 0	22152	
G	U	S3	B	G	1, 2.5	0.07271, 0.11333	1.46777, 0.74135	2.1782, 1.56351	3.70773, 0.981	3.01858, 0.00935	3.00001, 0	25329	
G	U	S1	B	B	1.7, 1.1	1.0467, 1.80331	0.97089, 0.62481	2.06289, 2.16346	3.00119, 0.0004	2.99975, 0	2.99994, 0	22321	
G	U	S2	B	B	1.5, 2	0.84924, 0.64234	1.9328, 1.93909	0.5835, 0.66561	3.00109, 0.00037	2.99999, 0	2.99998, 0	22245	
G	U	S3	B	B	1, 2.5	0.07271, 0.11333	1.46777, 0.74135	2.1782, 1.56351	3.70348, 0.2375	3.00006, 0.00002	3, 0	25329	
G	G	*	U	B								*	Infeasible
G	G	*	G	U								*	Infeasible
G	G	S1	G	G	0.2, 0.9	0.0199, 0.02203	0.00014, 0.00015	0, 0	5, 1	5, 1	5, 1	12857	

Table 7 (continued)

Payoff at time = 1		Initial	Payoff at time = 15001		Average population payoff at time (in 1000s)							Earliest time	Remarks
Provider	Consumer	State	Provider	Consumer	1	5	10	15	20	25	30	To reach ESS	
G	G	S2	G	G	1, 1	0.02804, 0.01494	0.0002, 0.00011	0, 0	5, 1	5, 1	5, 1	13053	
G	G	S3	G	G	1.6, 0.9	0.0571, 0.02858	0.0004, 0.0002	0, 0	5, 1	5, 1	5, 1	13769	
G	G	S1	G	B	0.2, 0.9	0.0199, 0.02203	0.00014, 0.00015	0, 0	6, 1	6, 1	6, 1	12857	
G	G	S2	G	B	1, 1	0.02804, 0.01494	0.0002, 0.00011	0, 0	6, 1	6, 1	6, 1	13053	
G	G	S3	G	B	1.6, 0.9	0.0571, 0.02858	0.0004, 0.0002	0, 0	6, 1	6, 1	6, 1	13769	
G	G	*	B	U								*	Infeasible
G	G	S1	B	G	1.7, 1.1	0.03411, 0.04192	0.00023, 0.00029	0, 0	4.97892, 0.98946	3.00841, 0.0042	3, 0	12857	
G	G	S2	B	G	1.5, 2	0.01586, 0.04299	0.00011, 0.00031	0, 0	4.97071, 0.98535	3.00603, 0.00301	3, 0	13053	
G	G	S3	B	G	1, 2.5	0.02861, 0.08568	0.0002, 0.00059	0, 0	4.93954, 0.96977	3.00291, 0.00146	3, 0	13769	
G	G	S1	B	B	1.7, 1.1	0.03411, 0.04192	0.00023, 0.00029	0, 0	5.96838, 0.98946	3.01261, 0.0042	3.00001, 0	12857	
G	G	S2	B	B	1.5, 2	0.01586, 0.04299	0.00011, 0.00031	0, 0	5.95606, 0.98535	3.00904, 0.00301	3.00001, 0	13053	
G	G	S3	B	B	1, 2.5	0.02861, 0.08568	0.0002, 0.00059	0, 0	5.90931, 0.96977	3.00437, 0.00146	3, 0	13769	
G	B	*	U	B								*	Infeasible
G	B	*	G	U								*	Infeasible
G	B	*	G	G								*	Infeasible
G	B	S1	G	B	0.2, 0.9	0.00687, 0.00357	0.00005, 0.00002	0, 0	6, 1	6, 1	5.99996, 1	11592	
G	B	S2	G	B	1, 1	0.02345, 0.01174	0.00016, 0.00008	0, 0	6, 1	6, 1	5.99996, 1	12840	
G	B	S3	G	B	1.6, 0.9	0.05661, 0.02831	0.0004, 0.0002	0, 0	6, 1	6, 1	6, 1	13752	

Table 7 (continued)

Payoff at time = 1		Initial	Payoff at time = 15001		Average population payoff at time (in 1000s)							Earliest time	Remarks
Provider	Consumer	State	Provider	Consumer	1	5	10	15	20	25	30	To reach ESS	
G	B	*	B	U								*	Infeasible
G	B	*	B	G								*	Infeasible
G	B	S1	B	B	1.7, 1.1	0.00371, 0.01045	0.00002, 0.00007	0, 0	5.98916, 0.99639	3.03679, 0.01226	3.00001, 0	11592	
G	B	S2	B	B	1.5, 2	0.01176, 0.0352	0.00008, 0.00024	0, 0	5.96327, 0.98776	3.01083, 0.00361	3.00001, 0	12840	
G	B	S3	B	B	1, 2.5	0.02831, 0.08492	0.0002, 0.00059	0, 0	5.91121, 0.9704	3.00447, 0.00149	3, 0	13752	
B	U	*	U	B								*	Infeasible
B	U	*	G	B								*	Infeasible
B	U	S1	B	U	1.7, 1.1	1.99554, 1.00219	1.99997, 1.00002	2, 1	1, 2	1, 2	1, 2	11544	
B	U	S2	B	U	1.5, 2	0.99983, 2.99949	1.00003, 3.00001	1.00003, 3.00001	2.99999, 0	2.99997, 0	2.99999, 0	7595	
B	U	S3	B	U	1, 2.5	0.99998, 2.99995	0.99997, 2.99998	1.00001, 3.00001	2.99999, 0	3.00004, 0	2.99969, 0	5707	
B	U	S1	B	G	1.7, 1.1	1.99554, 1.00219	1.99997, 1.00002	2, 1	2.00007, 1	2.01091, 1	3.05428, 1	11544	
B	U	S2	B	G	1.5, 2	0.99983, 2.99949	1.00003, 3.00001	1.00003, 3.00001	2.99999, 0	2.99997, 0	2.99999, 0	7595	
B	U	S3	B	G	1, 2.5	0.99998, 2.99995	0.99997, 2.99998	1.00001, 3.00001	2.99999, 0	3.00004, 0	2.99969, 0	5707	
B	U	S1	B	B	1.7, 1.1	1.99554, 1.00219	1.99997, 1.00002	2, 1	3.01007, 1	5.96043, 1	5.9998, 0.99993	11544	
B	U	S2	B	B	1.5, 2	0.99983, 2.99949	1.00003, 3.00001	1.00003, 3.00001	2.99999, 0	2.99997, 0	2.99999, 0	7595	

Table 7 (continued)

Payoff at time = 1		Initial	Payoff at time = 15001		Average population payoff at time (in 1000s)							Earliest time	Remarks
Provider	Consumer	State	Provider	Consumer	1	5	10	15	20	25	30	To reach ESS	
B	U	S3	B	B	1, 2.5	0.99998, 2.99995	0.99997, 2.99998	1.00001, 3.00001	2.99999, 0	3.00004, 0	2.99969, 0	5707	
B	G	*	U	B								*	Infeasible
B	G	*	G	B								*	Infeasible
B	G	*	B	U								*	Infeasible
B	G	S1	B	G	1.7, 1.1	0.98305, 2.94758	1, 2.99999	1.00002, 3.00001	2.99999, 0	2.99997, 0	2.99999, 0	9095	
B	G	S2	B	G	1.5, 2	0.99989, 2.99965	1.00004, 3.00002	1.00002, 3.00001	2.99999, 0	3.00004, 0	2.99968, 0	7305	
B	G	S3	B	G	1, 2.5	0.99998, 2.99997	0.99997, 2.99998	1.00002, 3.00001	2.99999, 0	3.00004, 0	2.99996, 0	5118	
B	G	S1	B	B	1.7, 1.1	0.98305, 2.94758	1, 2.99999	1.00002, 3.00001	2.99999, 0	2.99997, 0	2.99999, 0	9095	
B	G	S2	B	B	1.5, 2	0.99989, 2.99965	1.00004, 3.00002	1.00002, 3.00001	2.99999, 0	3.00004, 0	2.99968, 0	7305	
B	G	S3	B	B	1, 2.5	0.99998, 2.99997	0.99997, 2.99998	1.00002, 3.00001	2.99999, 0	3.00004, 0	2.99996, 0	5118	
B	B	*	U	B								*	Infeasible
B	B	*	G	B								*	Infeasible
B	B	*	B	U								*	Infeasible
B	B	*	B	G								*	Infeasible
B	B	S1	B	B	1.7, 1.1	0.9961, 2.98825	1, 3	1.00001, 3.00001	2.99998, 0	3.00004, 0	2.99971, 0	8003	
B	B	S2	B	B	1.5, 2	0.9999, 2.99968	1.00004, 3.00002	1.00007, 3.00003	3, 0	2.99999, 0	2.99999, 0	6361	
B	B	S3	B	B	1, 2.5	0.99999, 2.99997	0.99999, 2.99999	1.00002, 3.00001	2.99999, 0	2.99997, 0	2.99999, 0	5109	

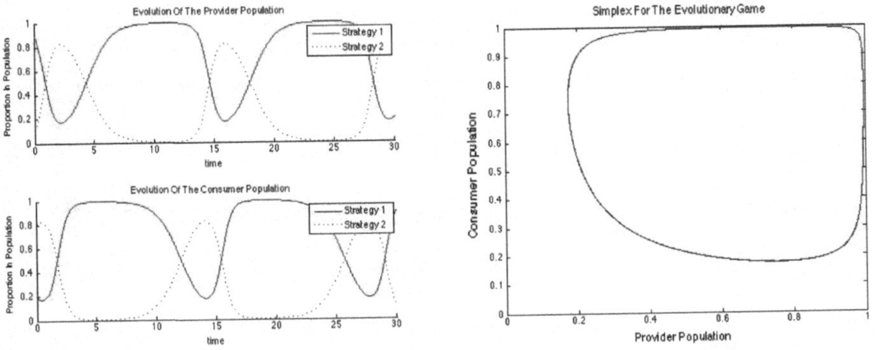

Fig. 7 Replicator dynamics: has no ESS (NHS)

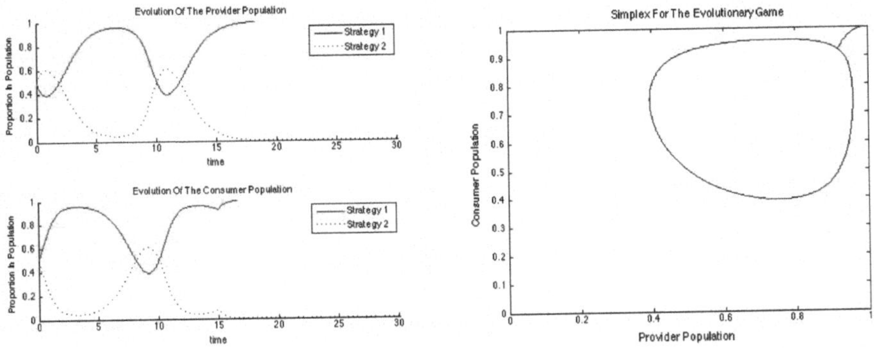

Fig. 8 Replicator dynamics: has ESS (NHS)

5.3.3 Influence in Policy-Making

The payoff matrix for the consumers and the providers and the starting state provide an indication of how far the supply chain is away from sustainability at any point of time. The three charts in Fig. 9 indicate the time taken to reach ESS for various combinations of provider and consumer payoff for starting states S_1, S_2 and S_3, respectively. The broken bar indicates that there is no ESS for the specific combination. Some of the combinations listed may lead to the entire population of consumers not using NHS. Thus the system does have an ESS, but the scheme is not sustainable as the consumers are no longer using the NHS network. In other cases, the entire population of the consumers use the NHS network. In this case, the system does have an ESS and the scheme is sustainable. Figure 9 can be used by the government and regulatory bodies to bring in policy changes and governance rules so as to reach stability earlier. For instance, let the system start in state S_1 where no favourable policies are in place for the providers and consumers to use NHS (i.e. providers and consumers have undesirable (U) payoffs). To make the system sustainable, the

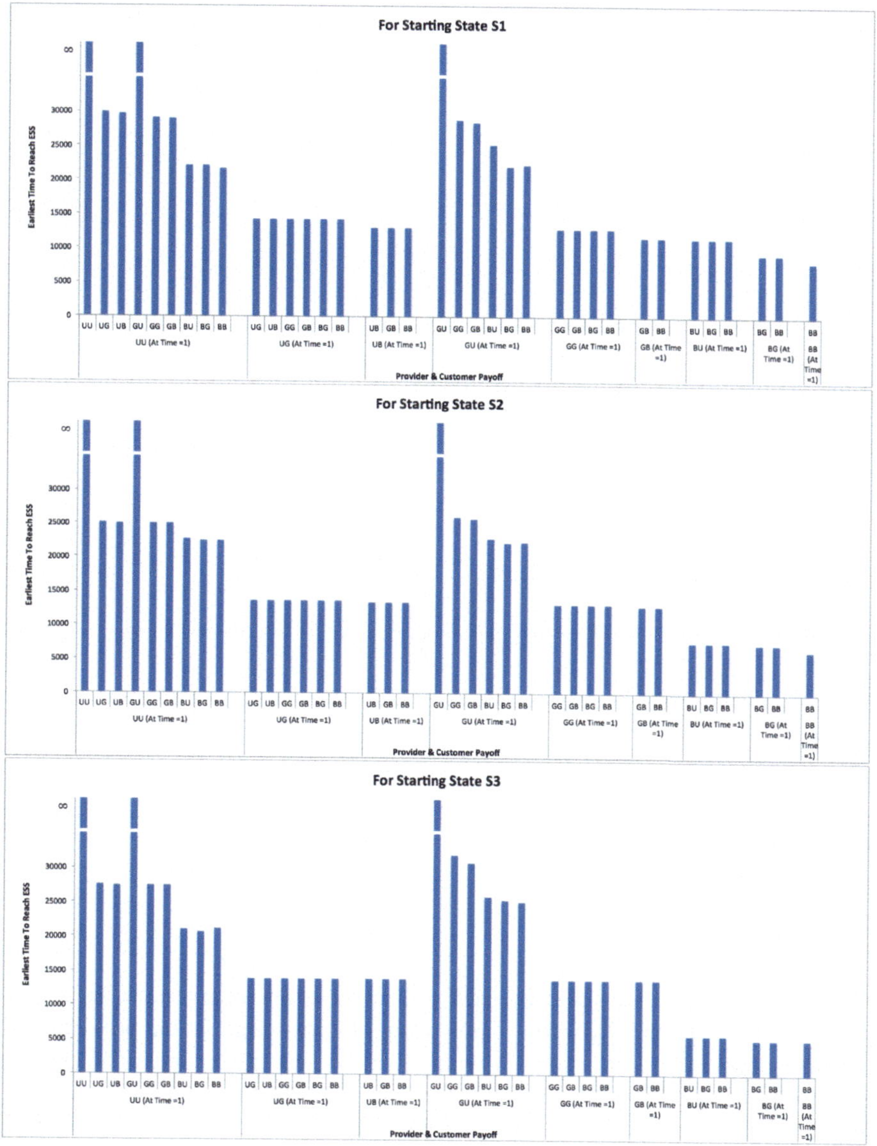

Fig. 9 Time to reach ESS for different starting states (NHS)

policy-maker must create a positive policy aimed at benefiting consumers. That is, they must design a policy that will change the payoff matrix for the consumer to G or B. In addition, the policy-maker may also create positive policies benefiting providers to make the system reach ESS (and hence for NHS to be sustainable) faster. The figures for each of the three states represent how fast the system can reach ESS.

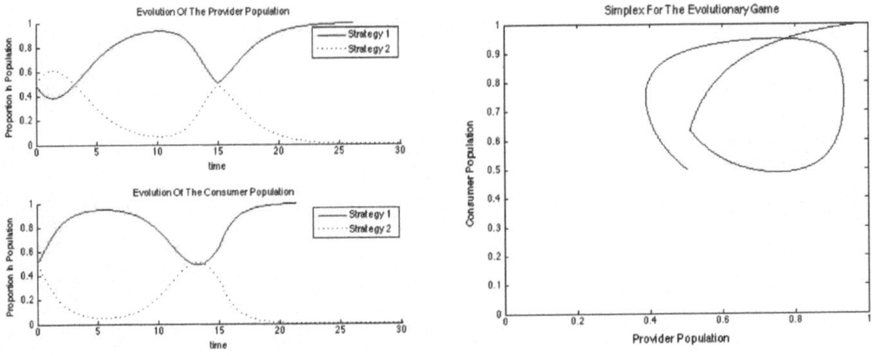

Fig. 10 Maynard dynamics (Sandholm 2011) for NHS

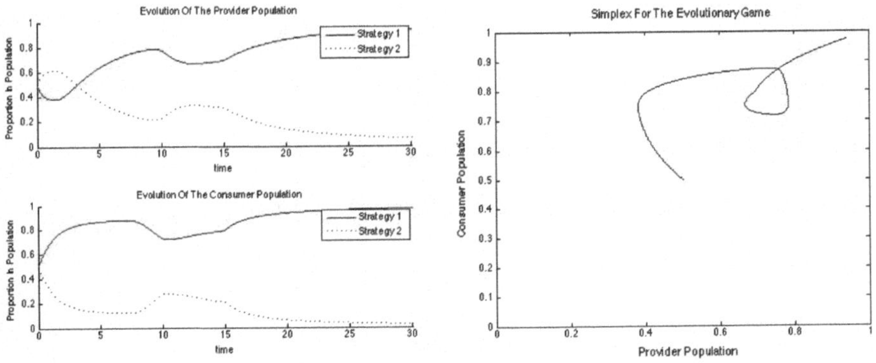

Fig. 11 Brown–von Neumann–Nash (BNN) dynamics (Sandholm 2011) for NHS

5.3.4 Other Game Dynamics

Other game dynamics can also be used to analyse evolutionary stable strategies and hence the sustainability of the supply chain. For instance, let the provider and consumer population start in state S_2 with payoff U, and move to payoff G and B, respectively, at time 15001. As shown earlier, Fig. 8 depicts the population evolution and the simplex for the game using replicator dynamics. Figures 10, 11, 12 and 13 depict the system evolution when other game dynamics are applied.

6 Modelling the Stochastic Aspect of Evolutionary Game

We had introduced a "kind-of" stochastic component in the earlier example in the following manner. Initially, the payoff and the proportion of the players in each of the population choosing either of the action was fixed. The game was played for a

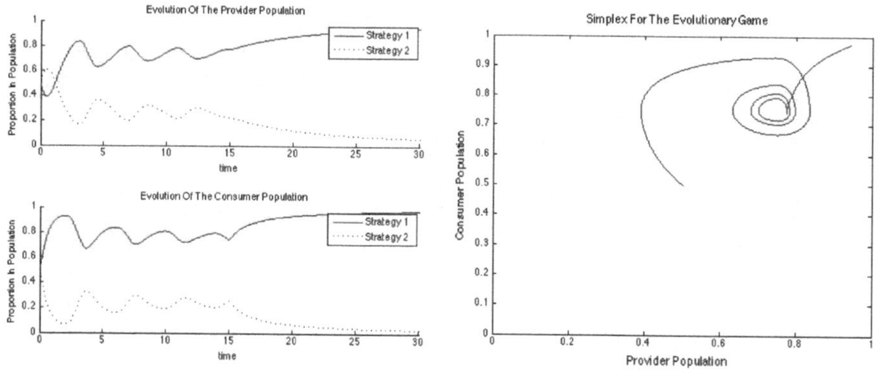

Fig. 12 Smith dynamics (Sandholm 2011) for NHS

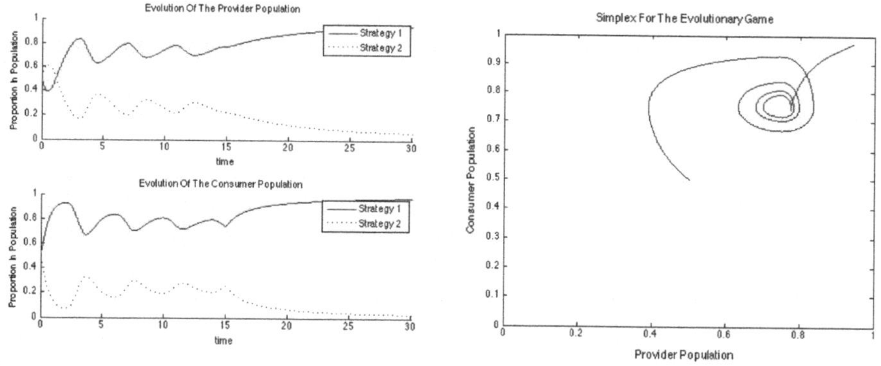

Fig. 13 Logit dynamics (Sandholm 2011) for NHS

predefined t units of time. At the end of time t, the game continued but with a new payoff. This is akin to a stochastic evolutionary game where the game transitioned to an absorbing state (the new state) with a probability of 1 at time t (though technically you cannot specify the time at which the transition will happen).

In real-life scenarios, it is natural that the actions chosen by the players would cause the game to move to a new state. For example, consider a manufacturing supply chain with multiple suppliers. If one or more supplier chooses to invest in a new technology while the other suppliers continue to use the old technology, the game may move to a new state with a new payoff matrix and maybe even a new set of actions. Thus, it is a natural extension to model sustainability of the supply chain as ESG at least for a single population scenario. There are two ways of demonstrating this.

(1) Consider a public healthcare system where patients can avail of services in either private hospitals or government hospitals under the scheme. The scheme may allow for either some or all treatment procedures to be available in both pri-

vate and government hospitals. It may happen that some treatment procedures are available only in private or only in government hospitals. Let us consider the scenario where the patient has a choice of availing a set of treatments in either private or government hospitals. So the government and private hospitals are vying for the same market share of patients. As doctors may need to refer patients to other doctors or facilities in other hospitals, each hospital has two choices of actions-refer patient to private hospital and refer patient to government hospital. As the government is trying to increase usage of government hospitals, they may choose to reward government hospitals when their usage number increases; or the government may choose to bring more treatment options under government hospitals. This can be modelled as a new state with a new payoff function. Similarly, there may be a state where the rewards to the government hospitals fall due to lesser usage. So there are three states: current state s_0 with current payoff; s_1 where the reward to the government hospital is more; and s_2 where the reward to the government hospital is less. Transition probabilities can be defined accordingly. The payoffs and transition probabilities will be symmetric in this case. The initial population for private hospitals could indicate what % of the private hospital doctors refer patients to private hospitals. The initial population for government hospitals can be defined similarly.

(2) Consider the single population to be the set of all patients. Let us go back to our example of NHS. Currently, the patients can either use the scheme or use private insurance in case their employer offers it. It is now possible for the game to have two states-current state where patients prefer to use NHS and new state where the benefit of using private insurance outweighs the long waiting time under the NHS scheme.

7 Summary

This chapter contributes to a much needed quantitative framework to assess sustainability in supply chains through game-theoretic methods, specifically evolutionary game theory. The work first recognizes the gap in existing sustainability literature which does not provide with a framework to analyse sustainability across all dimensions: economic, environmental and social. Equating the concept of stability in an evolutionary game to the concept of sustainability, the framework provides to identify the conditions under which sustainability is reached. Further, the impact of all the stakeholders on the trajectory of the supply chain in order to lead it towards sustainability is demonstrated. As an example, we demonstrate sustainability of a health insurance supply chain.

The work can be extended in many directions and domains. The framework can be applied to various supply chains to analyse their sustainability. It will also be interesting to see how the ESS will be reached. It differs from our earlier representation because there is no predefined time t when the game moves to a new state. Instead, the transition probability for the set of actions chosen by the players determines when the game moves to the new state. In the theoretical development domain, the work can be extended to develop efficient algorithms to achieve sustainability. Finally, it would be interesting to develop incentives for supply chains to attain sustainability, thus contributing to the mechanism design literature.

Acknowledgements The data in the tables and figures in Sects. 4 and 5 are reprinted from Babu and Mohan (2017) with the permission from Elsevier

References

Appleby, J., & Phillips, M. (2009). The NHS: Satisfied now? British Social Attitudes: The 25th Report, SAGE Publications (pp. 25–54).

Aydin, G., & Porteus, E. L. (2009). Manufacturer to retailer versus manufacturer to consumer rebates in a supply chain. In N. Agrawal & S. A. Smith (Eds.), *Retail supply chain management* (Vol. 122, pp. 237–270)., International Series in Operations Research and Management Science US: Springer.

Babu, S., & Mohan, U. (2017). An integrated approach to evaluating sustainability in supply chains using evolutionary game theroy. *Computer & Operations Research, 89*, 269–283.

Barari, S., Agarwal, G., Zhang, W. J. C., Mahanty, B., & Tiwari, M. K. (2012). A decision framework for the analysis of green supply chain contracts: An evolutionary game approach. *Expert systems with applications, 39*(3), 2965–2976.

Brandenburg, M., Govindan, K., Sarkis, J., & Seuring, S. (2014). Quantitative models for sustainable supply chain management: Developments and directions. *European Journal of Operational Research, 233*, 299–312.

Burns, L. R. (2002). *The health care value chain: Producers, purchasers and providers—Chapter 1*. A Wiley Company, San Francisco: Jossey-Bass.

Cachon, G. P., & Netessine, S. (2006). Game theory in supply chain analysis. *In Models, Methods, and Applications for Innovative Decision Making, INFORMS*, 200–233.

Cachon, G. P., & Zipkin, P. H. (1999). Competitive and cooperative inventory policies in a two-stage supply chain. *Management science, 45*(7), 936–953.

Carter, C. R., & Rogers, D. S. (2008). A framework of sustainable supply chain management: Moving toward new theory. *International Journal of Physical Distribution & Logistics Managemen, 38*(5), 360–387.

Chakravarty, A. K. (2014). *Supply chain transformation—Chapter 10*, Springer Texts in Business.

Chen, F. (2000). Sales-force incentives and inventory management. *Manufacturing & Service Operations Management, 2*, 186–202. April.

Chen, F. (2005). Salesforce incentives, market information, and production/inventory planning. *Management Science, 51*, 60–75. January.

Cho, S., McCardle, K. F., & Tang, C. S. (2009). Optimal pricing and rebate strategies in a two-level supply chain. *Production and Operations Management, 18*(4), 426–446.

Department of Health. (2015). NHS hospital waiting times statistics, 2013.

Flesch, J., Parthasarathy, T., Thuijsman, F., & Uyttendaele, P. (2013). Evolutionary stochastic games. *Dynamic Games and Applications, 3*, 207–219.

Ghatak, A., Rao, K. S. M., & Shaiju, A. J. (2012). Evolutionary stability against multiple mutations. *Dynamic Games and Applications, 4*, 376–384.

Hofbauer, J., & Sigmund, K. (2003). Evolutionary game dynamics. *Bulletin of the American Mathematical Society, 40–4*, 479–519.

Leng, M., & Parlar, M. (2005). Game theoretic applications in supply chain management: A review. *INFOR: Information Systems and Operational Research*, *43*(3), 187–220.
Li, J., Du, W., Yang, F., & Hua, G. (2013). The research on evolutionary game of remanufacturing closed-loop supply chain under asymmetric situation. In *LTLGB 2012* (pp. 473–479), Springer.
Nagarajan, M., & Sošić, G. (2008). Game-theoretic analysis of cooperation among supply chain agents: Review and extensions. *European Journal of Operational Research*, *187*(3), 719–745.
Naini, S. G. J., Aliahmadi, A. R., & Jafari-Eskandari, M. (2011). Designing a mixed performance measurement system for environmental supply chain management using evolutionary game theory and balanced scorecard: A case study of an auto industry supply chain. *Resources, conservation and recycling*, *55*(6), 593–603.
NHS. (2015). NHS patient safety, 2011.
Office For National Statistics. (2015). NHS hospital productivity efficiency, 2010. Retrieved April, 2015.
Sandholm, W. (2011). *Population games and evolutionary dynamics (economic learning and social evolution)* (1st edn.), The MIT Press.
Schlag, K. (1994). *Evolution in partnership games, an equivalence result*. Discussion Paper Serie B (p. 298). Germany: University of Bonn.
Seuring, S. (2013). A review of modelling approaches for sustainable supply chain management. *Decision Support Systems*, *54–4*, 1513–1520.
Shapley, L. (1953). Stochastic games. *Proceedings of the National Academy of Sciences*, *39*, 1095–1100.
Smith, J. M., & Price, G. R. (1973). The logic of animal conflict. *Nature*, *246*, 15–18.
Taylor, P. D., & Jonker, L. (1978). Evolutionarily stable strategies and game dynamics. *Mathematical Biosciences*, *40*, 145–156.
Zhu, Q.-H., & Dou, Y.-J. (2007). Evolutionary game model between governments and core enterprises in greening supply chains. *Systems engineering-theory & practice*, *27*(12), 85–89.

Part II
Applications of SSC in Specific Sectors

Chapter 7
Applications of Green Supply Chain Management in the U.K. Restaurant Industry

Vinaya Shukla, Arvind Upadhyay, and Bhushan Khandve

Abstract Green Supply Chain Management (GSCM), which involves incorporating ecological considerations into supply chain management, has gained prominence in recent times.

1 Introduction

Green Supply Chain Management (GSCM), which involves incorporating ecological considerations into supply chain management, has gained prominence in recent times. Large-scale resource depletion, environmental degradation and climate change have triggered this, with companies across a range of sectors including manufacturing, construction and services now actively engaged in embedding environmental aspects into their design, procurement, manufacturing, packaging and logistic activities (Zhu et al. 2012). While earlier, the focus was primarily on the economic dimension, environmental behaviour/credentials have also become necessary for success now (Ahi and Searcy 2013). Appropriate green products and processes are consequently being developed, with organisations seeking to exploit the marketing potential of 'green' by targeting environmentally conscious consumers (Schubert et al. 2010).

A key sector from a GSCM perspective is restaurants. Restaurants consume an enormous amount of energy (from the heating, ventilation, air conditioning, refrigeration and lighting that is involved) and water (for washing/cleaning), and also generate significant amounts of waste of plastic, aluminium, paper and glass packaging/tableware, leftover food/ingredients and used cooking oil. This, coupled with

V. Shukla · B. Khandve
Business School, Middlesex University London, London NW4 4BT, UK

A. Upadhyay (✉)
Brighton Business School, University of Brighton, Mithras House, Lewes Road, Brighton BN2 4T, UK
e-mail: a.upadhyay@brighton.ac.uk

U. Ramanathan and R. Ramanathan (eds.), *Sustainable Supply Chains: Strategies, Issues, and Models*, https://doi.org/10.1007/978-3-030-48876-5_7

the fact that a large number of restaurants are in operation, makes direct environmental impact from this sector, significant. Restaurants also have an indirect environmental impact which is from upstream via suppliers/farmlands, manufacturers, pesticide residues and animal wastes. Applying GSCM to restaurants therefore provides several benefits including conservation of resources, effective waste management (of large quantities of waste) and sustainable/environmentally friendly food (Namkung and Jang 2013; Sarkis and Dou 2017); the benefits are not just in financial terms such as through cost savings and premium pricing (on account of enhanced (environmental) image and development of new environmentally friendly products and services), but also environmental (Achillas 2018).

Appreciating the role and importance of GSCM, several restaurants across both the developed and developing world have implemented green practices. Some related research has also been done: some researchers have looked at green practices (e.g. Wang et al. 2013 in the Taiwanese restaurant context), while others have looked at both the practices and the drivers and barriers to those practices (e.g. Kasim 2009 in the Malaysian restaurant context). However, researchers still lament the lack of comprehensive knowledge of this subject (Kasim and Ismail 2012; Jackson 2010). Another lacuna of previous research is that it is mostly set in Asia with only a few studies based out of the Western World.

The UK has a large and thriving restaurant sector where small and medium (SM) restaurants dominate, and whose environmental impact therefore is significant: SM restaurants are responsible for 60% of the carbon dioxide emissions and 40% of the commercial waste in the UK (Revell and Blackburn 2007). However, the nature and the extent to which they implement green practices as also the associated motivations and hindrances are unclear from previous research: only one study (by Revell and Blackburn 2007) has been done, which found the primary focus of SM restaurants in the UK to be on quality of food and service rather than on environmental aspects. This gap in knowledge (about green practices implementation as well as drivers and barriers affecting it for small- and medium-scale restaurants in the UK) therefore needs to be filled up, which constitutes the focus of our work. Findings from it will be useful to practitioners as they will learn about the variety of different green practices being implemented in restaurants. Similarly, from the policymakers' perspective, learnings about drivers and barriers would help them in developing policies/strategies that amplified the role of key drivers and dampened those of key barriers to enable greater green practice implementation.

The rest of the paper is structured as follows. In Sect. 2, previous literature on GSCM in the restaurant sector including that on green practices, drivers and barriers is discussed, while the research methodology used for the investigation is covered in Sect. 3. The findings are discussed and analysed in Sect. 4. Finally, we conclude in Sect. 5, where the research contribution, limitations and suggestions for further work are covered.

2 Literature Review

2.1 Literature on GSCM in the Restaurant Industry

Green Supply Chain Management (GSCM) has emerged as a systematic and integrated approach to tackling environmental concerns across supply chains (Malviya and Kant 2015). It refers to the application of environmental management principles to the entire set of activities across the whole customer order cycle, including design, procurement, manufacturing and assembly, packaging, logistics and distribution (Handfield et al. 1997). Seuring (2004) defines GSCM as the managerial integration of material and information flow throughout the supply chain to satisfy the demand of customers for green products and services produced by green processes. The main goal of GSCM is to ensure efficient, effective and extensive implementation of green practices (or activities/initiatives to reduce environmental footprint) across the different supply chain stages (Perotti et al. 2012) by managing the 'antecedents', i.e. drivers and barriers affecting that implementation (Luthra et al. 2015). Green practices are typically referred to in terms of the supply chain stages where they belong as green design, green procurement, green manufacturing, green packaging and green logistics. These have been adapted for restaurants and framed in terms of (their) upstream and downstream activities as green design, green purchasing, green menu design and green cooking (Wang et al. 2013).

2.1.1 Green Design

Green design in the restaurant's case means their being designed (in a physical sense) to be environmentally sustainable, i.e. where less material, energy and water are consumed (Lewis et al. 2005). This requires eco-friendly materials to be used and environmental aspects to be considered during the (restaurant's) design and construction phases (Alcorn 2009). As per Bartlett and Howard (2000), green-designed restaurants have natural lighting and ventilation, occupant sensitive controls and stable temperatures, while Jernigan (2012) suggests such restaurants to reuse existing building materials so that the requirement of virgin materials is lower, and also to use low volatile organic compound (VOC) sealants, adhesives, paints and carpets. From a performance perspective, Alcorn (2009) note that green-designed restaurants lower resource depletion and show better environmental and economic performance.

2.1.2 Green Purchasing

This refers to restaurants following a sustainable protocol in their choice of supplies and suppliers. For instance, as per the guide on sustainable purchasing for the UK public sector, five principles need to be considered: using local supplies over imports, promoting organic or sustainable purchases, restricting purchases that damage the environment, utilising centralised purchasing systems and inspecting purchases to ensure that their processing has been eco-friendly (Rimmington et al. 2006). With regard to local sourcing, it not only helps support the local community/economy but also helps lower vehicle emissions. For it to work effectively though, restaurants need to offer seasonal dishes as per the local produce.

While the criticality of purchasing to restaurants meeting their sustainability objectives is well recognised (Zsidisin and Siferd 2001), the fact that many of them focus on price, quality and service in purchases rather than eco-friendliness is also noted (Bergstorm et al. 2005). Researchers (e.g. Chiu and Hsieh 2016) have suggested supplier development programmes on environmental aspects as well as collaboration with suppliers on green-related investments and related risk sharing (e.g. Large and Thomsen 2011) as the way forward.

2.1.3 Green Menu Design

Green menu design is offering food on the menu that is processed considering environmental aspects, for example, offering organic dishes that are cultivated using non-toxic fertilisers and pesticides, and without genetic engineering. Also, menus are designed considering seasonal produce from local supplies (with lower emissions from transportation) (Energy Star 2007). It can also mean taking away meats as an option as they are known to be more environmentally damaging than the vegetarian alternatives. Offering fish/seafood that is sustainably harvested and free from hazardous pollutants can also be an option (Jeong and Jang 2010). Designing a green menu though is not easy as the restaurant has to ensure that the taste, nutrition and look/feel are not compromised in the process.

Given that many restaurants tend to serve specialist ethnic cuisines (e.g. French/Indian/Mexican), an important question is whether some cuisines are more green than others. We came across no study that has done such a comparison. One reason for this could be the challenge of doing so, given that a large number of variables with differential green impacts are involved. For example, even though anecdotally, one may say that Indian cuisine is more green given the greater use of vegetables, the fact that there is more frying and more spices are used may make it less green overall.

2.1.4 Green Cooking

Cooking and related activities consume a significant amount of energy such as on food preparation, ventilation, cleaning/dishwashing and refrigeration; improving the energy efficiency of these activities/equipment, referred to as green cooking by some authors, is therefore not only useful in improving environmental performance, but also helps lower energy costs (Energy Star 2007). Some improvement activities in this regard include cooking with fully loaded oven, or keeping the lids closed on kettles and braising pans during extended use, and also, using demand control ventilation for kitchen makeup air units and exhaust hoods and ensuring that the size of hoods is appropriate (ASHRAE 2003). With regard to used cooking oil, reprocess it into biodiesel, an environmentally friendly fuel for trucking, heating, etc.; in case of in-house reprocessing, use biodiesel to generate (environmentally friendly) electricity for use.

Some authors (e.g. Tran et al. 2016) consider green cooking to be analogous to green manufacturing and take a broader view; according to them, not just energy efficiency, but efficiency in the use of other resources (used in cooking) and related waste management as well as reducing carbon emissions, all constitute a part of green cooking. With regard to efficiency of resources, Lion et al. (2018) consider those of cooking equipment and chefs/employees in the restaurant; hygiene in the restaurant kitchen, and taste and presentation style of cooked food need to be considered when assessing efficiency. On food waste and its management, Tian et al. (2017) consider the waste incurred in cooking as well as that which is leftover because of the serving size being excessive. Finally, monitor/manage the gases used in cooking to reduce their wastage/excessive use, and thereby ensure healthier kitchen environment, lower carbon emissions and lower costs (Govindan et al. 2018).

2.1.5 Green Transport/Logistics

Green transport refers to minimising consumption of fossil fuels and associated carbon emissions during transportation of inputs/products (which in the UK's case is predominantly via trucks). Key green transportation practices for restaurants include monitoring and collaborating with suppliers to ensure (their) use of fuel-efficient vehicles, and alternative (more environmentally friendly) fuels for delivery (Fernie and Sparks 2014), and also sourcing locally to reduce transport distances. On the outbound side for restaurants offering food delivery services to customers, ensure the use of fuel-efficient vehicles and bulk deliveries to minimise number of trips. Some authors (e.g. Chan et al. 2016) consider the use of recyclable as well as edible packaging to reduce waste in distribution under green transportation; transport of food waste (to ensure its effective management) is also considered.

2.1.6 Green Packaging

With more and more restaurants offering takeout/and or delivery options for meals which requires packaging (e.g. food containers, cutlery and utensils), packaging's environmental impact has been going up. This is particularly so because of the large availability (and use) of cheap but environmentally unfriendly disposable packaging options. Green packaging's focus is to ensure that the packaging is such that it uses less material (e.g. through better design, alternative material choice), uses more recycled material, avoids hazardous material and can be biodegraded/composted. Many large fast-food restaurants such as McDonald's (Bright 2018) and KFC (Cottom 2019) are aggressively taking green packaging-related initiatives spurred by consumer and government pressure. However, this picture is unclear for small and medium restaurants where little work has been done. For example, Wang et al (2013) discuss green packaging (along with other green practices); however, the restaurants covered in the investigation are mostly large ones.

Overall, there appears to be some understanding of the different green practices used in restaurants. However, a comprehensive understanding is missing, which has also been noted by some authors (e.g. Kasim and Ismail 2012; Wang et al. 2013). One reason for this is that researchers have mostly focussed on specific green practices: for example, while some have focussed on the downstream segment, i.e. those in relation to customers (e.g. Namkung and Jang 2013), others (e.g. Chiu and Hsieh 2016) have focussed upstream. A similar gap in knowledge exists from a country perspective: most GSCM studies on restaurants have been Asia focussed rather than the Western world. The situation is particularly acute for the UK where only one study by Revell and Blackburn (2007) has been done; this study is also more than a decade old, and which noted restaurants to be indifferent about green practices. As such a comprehensive, present-day understanding of green practices covering all restaurant operation stages is needed for the UK context.

2.2 Drivers and Barriers of GSCM in the Restaurant Industry

The generic GSCM literature suggests organisations being motivated to implement green practices on account of government regulatory pressure and pressure from consumers, competitors, non-governmental organisations and other stakeholders (Balasubramaniam and Shukla 2017); they could also be influenced by internal drivers such as business benefits (from implementation) as well as the environmental commitment of business owners/management. On the other hand, the key barriers or impediments to GSCM noted in the generic literature include shortage of green professionals and suppliers, lack of knowledge/awareness of green aspects and high cost of implementation.

Studies on greening of the restaurant sector have highlighted similar drivers and barriers as in the other sectors. For instance, Kasim and Ismail (2012) identified government regulations, competitive advantage and stakeholder demands as the green drivers for restaurants; high cost and lack of information of green practices were highlighted as the green barriers. Chou et al. (2012) and Wang et al (2013) highlight the role of Green Restaurant Association (in America) and Japan Environment Association in encouraging restaurants to go green. Environmental risk reduction and meeting legal requirements were highlighted as the green drivers by Bonilla-Priego et al. (2011), while the same in the cases of Kasim (2009) and Tsai et al. (2010) were noted to be consumer and stakeholder demand. With regards to barriers or challenges/impediments to GSCM implementation, lack of awareness of green practices, lack of knowledge of customer attitudes to green practices and the fear of increased costs have been suggested (Schubert et al. 2010). Namkung and Jang

(2013) emphasise lack of clarity on customers' perceptions (on green) to be a key barrier.

In summary, there is a reasonable generic understanding of green drivers and barriers for restaurants, although some authors (e.g. Jackson 2010) have highlighted the need for more clarity. The knowledge of UK restaurants from this perspective though is quite limited. Given the contextual nature of drivers and barriers (e.g. government green regulations, stakeholder green-related demands and customers' green perceptions could differ across locations), transplanting them across geographies is difficult. A separate, comprehensive understanding of green drivers and barriers for the UK restaurants case is therefore needed.

3 Methodology

An explorative approach based on qualitative research methodology was considered. Exploratory research is appropriate where there is a lack of published literature/knowledge (Wilson 2014); such an approach also gives better insights into a particular topic and helps to set future research directions. Twenty restaurants in the UK (who agreed to participate out of the 32 contacted) were considered for the investigation. The potential restaurants were first researched to ensure their environmental activeness before being contacted (via telephone in most cases) to set up interview meetings. To avoid cuisine-related biases, care was taken to ensure that a wide variety of different cuisines are considered when making the restaurant choices.

Face-to-face semi-structured interviews were conducted with senior staff at the restaurants (restaurant owners (5 cases), general managers (11 cases) and operations managers (4 cases)) to gather information. Semi-structured interviews were considered because they provide flexibility to explore new concepts and allow comparison of responses while avoiding information overload (Weller and Romney 1988). The interview questions were developed on the basis of the review of the literature on GSCM in restaurants and were of the 'what', 'how' and 'why' types aimed at understanding green practices, drivers and barriers; the interview protocol used is given in Appendix. Interviews were tape-recorded and transcribed, and where this was not possible, detailed notes were taken and then directly transcribed. Secondary data such as annual and environmental reports, official website blogs and newsletters were used to improve understanding.

The interview transcripts were coded using various approaches discussed in the literature and from terminologies used by the interviewees. Quotes which best illustrated a particular situation were chosen to highlight key points. To keep the restaurants anonymous, alphabetical codes (A, B, C.....T) were assigned to individual restaurants. The data drawn from interview transcripts and supporting literature in the case of green practices was further classified into the following sub-categories, i.e. green design, green menu design, green purchasing and packaging, green cooking, green transportation, green drivers and green barriers.

4 Findings and Discussion

Table 1 gives the overall extent of implementation of green practices (in terms of high medium and low), the nature of green practices implemented, and the green drivers and barriers (as perceived) for each of the 20 restaurants, including the type of cuisine they serve.

Examining Table 1, it is clear that a wide variety of cuisines are covered in the restaurant choices: there is world cuisine, pan Asian cuisine, Indian cuisine and Mexican cuisine (1 restaurant each), and also, European cuisine (3 restaurants), Japanese cuisine (3 restaurants), British cuisine (4 restaurants) and generic vegan/vegetarian cuisine (4 restaurants). Such a (wide) selection ensures that any biases in GSCM findings with regard to cuisine types are avoided.

Next, looking at the pattern of green practices implementation, we find that only a few restaurants (5 out of 20 or 25%) have implemented these practices across all/most operational areas (spanning green design to green transportation), and covering most environmental aspects (from conserving resources (incl. energy), to maximising use of recyclable/sustainable/healthful inputs, to minimising carbon emissions and waste). These restaurants (refer restaurants A, C, M, N and S in the table) are consequently rated as high in terms of green practices implementation. At the other extreme, there are other restaurants (7 out of 20) that have covered only a few environmental aspects in a few operational areas: they have therefore been rated as low. The remaining eight restaurants with moderate levels of implementation in terms of operational areas and environmental aspects covered are rated as medium. Many of these low- and medium-rated restaurants though did highlight the need to implement environmental practices in the missing areas so as to have a more comprehensive implementation. This is captured in the words of one Restaurant Manager as: "*When talking about sustainability, it's not just about the final dish. It's the whole back story*".

Analysing the green practice implementation ratings vis-à-vis cuisine type for restaurants in Table 1, no pattern can be seen. For example, the high-rated restaurants are seen to serve different kinds of cuisines. Similarly, restaurants serving the same cuisine are not all rated high or low but have a mix of ratings; for example, two Japanese restaurants are rated medium and one low; similarly, among the four vegan/vegetarian restaurants, one is rated high, two medium and one low. What this shows is that we cannot associate a particular cuisine type with a particular level of green practice implementation.

With regard to green practices implementation in different operational areas, we see green purchasing and green cooking being the focus for most restaurants. This is not surprising for green cooking given that it provides not just environmental but also significant economic benefits. On the other hand, green purchasing is generally more expensive; its extensive implementation therefore comes as a surprise (although, as we can see from the table, the implementation is superficial in many cases). Finally, green transportation can be seen to be considered by only a few restaurants. This is logical given that transportation, and especially on the inbound side, is generally

Table 1 Restaurant-wise summarised findings on green practices, green drivers and barriers

Restaurant (cuisine) and Impl. extent[a]	Green design	Green menu design	Green purchasing and green packaging	Green cooking	Green transportation	Green drivers	Green barriers
A (Wholly organic pub) *high*	Only second-hand furniture used	All dishes are organic	Sourcing from family-run organic farm; Separate sustainable fish sourcing policy	Bread and pickles made in-house; Food waste used to create electricity via anaerobic digester	–	Organisation commitment, local community, government regulations	Cost of inputs, supplier commitment
B (Casual Indian dining) *low*	–	–	Quality preferred vis-à-vis eco-friendliness	Normal waste management; no special attention to carbon emissions	Fuel-efficient mobiles used for deliveries	Government regulations	Cost of inputs, lack of knowledge
C (Modern European eatery) *high*	Hydroelectric (renewable) energy provider used	No bottled mineral water; No beef and chicken dishes; Quality rather than organic label focus	Ethical suppliers and suppliers with green credentials used; local sourcing; sustainable fish supplies used	Energy-efficient induction cookers used; ozone rather than chemicals used in dishwasher; energy and waste reduction focus	–	Local community, customer demands	Cost of inputs, supplier commitment

(continued)

Table 1 (continued)

Restaurant (cuisine) and Impl. extent[a]	Green design	Green menu design	Green purchasing and green packaging	Green cooking	Green transportation	Green drivers	Green barriers
D (European breakfast and lunch) *medium*	Energy-efficient lighting	No bottled mineral water; most dishes organic and all meat free range	Biodegradable packaging used; focus on suppliers that can supply organic and free-range produce	Waste is recycled	–	Organisation commitment, Competitive advantage	Cost of inputs, lack of knowledge
E (Japanese sushi chain) *medium*	–	Environmentally unfriendly fishes like Bluefin tuna and eel not on menu	Biodegradable packaging, sustainable bamboo chopsticks; key vegetable purchases from UK	Leftover oil recycled into biodiesel	–	Government regulations, Competitive advantage	Cost of inputs, supplier commitment
F (Seasonal British dishes) *low*	Small onsite farm shop for sustainable eating	–	Select sustainable farms used for meat and fish sourcing	Bread made in-house	Livestock bred & slaughtered in one farm to minimise transportation	Local community, competitive advantage	Cost of inputs, supplier commitment
G (Mostly vegetarian dishes) *medium*	Concern about energy savings	Eco-friendly menu; predominantly vegetarian options	Selective organic vegetables and free-range meat suppliers	State of the art energy monitor in kitchen; emphasis on sustainable waste management	–	Organisation commitment, local community, regulations	Cost of inputs, lack of knowledge

(continued)

Table 1 (continued)

Restaurant (cuisine) and Impl. extent[a]	Green design	Green menu design	Green purchasing and green packaging	Green cooking	Green transportation	Green drivers	Green barriers
H (Eatery with world flavours) *low*	–	–	Sourcing from mostly local suppliers	Pickles, jams and chutneys made in-house; only seasonal ingredients used (to minimise waste)	Hand or bicycle deliveries as suppliers nearby	Organisation commitment, reduce cost, competitive advantage	Cost of inputs
I (Traditional British pub food) *medium*	Actively raise awareness of deforestation, sustainability	No bottled mineral water (to minimise waste)	Sourcing from sustainable fish suppliers only; own beehive on rooftop for honey supply	Recycle all waste	–	Government regulations, local community	Cost of inputs
J (Pan Asian dishes) *low*	Energy-efficient lighting	–	Focus on suppliers that can provide more variety of products	–	Fuel efficiency motorcycles used for deliveries	Government regulations	Cost of inputs, lack of knowledge
K (Vegan dishes) *medium*	Reclaimed wood and coffee bean sack furniture	Only vegan and organic recipes; focus on a healthy optimum diet	Only sustainable ingredients purchased	No cleaning products used, use of low-energy, low-waste cooking methods preferred	–	Stakeholders, reduce cost, local community	Cost of inputs

(continued)

Table 1 (continued)

Restaurant (cuisine) and Impl. extent[a]	Green design	Green menu design	Green purchasing and green packaging	Green cooking	Green transportation	Green drivers	Green barriers
L (Vegetarian dishes) *low*	–	Only organic recipes	Focus on sourcing local produce as much as possible	–	–	Stakeholders, customer demand	Cost of inputs
M (Vegetarian dishes) *high*	Maximum natural lighting; max usage of recycled and organic material	Purified tap water (to reduce plastic bottle usage); organic recipes; customers able to specify dish sizes (so less waste)	Local sourcing; suppliers providing organic produce preferred; used packages returned to suppliers for reuse	All waste recycled; local youths trained to become eco-friendly chefs	Pick up and deliveries in biodiesel car	Stakeholders, local community, government regulations	Cost of inputs, lack of knowledge
N (Mexican mini-chain) *high*	Carbon neutral rating; Building made more energy-efficient; motion sensor lights, recycled material used	–	Chillies in sauces and chutneys grown in select UK farms (to minimise imports); Sustainable fish sourcing	Heat from fridges and freezers reused to heat hot water (to minimise burning additional gas). Used oil taken away and processed into biofuel	Suppliers manage the deliveries with fuel-efficient vehicles	Organisation commitment, local community, government regulations	Cost of inputs
O (Japanese sushi chain) *low*	Front side covered with glass to maximise natural lighting	–	–	–	–	Government regulations	Cost of inputs

(continued)

Table 1 (continued)

Restaurant (cuisine) and Impl. extent[a]	Green design	Green menu design	Green purchasing and green packaging	Green cooking	Green transportation	Green drivers	Green barriers
P (British steak house) *low*	–	Free-range meat and vegetarian option offered	All eggs, dairy and meat are free-range and from sustainable British farms	–	Suppliers manage deliveries with fuel-efficient vehicles	Stakeholder, local community	Cost of inputs, lack of knowledge
Q (Bar cum restaurant serving European dishes) *medium*	Cross ventilation system and LED lights; rainwater harvested for hydrating herb garden	Dishes as per seasonality	Quality first but within that preference for local ingredients, beers, spirits, beef and coffee. Some inconsistency in local supply (a problem)	Eco-friendly products used for cleaning and washing; all waste recycled; focus on efficiency	–	Stakeholders, reduce cost, local community, customers	Cost of inputs, supplier commitment
R (Simple fish dishes) *medium*	–	Mostly organic fish recipes	Sustainable fish and biodegradable products purchased; 90% of purchases local	Root-to-fruit (waste minimising) cooking strategy used; Emphasis on recycling all waste	Suppliers manage deliveries with fuel-efficient vehicles	Stakeholders, reduce cost, competitive advantage	Cost of inputs
S (Staff canteen for breakfast, lunch) *high*	Solar energy used to power the whole facility	Dynamic menu as per vegetable supplies from suppliers (focus on what is freshly available)	Significant efforts to develop network of local, sustainable suppliers; dairy and meat from one sustainable farm	Nose-to-tail cooking strategy used for meat (to minimise waste and for economic pricing)	–	Stakeholder, government regulations, competitive advantage	Cost of inputs, supplier commitment

(continued)

Table 1 (continued)

Restaurant (cuisine) and Impl. extent[a]	Green design	Green menu design	Green purchasing and green packaging	Green cooking	Green transportation	Green drivers	Green barriers
T (Japanese restaurant) *medium*	Unable to have sustainable design due to building constraints (though keen)	–	Farmed salmon & yellow (not blue) fin tuna sourced as more sustainable; meat sourced locally; local vegetables more expensive; recyclable containers, bags	Focuses on recycling of all waste	Supplier manage the deliveries with fuel-efficient vehicles	Government regulations	Cost of inputs, supplier commitment

[a]Refers to extent of green practices implementation in high, medium, low terms

managed by the supplier, where small and medium restaurants of the kind considered have limited influence/bargaining power.

We now look at the findings (with reference to Table 1) on green practices implementation within each operational area separately.

4.1 Findings on Green Practices

4.1.1 Green Design

Most restaurants were found to have not considered infrastructural alterations to convert their facility into a green one. Uncertainty about returns from related investments was highlighted as a factor, with the (large) investments itself being a deterrent in some cases. However, many restaurants have considered energy-related options such as natural and low-energy (LED) lighting, which was because they provided economic benefits also. Some were found to have gone even further and using renewable energy (such as solar and hydroelectric) sources/suppliers; natural ventilation for energy savings is also being used selectively. With regard to other resources and their conservation, one restaurant was found to have considered rainwater harvesting (for water), with a couple of others emphasising wood conservation through the use of second-hand/recycled material furniture. Overall, we find a lack of uniformity in the nature of green design practices implemented. Also, an important factor in green designing, as pointed out by some respondents, was that the attractiveness/ambience (of the restaurant) should not get compromised in the process.

4.1.2 Green Menu Design

Green menu (although to varying degrees) was observed in two-thirds of the restaurants investigated. Many of these restaurants were found to offer organic dishes and predominantly vegetarian ones. This was partly in response to (health-conscious) customers' demands and partly to environmental concerns on the part of the restaurants. The fact that relevant (environmentally friendly) organic and free-range ingredients could be locally sourced to support this was also a factor. In some cases, the vegetarian menu choice was found to be driven by cost factors also (with chicken and beef options being more expensive to cook). The strong vegetarian focus of some restaurants can be gauged by the response of one Restaurant manager, as per whom: "*We always look at vegetables, not as a garnish, but as the core of a dish*".

Another feature observed for many restaurants, and entirely for environmental reasons, was offering tap water rather than the bottled one (so that associated plastic/glass waste could be reduced). Some other waste reduction options, which were observed for a few restaurants, include offering choice in terms of dish size, having a dynamic menu as per what was freshly available.

4.1.3 Green Purchasing and Green Packaging

Almost all the investigated restaurants were found to follow environmental principles in purchasing. Most were found to source organic and free-range vegetables, eggs, meat and fish, though in some cases this was partly driven by customer demands also. In the meat and fish's case, this (sustainable purchasing) was also indirectly influenced by government regulations. The preference of most restaurants was also found to be for local sourcing as it yielded both environmental benefits (lesser emissions) and economic ones (lower transport costs). However, this is not at the cost of quality, taste or other commercial considerations. With regard to packaging and customer (carry home) containers and boxes, purchases of only the biodegradable and recyclable kinds are allowed by some restaurants. Finally, some of the large restaurants were found to be vertically integrated into select areas, for example, growing some of their vegetable requirements themselves or maintaining their own beehives for honey; they also spend time and effort in developing and managing supplier networks.

4.1.4 Green Cooking

All the different environmental aspects such as minimising use of resources inc. energy, reducing waste and maximising waste recycling, and reducing the use of hazardous substances were found to be considered by the restaurants in their (green) cooking practices, although to varying degrees. Customer and commercial considerations (besides environmental ones) were found to be important for implementing some of the practices.

Many restaurants were found to be making/cooking certain products such as breads, jams, pickles and chutneys in-house rather than sourcing them from outside. The reasoning given was that this not only provided better quality (fresher, tastier and as per customer requirements) and cost efficiency, but also made the products more environmentally friendly as less transportation and storage is involved. Energy-saving cooking, both through choice of related equipment and their use, was also found to be popular among restaurants (due to economic as well as environmental reasons). Energy-efficient induction cooking hobs (which only use power when the pot sits on the hob), energy monitors, reusing heat from fridge/freezers to heat water are some of the approaches being used. On waste, both waste reduction and waste recycling were observed in many restaurants, which is driven partly by environmental commitment and partly by waste/environmental regulations. While root-to-fruit and nose-to-tail cooking strategies are being used by some (to ensure the entire plant/vegetable/animal/fish is used), some others are using seasonal ingredients to minimise waste. On waste recycling, used oil is recycled to biofuel (via third party), with most organisations focused on recycling other wastes also. Finally, some of the restaurants were found to use more eco-friendly chemicals, or avoid chemicals altogether (by using ozone) for cleaning. Here again, it is difficult to say whether the motives are environmental or partly food hygiene/health/safety (and related regulations) related.

4.1.5 Green Transportation

Only a few restaurants were found to be active in environmentally sustainable transportation of goods. This is not surprising given that on the more significant inbound transportation part, restaurants have little say/control: this is managed entirely by the suppliers (of inputs). Suppliers generally use large trucks and bulk deliveries with consequential low environmental impact; restaurants also therefore do not see any reason to be involved. On the outbound transportation side though, when delivering to customer locations, some restaurants use fuel-efficient vehicles/motorcycles; the motivations are both economic and environmental.

4.2 Findings on Green Drivers and Barriers

4.2.1 Green Drivers

The green drivers were found to be mostly in line with those suggested by previous researchers for restaurants. Government regulations, especially those on (greater) recycling and waste management, were identified as the principal driver. As appropriately captured by one Restaurant Operations Manager: "*The only way it is going to impact is if the government steps in*". There is indirect influence from other regulations too, such as the health, safety and sanitation/hygiene ones involving restaurants.

Pressure from local community was identified as the next important driver, where local community refers to the local council (or government machinery) and local businesses. Local businesses operate in close proximity with one another; they therefore exert pressure on each other to project an environmentally friendly image so that the community as a whole can acquire such an image. Local council also exerts pressure via different environmental initiatives at the local level.

Many restaurants highlighted gaining competitive advantage as an important green driver; their focus is on repositioning and acquiring a green image by offering organic vegan/vegetarian options and attracting more customers, especially the increasing number of environment/health-focussed ones. Customers' attractiveness towards green restaurants and their driving green practices was in fact observed for some of the restaurants.

Stakeholder pressure was identified as a green driver by a moderate number of restaurants, specifically the ones which are part of a large chain. This refers to the restaurant's parent company or stakeholder (generally with stringent environmental requirements and systems) pressuring/mandating it to follow those requirements.

A moderate number of restaurants also highlighted organisation commitment (for greening) and reducing costs (through greening) as drivers. The role of senior management's green-related knowledge, attitude and commitment on the extent and nature of green practices implementation is well recognised across sectors incl. for restaurants; therefore, this finding is not surprising. Similarly, many green initiatives such as those which lead to reduction of resource and energy requirements also consequently lower the cost of operations; hence, it makes sense for cost reduction to be considered as a motivation/driver for those initiatives.

Most of the above drivers can be explained from the perspective of the institutional theory, which examines how external pressures influence organisational actions (Hirsch 1975). According to this theory, firms are under constant coercive, normative and mimetic pressure to adapt to and be consistent with their external institutional environment (in this case implementing green practices) (DiMaggio and Powell 1983). Exerted by those in power, coercive pressures are powerful institutional forces that influence organisations to change their behaviour, while mimetic pressures are faced by firms to imitate/mimic the actions of their successful competitors in the industry in order to follow their success or to avoid losing their competitive advantages. Normative pressure arises from end consumers/owners to ensure the implementation of what constitutes appropriate and legitimate behaviour.

Here, we can see coercive pressure coming from government regulations, non-compliance to which could mean penalties/termination of business permits; in the case of restaurants which are part of a large chain, the coercive pressure also comes from their parent companies with associated infringement penalties. There is some element of mimetic pressure to go green as per nearby competitors, as also a lot of normative pressure from health/environmental conscious customers. Finally, the environmental commitment of restaurants can also be viewed from an institutional theoretic perspective (Scott 2001), namely, mimetic cultural cognitive isomorphism (socio-cultural responsibility), a rational desire to embrace environmental practices that are consistent with the obligations and values of society where they operate (Hsu et al. 2013).

4.2.2 Green Barriers

The green barriers highlighted by the restaurants are also largely in line with the literature. Green inputs (ingredients and equipment) are more expensive than normal ones; cost of these inputs was therefore identified as a barrier by all the restaurants. As per the respondents, the unpredictable nature of the restaurant business and changing customer habits makes it difficult to be confident about recovering these (higher) green input costs/investments. In the words of one restaurant General Manager: "*It is not easy being green*".

The second important barrier, which was identified by a large proportion of restaurants, is the lack of knowledge (of GSCM). Application of GSCM in restaurants has started not that long ago, and therefore, it was unsurprising to find many restaurants unaware of the range of green practices they could implement, the significance of those practices and their potential benefits.

Finally, many restaurants were found to be struggling to source eco-friendly and organic products locally. The difficulty was sometimes on the quality front, and sometimes the requisite quantity was not available. This lack of consistent supply or supplier commitment (to provide green products) was therefore identified as the third key barrier.

Theoretically, the findings on barriers can be explained from the perspective of the resource-based view (RBV) (Barney 1991), which considers firms to be a bundle of resources and capabilities (both tangible, such as buildings, and money and intangible, such as knowledge) that need to be distinctive if they are to be competitive. On the other hand, a lack of resources and knowledge could mean being less/non-competitive. The restaurants considered here are all small and medium restaurants with limited resources, which constrains their ability to invest in expensive green equipment, in training themselves and upgrading their green knowledge, in hiring and training green chefs, and in sourcing eco-friendly/organic products from distant suppliers (if not available locally). This in turn is reducing their competitiveness on the critical (and emerging) green dimension, and leading to their replacement with more resourceful, large organised restaurant chains on the high street. In Table 1, we can see, restaurants, which are part of a chain such as E, T and N, and therefore with greater resources implementing green practices to a greater extent than the others.

5 Conclusions

This study is arguably among the first comprehensive studies to understand GSCM application in restaurants, both in general and for the United Kingdom. A good understanding of green practices both in terms of their nature and extent of implementation is now available for individual restaurant operational stages: specifically, the green design, green menu design, green purchasing and packaging, green cooking and green transportation practices. We also have a good understanding of the drivers and barriers that motivate and impede restaurants to implement these practices. The findings have several managerial implications.

Firstly, restaurant owners and other stakeholders now know all the green practices and sub-practices they could implement at each operational stage. Also, the challenges they need to overcome such as on keeping green costs/investments down and being able to recover these, ensuring regularity of green/sustainable supplies and the need to transfer green-related knowledge to staff. This knowledge itself would enable more green practices and sub-practices to be tried as well as successfully implemented.

On their part, policymakers can note the disparity in green practices implementation across restaurants, both in terms of the number of operational stages covered and the variety/range of sub-practices within each. They can work towards reducing this disparity (either across the board or selectively) by strengthening some of the identified drivers and/or dampening the barriers. For example, they could strengthen the regulations with regard to green menu design (mandating a certain percentage of sales to be green or banning provision of bottled plastic/glass bottle), or green cooking (penalising energy consumption beyond specified limits based on restaurant size). Other options could be: (i) Providing tax relief and financial support to restaurants implementing green practices; (ii) Facilitating training on GSCM to restaurants; and (iii) Promoting collaborative relationships and GSCM knowledge transfer between high and low green practice implementing restaurants.

Despite the novelty and comprehensiveness of the study, it has some limitations. Firstly, it is based on only 20 restaurants; a more exhaustive study through a survey could serve to enhance the generalizability of the findings. Secondly, only small and medium restaurants are covered. Given that nature of GSCM could differ for large/organised sector restaurants, a separate study of a similar kind is needed for them. Performance implications (incl. environmental, economic and organisations) from GSCM is another area which has been discussed quite superficially in this work. This needs to be explicitly discussed in future studies. Finally, this work can be repeated for other countries and the differences in findings across them compared and contrasted.

Appendix

Interview protocol

- What does the restaurant offer?
- What sustainable/ eco-friendly practices are carried out at the restaurant?
- How familiar are you with the concept of green design? Were there any eco-friendly aspects considered while designing the restaurant?
- Are there any environmental-friendly factors considered such as energy consumption and carbon emissions during cooking operations?
- What kind of dishes is offered on the menu? Are there any dishes which are organic?
- How do you select your suppliers? Do you prefer to choose suppliers that implement eco-friendly practices while manufacturing the products?
- Do suppliers use fuel-efficient vehicles while transporting the goods to the restaurant?
- Does your restaurant offer delivery service? If yes, do you use fuel-efficient medium for transportation?
- What do you think are the drivers to implement green practices? For example, government regulations, customers?
- What do you think are the barriers to implement green practices? For example, costs, lack of knowledge?
- To what extent do environmental practices affect the economic performance of the restaurant?

References

Ahi, P., & Searcy, C. (2013). A comparative literature analysis of definitions for green and sustainable supply chain management. *Journal of Cleaner Production, 52,* 329–341.

Achillas, C. (2018). *Green supply chain management* (2nd edn., pp. 253–258).

Alcorn, M. (2009). *The green restaurant in practice: Employee attitude and behavior towards environmental sustainability*. MSc thesis. Stillwater: Oklahoma State University.

ASHRAE. (2003). Ashrae handbook: Heating, ventilating, and air-conditioning applications. In *American society of heating, refrigerating and air conditioning engineers*. Atlanta, Georgia.

Balasubramanian, S., & Shukla, V. (2017). Green supply chain management: An empirical investigation on the construction sector. *Supply Chain Management: An International Journal, 22*(1), 58–81.

Barney, J. (1991). Firm resources and sustained competitive advantage. *Journal of Management, 17*(1), 99–120.

Bartlett, E., & Howard, N. (2000). Informing the decision makers on the cost and value of green building. *Building Research & Information, 28*(5–6), 315–324.

Bergström, K., Solér, C., & Shanahan, H. (2005). Professional food purchasers' practice in using environmental information. *British Food Journal, 107*(5), 306–319.

Bright, E. (2018). McDonald's makes environmentally sustainable packaging pledge. Retrieved September 3, 2019, from https://www.thegrocer.co.uk/high-street/mcdonalds-makes-environmentally-sustainable-packaging-pledge/562230.article.

Bonilla Priego, M. J., Najera, J. J., & Font, X. (2011). Environmental management decision-making in certified hotels. *Journal of Sustainable Tourism, 19*(3), 361–381.

Chan, H. K., Yee, R. W., Dai, J., & Lim, M. K. (2016). The moderating effect of environmental dynamism on green product innovation and performance. *International Journal of Production Economics, 181,* 384–391.

Chiu, J. Z., & Hsieh, C. C. (2016). The impact of restaurants' green supply chain practices on firm performance. *Sustainability, 8*(1), 42.

Chou, C. J., Chen, K. S., & Wang, Y. Y. (2012). Green practices in the restaurant industry from an innovation adoption perspective: Evidence from Taiwan. *International Journal of Hospitality Management, 31*(3), 703–711.

Cottom, T. (2019). KFC commits to sustainable packaging globally. Retrieved September 3, 2019, from https://www.recyclingtoday.com/article/kfc-commits-sustainable-packaging/.

DiMaggio, P., & Powell, W. W. (1983). The iron cage revisited: Collective rationality and institutional isomorphism in organizational fields. *American Sociological Review, 48*(2), 147–160.

Energy Star. (2007). Guide for Cafés, Restaurants, and Institutional Kitchens. Retrieved September 6, 2017, from www.energystar.gov/sites/default/files/asset/document/CR%20ES%20Restaurant%20Guide%202015%20v8_0.pdf.

Fernie, J., & Sparks, L. (2014). *Logistics and retail management: Emerging issues and new challenges in the retail supply chain*. Philadelphia: Kogan page.

Govindan, K., Rajendran, S., Sarkis, J., & Murugesan, P. (2018). Multi criteria decision making approaches for green supplier evaluation and selection: A literature review. *Journal of Cleaner Production, 98,* 66–83.

Handfield, R., Walton, S., Seegers, L., & Melnyk, S. (1997). 'Green' value chain practices in the furniture industry. *Journal of Operations Management, 15*(4), 293–315.

Hirsch, P. M. (1975). Organizational effectiveness and the institutional environment. *Administrative Science Quarterly, 20*(3), 327–344.

Hsu, C. C., Choon, T. K., Zailani, H. M. S., & Jayaraman, V. (2013). Supply chain drivers that foster the development of green initiatives in an emerging economy. *International Journal of Operations & Production Management, 33*(6), 656–688.

Jackson, L. A. (2010). Toward a framework for the components of green lodging. *Journal of retail & Leisure property, 9*(3), 211–230.

Jeong, E., & Jang, S. (2010). *Effects of restaurant green practices: Which practices are important and effective?* (p. 13). Paper: Caesars Hospitality Research Summit.
Jernigan, D. A. (2012). *International Facility Management Association (IFMA) conference* (No. SAND2012-3403C). Albuquerque, NM (United States): Sandia National Laboratories (SNL-NM).
Kasim, A. (2009). Managerial attitudes towards environmental management among small and medium hotels in Kuala Lumpur. *Journal of Sustainable Tourism, 17*(6), 709–725.
Kasim, A., & Ismail, A. (2012). Environmentally friendly practices among restaurants: drivers and barriers to change. *Journal of Sustainable Tourism, 20*(4), 551–570.
Large, R. O., & Thomsen, C. G. (2011). Drivers of green supply management performance: Evidence from Germany. *Journal of Purchasing and Supply Management, 17*(3), 176–184.
Lion, R., Arulogun, O., Titiloye, M., Shaver, D., Jain, A., Godwin, B., et al. (2018). The effect of the "Follow in my Green Food Steps" programme on cooking behaviours for improved iron intake: A quasi-experimental randomized community study. *International Journal of Behavioral Nutrition and Physical Activity, 15*(1), 79.
Lewis, L. B., Sloane, D. C., Nascimento, L. M., Diamant, A. L., Guinyard, J. J., Yancey, A. K., et al. (2005). African Americans' access to healthy food options in South Los Angeles restaurants. *American Journal of Public Health, 95*(4), 668–673.
Luthra, S., Garg, D., & Haleem, A. (2015). Critical success factors of green supply chain management for achieving sustainability in Indian automobile industry. *Production Planning & Control, 26*(5), 339–362.
Malviya, R., & Kant, R. (2015). Green supply chain management (GSCM): A structured literature review and research implications. *Benchmarking: An International Journal, 22*(7), 1360–1394.
Namkung, Y., & Jang, S. S. (2013). Effects of restaurant green practices on brand equity formation: Do green practices really matter? *International Journal of Hospitality Management, 33,* 85–95.
Perotti, S., Zorzini, M., Cagno, E., & Micheli, G. J. (2012). Green supply chain practices and company performance: The case of 3PLs in Italy. *International Journal of Physical Distribution & Logistics Management, 42*(7), 640–672.
Revell, A., & Blackburn, R. (2007). The business case for sustainability? An examination of small firms in the UK's construction and restaurant sectors. *Business Strategy and the Environment, 16*(6), 404–420.
Rimmington, M., Smith, J. C., & Hawkins, R. (2006). Corporate social responsibility and sustainable food procurement. *British Food Journal, 108*(10/11), 824–837.
Sarkis, J., & Dou, Y. (2017). *Green supply chain management* (3rd edn., pp. 423–430).
Schubert, F., Kandampully, J., Solnet, D., & Kralj, A. (2010). Exploring consumer perceptions of green restaurants in the US. *Tourism and Hospitality Research, 10*(4), 286–300.
Scott, W. R. (2001). *Institutions and organizations*, Sage.
Seuring, S. (2004). Integrated chain management and supply chain management comparative analysis and illustrative cases. *Journal of Cleaner Production, 12*(8–10), 1059–1071.
Tian, Z. P., Wang, J., Wang, J. Q., & Zhang, H. Y. (2017). Simplified neutrosophic linguistic multi-criteria group decision-making approach to green product development. *Group Decision and Negotiation, 26*(3), 597–627.
Tran, T. T. V., Kaiprommarat, S., Kongparakul, S., Reubroycharoen, P., Guan, G., Nguyen, M. H., et al. (2016). Green biodiesel production from waste cooking oil using an environmentally benign acid catalyst. *Waste Management, 52,* 367–374.
Tsai, M. C., Cheng, C. C., & Chang, Y. Y. (2010). Drivers of hospitality industry employees' job satisfaction, organizational commitment and job performance. *African Journal of Business Management, 4*(18), 4118–4134.
Wang, Y. F., Chen, S. P., Lee, Y. C., & Tsai, C. T. S. (2013). Developing green management standards for restaurants: An application of green supply chain management. *International Journal of Hospitality Management, 34,* 263–273.
Weller, S. C., & Romney, A. K. (1988). *Systematic data collection* (Vol. 10), Sage publications.
Wilson, J. (2014). *Essentials of business research: A guide to doing your research project*, Sage.

Zhu, Q., Sarkis, J., & Lai, K.-H. (2012). Examining the effects of green supply chain management practices and their mediations on performance improvements. *International Journal of Production Research, 50*(5), 1377–1394.

Zsidisin, G. A., & Siferd, S. P. (2001). Environmental purchasing: A framework for theory development. *European Journal of Purchasing & Supply Management, 7*(1), 61–73.

Chapter 8
Factors Motivating Indian Manufacturing SME Employers in Adopting GSCM Practices

Manpreet K. Dhillon and Yongmei Bentley

Abstract The growth of manufacturing SMEs is vital as their contribution towards the national economy is significant. In this era of globalisation, SMEs are compelled to ensure sustainable profitability through cost saving while being environmentally conscious at the same time. It has been reported in the past empirical studies that adoption of green supply chain management (GSCM) practices by SMEs could enable such enterprises to improve their performance and succeed in their operations. Hence, to gain and maintain competitive advantage and succeed, SMEs need to change their practices and adapt their strategies to the dynamic environment of today. The factors motivating adoption of GSCM among Indian SMEs have not been thoroughly explored in the past studies. This sets the motivation for the present research. Thus, the purpose of this paper is to explore the factors motivating Indian manufacturing SME employers in adopting GSCM practices in their firms and to develop a GSCM framework based on the literature review and the empirical findings of this study. An interview-based qualitative research methodology is used in order to gather rich and rigorous information from experienced SME employers. The contents of the interviews were analysed using thematic analysis method and the recurring themes were identified and highlighted. The findings presented in this chapter clearly illustrate the low level of involvement of SME employers in adopting GSCM in their firms. The results reveal that financial benefits, saving cost, government pressures and awareness by government are the most important factors that are motivating Indian SME employers to adopt GSCM practices in their firms. The findings can help employers to focus on important factors to facilitate adoption of GSCM practices in SMEs with limited resources.

Keywords SME employers · GSCM practices · Motivations · India · Manufacturing SMEs

---

M. K. Dhillon (✉)
Coventry University, Coventry, UK
e-mail: manpreet.dhillon@coventry.ac.uk

Y. Bentley
University of Bedfordshire, Luton, UK

U. Ramanathan and R. Ramanathan (eds.), *Sustainable Supply Chains: Strategies, Issues, and Models*, https://doi.org/10.1007/978-3-030-48876-5_8

1 Introduction

Environmental issues are becoming noteworthy concerns of the government, society and business organisations (Rengarajan et al. 2015; Rozar et al. 2015). Business organisations are considered to be the source of different environmental problems such as ozone depletion, global warming, solid waste and water and air pollutions (Gandhi et al. 2018; Rozar et al. 2015; Worku and Virdi 2019). Due to the rapid economic growth, there has been increased consumption of energy and materials which have given rise to resource depletion problems and environmental issues (Soda et al. 2016). The growing number of companies worldwide has now recognised the importance of protecting the natural environment (Singh et al. 2015b). With an aim to reduce the environmental impact of their operations, companies globally are developing new environmental strategies and programs in order to gain higher benefits (Abdullah et al. 2018). Green supply chain management (GSCM) has emerged as a new approach for organisations to enhance profits by reducing the environmental risks (Oliveira et al. 2018). Ahi and Searcy (2013) in their paper identified 22 definitions of GSCM and observed that the definition by Srivastava (2007) is the most widely used one in the literature. According to Srivastava (2007), GSCM is "*integrating environmental thinking into supply-chain management, including product design, material sourcing and selection, manufacturing processes, delivery of the final product to the consumers as well as the end-of- life management of the product after its useful life*".

Globally, companies are adopting GSCM practices to improve their environmental performance and to achieve competitive advantage (Mathiyazhagan et al. 2018). Nowadays, Indian economy is getting integrated with the world economy (Mathiyazhagan et al. 2018). This globalisation is providing many opportunities to the Indian manufacturing industries to export their products (Singh et al. 2015b). However, according to the research conducted by Worku and Virdi (2019) and Gandhi et al. (2018), manufacturing organisations are the main source of generating air, water and solid waste emissions in the environment.

GSCM issues are now becoming significant in India due to the growing globalisation and penetration of international traders in the Indian market (Hu and Hsu 2010a, b; Jayaram and Avittathur 2015; Luthra et al. 2015). This has created many opportunities and given rise to environmental and social concerns (Luthra et al. 2015). According to the Environmental Performance Index (2018), India ranks 177th out of 180 countries which is worse than the Environment Performance Index (EPI) of 2014 (155th out of 178 countries). Hence, this indicates the miserable condition of India and its current awareness about GSCM and its practices.

SMEs account for 99% of the total industrial sector in India by contributing around 45% of the total manufacturing output and 40% to total export (Gandhi et al. 2018; Singh et al. 2015b). Furthermore, the total manufacturing sector in India contributes 16% towards the total GDP of India and out of this, 7% is contributed by SME manufacturing sectors of India (Gandhi et al. 2018). Hence, this indicates the significant contributions made by Indian manufacturing (IM) SMEs towards the economic growth. At the same time, Indian manufacturing SMEs generate 70% of the total industrial pollution of the nation (Singh et al. 2015b; TERI 2013). Most of these industrial wastes are created due to the lack of adequate waste disposal facilities in India (Joshi and Ahmed 2016; Singh et al. 2015b). According to the report by Brassaw (2017), 90% of the waste in India is dumped in unsatisfactory places and is not recycled due to the lack of waste treatment and disposal facilities. This explains the poor state of pollution and wastes that call for immediate adoption of GSCM practices across the Indian manufacturing SME sectors (Singh et al. 2015b; Soda et al. 2015). Thus, greening of SC in the manufacturing SME sector needs more attention in India.

Moreover, to succeed and gain a competitive advantage Indian manufacturing SMEs need to embrace change and adapt to the new environmental practices of today. The government of India is using different initiatives to help the Indian industries to adopt GSCM practices in their firms. Due to environmental changes, the world is experiencing a drastic change in the nature. This study will open up a new debate on the different motivational factors on adopting GSCM practices in Indian manufacturing SME sectors. More specifically, this study aims to explore various motivational factors that encourage SME employers to adopt GSCM in their firms and to develop a GSCM adoption framework based on existing literature and the empirical findings from this study.

2 Overview of Green Supply Chain Management

The evolution of the concept of SCM has been proved as an efficient and effective way to achieve competitive advantage (Matthew 2017; Rajeev et al. 2017). It has been perceived as one of the key aspects for long-term sustainability in terms of maximising revenue and profits for organisations as well as increasing the number of customers (Sambrani and Pol 2016). Over the past decade, it has also been proven to be successful in reducing cost by adding value to the supply chain by adopting green initiatives (Jaggernath 2015). Thus, the integration of environmental concerns within the traditional SCM has given rise to a new and growing field named green supply chain management (GSCM) (Ali et al. 2017; Jia et al. 2018; Laosirihongthong et al. 2013a; Sarkis 2012). GSCM has emerged as a significant strategy within the domain of sustainability, including initiatives from green purchasing to reverse logistics (Geng et al. 2017; Jia et al. 2018).

This concept was formally known in the mid-1990s and it was from then that investigation on GSCM became more established (Rajeev et al. 2017; Sarkis et al. 2011; Seuring and Muller 2008; Verma et al. 2018). Presently, adopting green agenda in the SCM has become an interesting focus area for practitioners and future academic research (Fang and Zhang 2018; Jayaram and Avittathur 2015; Sarkis 2003; Srivastava 2007; Verma et al. 2018). The deteriorating environment, diminishing natural resources, increased level of pollution and overflowing waste sites have increased the importance of GSCM (Mangla et al. 2017; Srivastava 2007). Thus, according to Batra and Chanana (2015) the rising attention towards greener SC is mainly due to

- Increased climatic changes caused by global warming;
- Increased depletion of natural resources caused by increased population and industrialisation;
- Increased waste and pollution caused by rapid development of industrialisation.

Thus, GSCM is important for all organisations as it aims to reduce the emissions of greenhouse gases and solid wastes (Côté et al. 2008). Consequently, global organisations have started implementing environmental practices with the intension to conserve and reduce the degradation of the environment (Epoh and Mafini 2018). Extending the concept of SCM to GSCM has led to the improvement of product and services, by making them environment friendly across its life cycle (Epoh and Mafini 2018; Gunasekaran et al. 2015). Thus, integrating green practices in the SC of the organisation helps in increasing the overall performance of the firm (Epoh and Mafini 2018).

The term GSCM has been widely defined by many scholars as follows:

Srivastava (2007, p. 54) defines GSCM as

> integrating environmental thinking into supply chain management, including product design, material sourcing and selection, manufacturing process, delivery of the final product to the consumers as well as end-of-life management of the product after its useful life.

Gardas and Narkhede (2013, p. 442) define GSCM as

> the process of using environmentally friendly inputs and transforming these inputs into outputs that can be reclaimed and re-used at the end of their lifecycle thus, creating a sustainable supply chain.

Khushbu and Shah (2014, p. 1657) define GSCM as

> a modern management approach where supply chain is a combination of economy and ecology.

According to Govindan et al. (2016, p.186),

> GSCM is one such idea that takes environmental elements into consideration when managing the supply chain, enabling industries to enhance their economical as well as ecological performance and so is seen by many as a promising organizing concept.

Sarkis (2012) defined GSCM as "closing the supply chain loop", which involves reverse supply chain activities. Sarkis diagrammatically represents GSCM as Fig. 1.

Figure 1 exhibits the extension of the traditional SCM activities by including reverse supply chains. The concept of closing the loop represents the end-of-life of the materials that are absorbed back into the system through activities like recycling, remanufacturing, reclamation and reverse logistics at all levels of an SC (Sarkis 2012).

In conclusion, the field of GSCM is gaining increased popularity among the industrialised countries. Despite the increasing popularity, there still remains gaps in the literature of GSCM that require further research (Laosirihongthong et al. 2013a, b), particularly, on identifying different motivational factors of adopting GSCM practices in the organisations. Thus, the next section discusses this gap in the body of literature in Indian context.

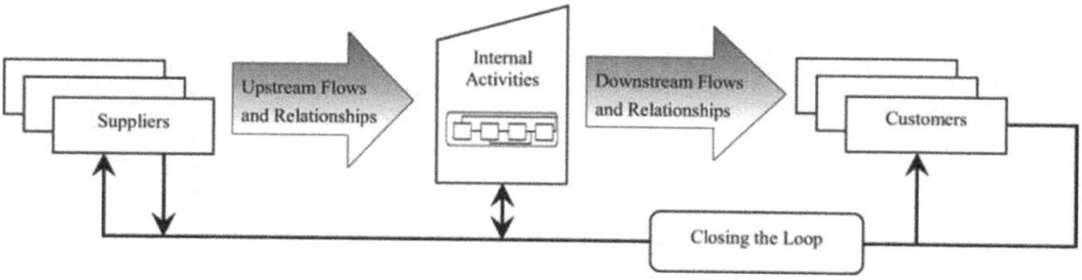

Fig. 1 Diagram of green supply chain management. *Sources* Sarkis (2012)

3 Problem Statement

As discussed earlier, GSCM issues are becoming noteworthy issues in India because of the shift of the world's manufacturing operations being carried out in Asian countries (Hu and Hsu 2010a, b; Luthra et al. 2015). Increased environmental concerns in these countries have drawn the attention of many researchers to investigate the adoption and implementation of GSCM practices (Mathiyazhagan et al. 2018; Seman et al. 2012). However, the concept of GSCM is relatively new in India and recent literature show that there is a lack of research on identifying the factors that are motivating the adoption and implementation of GSCM in Indian context (Jayaram and Avittathur 2015; Vijayvargy et al. 2017).

According to The Economic Times (2015), India is found to be a slow adopter of the concept green. Rapid industrialisation in India has led to enormous consumption of resources which is generating immense waste and only 3% is recycled in authorised recycling facilities (Soda et al. 2015). Again, in India, 50% of the total pollution is caused by industrial wastes (Soda et al. 2015). The status of pollution and wastage created by the industries in India calls for an immediate adoption of GSCM measures (Singh et al. 2015b). The importance of adopting green practices in India is greater than anywhere else (Soda et al. 2016).

Due to the increasing popularity of GSCM, many researchers have started undertaking research on this subject, but still there are many gaps in the literature in Indian context, for instance, Vijayvargy et al. (2017) in their research proposed to have more research focusing on different industries in order to gain information on the different motivational factors that are encouraging Indian industries to adopt GSCM practices. Again, Malviya and Kant (2015) in their research addressed an essential gap in the domain of GSCM and recommended to have in-depth and detailed studies on GSCM issues. Similarly, Kumar and Garg (2017) highlighted that there is a lack of qualitative research in Indian context. Finally, Gandhi et al. (2018), recommended that a GSCM framework for future extension be developed. Thus, the literature confirms that there is tremendous scope for research in this area in India and there are many aspects that still need to be addressed (Mathiyazhagan et al. 2018). Thus, based on the above gaps, this study aims to explore the various motivational factors of GSCM in Indian manufacturing SMEs by using qualitative research methods and finally develop a GSCM adoption framework.

4 Green Supply Chain Management Pressures on Indian SMEs

SMEs have always been the backbone of Indian economy as a highly vibrant and dynamic sector of India. Indian SMEs play an important role in economic development, employment generation and investments (Singh et al. 2015a; Uma 2013). As discussed earlier, SMEs constitute 99% of the total enterprise in India (Gandhi et al. 2018) and the total SME manufacturing sector in India contributes 16% towards the total GDP, of which 7% is contributed by the Indian manufacturing SMEs (Gandhi et al. 2018). Due to this, Indian manufacturing SMEs in India are under huge pressure to adopt GSCM practices in their firms because they account for 70% of industrial pollution (Singh et al. 2015b). The industrial waste issues are becoming alarming concerns in India as they are increasing the pollution of soil and water. These issues are increasingly creating pressure on the SME sectors (Singh et al. 2015b). Despite being an enormous sector and contributing significantly towards the GDP of India, this sector is not found progressing in GSCM due to various limitations faced by them (Gandhi et al. 2018). They face huge pressures from different sources in adopting GSCM practices. The government of India has taken up a lot of initiatives in promoting green practices among them (Small and Medium Business Development Chambers of India 2018). They announced various plans such as the National Action Plan on Climate Change (NAPCC) with an aim to reduce India's carbon consumptions by 20–25% by 2020 (Gandhi et al. 2018). Again, Swatch Bharat Abhiyan aims to clean India by 2019. Finally, the Indian government also aims to achieve the targets set in the Paris agreements made. Hence, to achieve these targets Indian manufacturing SMEs face lots of pressures that drive them to implement GSCM practice in their firms. The next section highlights the various drivers of GSCM from the existing literature.

5 Drivers/Motivational Factors of Green Supply Chain Management

Organisations are found to practice GSCM proactively or reactively (Ali et al. 2017; Laosirihongthong et al. 2013b). The driving forces that encourage GSCM are either internal (from within the organisation) or external (from outside the organisation). Many researchers (Ali et al. 2017; Lee et al. 2013; Walker et al. 2008; Zailani et al. 2012) believe that there are different kinds of internal and external pressures that enhance GSCM practice in the organisations.

According to Walker et al. (2008) the internal drivers (ID) that take an organisation towards GSCM practice are the personal commitments of the leaders, managers, policymakers and investors; whereas, the external drivers (ED) include increasing public awareness, customers demand and influence of NGO's. According to the research by Lee et al. (2013), internal pressures proactively manage any sort of environmental regulations and external pressures improve the flexibility of the organisation. Both the drivers are discussed in more detail below.

5.1 Internal Drivers (ID) for GSCM

The drivers that exist within the organisation itself are known as the ID. Organisational-related GSCM drivers such as top management commitments, employee health and safety, employee motivation, organisational policies for supporting GSCM, improvement of firm performance, and investor pressure are known as ID for GSCM.

5.2 External Drivers (ED) for GSCM

The drivers that exist from outside the organisation are the ED. There are a number of EDs such as regulatory bodies, customers, competition, marketing, suppliers and society.

Regulatory: Government rules and regulations form the strong driver especially if the firms have a positive attitude towards adoption of GSCM. This has been recognised by a number of studies (Dhull and Narwal 2016; Govindan et al. 2016; Luthra et al. 2016a). It includes factors such as ISO 14001 certification, government environmental regulations, stakeholders pressures etc.

Customers: Customers play a vital role in the implementation of GSCM in a firm. Increased awareness among the customers and their demand for eco-friendly products force a company to adopt GSCM practices. It includes factors like customers' demand, awareness among them, and so on.

Competition: Many researchers (Dashore and Sohani 2013; Dhull and Narwal 2016; Govindan et al. 2015, 2016; Luthra et al. 2016b) have acknowledged competition as an important driver. Competition with the competitors plays a strong role in the adoption of GSCM practice within the organisation.

Marketing: To receive publicity and enhance the company's image, marketing plays an important role in adopting GSCM practice. It includes factors like establishing companies' green image, etc.

Suppliers: Suppliers play an important role in the adoption of GSCM. Suppliers using green practices also encourage and support their co-companies to adopt such practices.

Society: The deterioration of the environment over the years has increased public awareness. The public is increasingly influenced by a company's green image. Their awareness about green products is making companies to adopt GSCM.

Huge pressures for implementing GSCM and its practices have motivated and encouraged many researchers to conduct research on this domain. Many scholars in their work had made it evident that adoption and implementation of GSCM and its practices are essential for the growth and development of an organisation. The research conducted by Drohomeretski et al. (2014) focused on identifying the main motivational factors of implementing GSCM and identified cost reduction as one of the major factors for implementing GSCM practices in the SC of the Brazilian automobile industry. Tachizawa et al. (2015) evaluated the interrelationships among different GSCM drivers, approaches and performance in Spanish firms. The results highlight coercive pressures (e.g. regulations) that are less effective than non-coercive (e.g. Customer, society) pressures. Walker et al. (2008) explored various internal and external drivers that motivated UK organisations to adopt GSCM practices and found external drivers affecting more than the internal drivers. Zhu et al. (2007) in their paper investigated the drivers, practices and performance of automobile enterprises within China and identified them facing high regulatory and market pressures. Again, Agi and Nishant (2017) evaluated different influential factors for implementing GSCM practice in Gulf countries (Middle east) and recognised the size of the company and top management commitment as the major factors in implementing green. Similarly, Balasubramanian and Shukla (2017) in their research revealed ISO 14001 and environmental auditing as the main drivers of GSCM of construction industries of UAE. Furthermore, Hu and Hsu (2010a, b) conducted a similar kind of research where they explored top management support as the critical factor for implementing GSCM practice in Taiwanese electrical and electronic industries using factor analysis. The study by Ariffin et al. (2015) focused on identifying drivers of implementing green manufacturing in Malaysian SMEs and the findings revealed improved company image, and improved competitiveness forms the top critical drivers.

Similarly, due to increased pollution and waste and diminishing resources, the importance of GSCM has also increased in India (Mathiyazhagan et al. 2013). There has been huge pressure on Indian industries to implement GSCM practices in their organisations (Soda et al. 2016). This has motivated many Indian scholars to conduct research on identifying the driving or motivational factors of implementing GSCM in India. Some of the studies conducted from the Indian point of view are summarised henceforth.

Diabat and Govindan (2011) focused on developing a model of drivers influencing the adoption of GSCM using ISM. The drivers were identified through extant literature review and industry experts. The model was confirmed by using a case study. Mathiyazhagan and Haq (2013) in their paper identified high penalties and customer demands as the most influential pressures for adoption of GSCM using ISM. A study by Jain and Sharma (2014) reviewed previous literature on GSCM drivers/pressures and identified government, customer pressure and competition as the most influential pressures. Sharma et al. (2015) analysed the key factors responsible for implementing GSCM in the dairy sector of India and the report highlighted environmental management and customer cooperation as the key factor.

The research by Gandhi et al. (2015) evaluated the key factors responsible for successful implementation of GSCM using DEMATEL approach. The proposed method evaluated top management commitment, human technical expertise and financial factor as the most influential factors. Luthra et al. (2015) used ISM technique to understand the relationship among the twenty-six critical success factors (CSF) and recognised enhanced brand image, economic benefits and firms competitiveness as the top-most CSF for Indian industries.

Dhull and Narwal (2016) focused on reviewing the literature on various barriers and drivers for implementing GSCM in India. Luthra et al. (2016b) conducted an empirical research to explore the CSF of Indian automobile industry to implement GSCM for sustainability. The study used multi-linear regression analysis method and recognised 'regulations' as the CSF for the automobile industries. Again, Seth et al. (2016) also conducted an empirical research to identify the CSF and the performance measures to implement green manufacturing practices in the Indian cement industry. Govindan et al. (2016) investigated the influential factors of GSCM in the context of mining industries in India and categorised 'top management realisation' as the most influential factor using DEMATEL technique.

More recent research conducted by Mathiyazhagan et al. (2018) focused on identifying and prioritizing the motivational factors of GSCM in the Indian construction industries using AHP. The government category was identified as the most important motivational factor. Since the main aim of this research is to identify the different motivational factors that are encouraging Indian manufacturing SMEs to adopt GSCM practices, this research aims to further summarise the studies performed by Indian researchers focusing especially on SME sector of India.

Very few researchers focused on identifying drivers or motivations for GSCM in Indian SME aspect except for the research conducted by Singh and Kumar (2015) who quantitatively analysed the drivers affecting the adoption of GSCM in the SMEs of Rajasthan. They used survey to identify the drivers which were further analysed using SPSS. The results showed a positive effect of drivers on adoption of GSCM. A review-based article by Sharma et al. (2015) identified nine major drivers and four output factors of implementing GSCM practices among SMEs through an extant literature review. Another study by Mathiyazhagan et al. (2015) investigated the different pressures for implementing GSCM in the mining and mineral SMEs of India and ranking the pressures based on the AHP technique. Kathiresan and Ragunathan (2017) identified different drivers of GSCM via literature review among the leather SMEs of Tamil Nadu and evaluated quantitative results using regression analysis, correlation and standard deviation. Gandhi et al. (2018) evaluated drivers for integrated lean and green manufacturing in Indian manufacturing SMEs with the support of the existing literature and expert opinions and further ranked the drivers using MCDM.

The research conducted by the above researchers discussed the drivers/motivational factors in different SME sectors of India. The key researchers and drivers identified are summarised in Table 1.

Table 1 Key drivers identified by Indian scholars focusing on Indian SMEs to adopt GSCM practice. Identification of such drivers was mostly based on literature review and expert opinions. Thus, this research aims to identify the key drivers that are motivating Indian manufacturing SME employers to adopt GSCM practice in their firms by interviewing them and presenting the actual status of GSCM in India.

Table 1 Key drivers of implementing GSCM in India SMEs

Source	Drivers identified
Singh and Kumar (2015)	Organisation related, Regulatory, Customers, Competition, Suppliers, Marketing, Society
Kumar et al. (2015)	Top management pressures, Government rules and regulations, ISO 14000 certification, Pressure from customers, Gaining competitive advantage, Improvement of firm performance, Environmental collaboration with suppliers, Collaboration between product designers and suppliers, Marketing
Mathiyazhagan et al. (2015)	Regulations
	Central government regulations, Regional environmental regulation, High penalty for environmental pollution, Environmental regulations for export countries
	External sources
	Foreign direct investment interest in green product, Pressure from non-governmental organisation for eco-friendly product, Media attention through environmental action groups
	Financial factors
	Scarcity of resources in India, Special tax exemption for ISO 14001 certified firms, Cost for disposal of hazardous materials, Carbon tax forcing fuel, cost reduction
	Production and operational factors
	Establishing company green image, Community pressure, Pressure from new economic, energy savings
Kathiresan and Ragunathan (2017)	Organisational commitments
	Top management commitment, Cross-functional team to minimize environmental impact, Environmental management system
	Eco-design related issues
	Eco product design, Designing product to sustain ecological framework
	Green procuring and marketing related issues
	Supplier collaboration, Environmental auditing, Supplier ISO certification, Customer demand, Customer pressures for green packaging
Gandhi et al. (2018)	Employee training, Top management commitment, Multi-skilled workers, Less machine breakdowns, Work standardisation, Technology upgradation, Employee empowerment, Organisation culture, Cost saving, Competitive advantage, Current legislation, Future legislation incentives, Green brand image, Public pressures

Fig. 2 Initial GSCM framework

Moreover, Gandhi et al. (2018) in their work highlighted that GSCM and its practices are still at its nascent stage in India especially among the SMEs due to various constraints faced by them. Nevertheless, according to them, there are many motivational factors that can lead to easy adoption of such practices. Similarly, the research by Mathiyazhagan et al. (2018) emphasises that while there is a vast scope of research in the domain of GSCM, a lot of drivers are yet to be explored that may play an important role in the adoption and implementation of GSCM in Indian industries. More research on identifying different drivers or motivating factors may help and encourage successful implementation of GSCM. Hence, this research aims to fill the gap in the literature by exploring various motivating factors that encourage Indian manufacturing SME employers to adopt GSCM practices by presenting a real picture of the status of GSCM in India. Based on the review of existing literature, this study develops an initial framework to highlight the findings from the literature. Figure 2 represents the initial framework developed by this study.

The next section discusses the research methods adopted to conduct this research.

6 Research Method

This study follows an exploratory approach in order to deeply investigate the existing motivational factors of Indian manufacturing SMEs employers to adopt GSCM practices in their organisations. Given the aim of this study, a qualitative method was used where the data were collected using semi-structured interviews. The interview method is used for an exploratory study when little is known about the area of research (Bryman and Bell 2011). Thus, exploratory design best suits this research as little is known about the motivational factors of the SMEs in India. This approach has helped the researcher discover what is actually happening in the organisations regarding GSCM, and to know more about the participants' experiences by acquiring multiple facets of the social world they are involved in.

Five semi-structured interviews were conducted over the phone with the SME employers located in India. The participants selected were the owners of the SMEs with manufacturing firms in operation for more than 2 years. According to Patton (2002) and Bryman and Bell (2011), there is no defined rule to sample size in a qualitative research. The research conducted by Hijaz et al. (2015) also used 5 interviews to conduct qualitative research on GSCM and SMEs. Again, Homburg et al. (2012) and Kumar et al. (1993) supported that a qualitative study can be conducted

Table 2 List of participants and fields of business

	Participants	Business
1	Participant 1 (P1)	Machinery manufacturer
2	Participant 2 (P2)	Zip manufacturer
3	Participant 3 (P3)	Paint manufacturer
4	Participant 4 (P4)	Shoe manufacturer
5	Participant 5 (P5)	Furniture manufacturer

using smaller sample size and single respondent in order to derive productive results. Research conducted by both authors focused on small SME samples and used single key SME informant as most SMEs are owned or managed by family members, selecting owners as respondents is crucial. Thus, based on the above discussion this study was conducted using small sample and single respondent to identify useful results.

Furthermore, all the interviewees had similar kind of businesses as demonstrated in Table 2. Since the interviewees were from manufacturing industry only, the SME owners were found to provide similar information. This is due to the fact that India is the most popular democratic country and is ruled by a single governing body with the same rules and regulations followed throughout the country. Similarly, the industrial laws are also the same throughout the country. Thus, the imposition of equal rules enforces all the industries to follow similar practices. Considering the time and budget, response rate and the kind of analysis used, the researcher decided to conduct this study using smaller sample size.

Since the researcher is located in the United Kingdom and taking into consideration the time and cost factors the researcher decided to conduct telephonic interviews with the participants. Each interview lasted for about 40 min. All the interviews were recorded with participants' consent and later were transcribed for analysis. The purpose of conducting the interviews was to capture the interpretative account of the participants to explore the different motivational factors for adopting GSCM practices in their firms.

All the participants were asked six questions during the interview process. The questions were related to the main aim of this study with the focus to explore the various motivational factors for adopting GSCM practices. Having all the transcribed interviews in hand, the data were analysed using Nvivo 11 plus software. This software helped the researcher organise the data and find different sub-themes under the major theme: Motivations. The emerging sub-themes form the key findings of this study that are discussed in the next section.

7 Findings and Discussion

The interview method undertaken for this study allowed the researcher to capture different factors that motivated SME employers to adopt GSCM practices in their organisations. When being asked about the different reasons or factors that motivated them to adopt GSCM practices, all the participants shared their stories and their motivations for practicing green. The term motivations in this study refer to the various reasons for an employers to adopt GSCM practices in their organisations. This study identified different reasons/factors that are encouraging SME employers to adopt GSCM practices. All the factors are segregated under two different groups—Internal factors and external factors. These are discussed in the sections below.

7.1 Internal Factors

(1) **Financial benefits**

Financial benefits are the key motivational factor for SMEs. All the employers interviewed claimed to have practised green with the aim of gaining some financial benefits out of it. This was apparent from the comments given by the participant below:

> …these practices are more or less useful as they help you economically. We save a bit like on paper usage and electricity. But operating on a small scale, I believe, that whatever is profitable is useful – (P4)

P4 saw GSCM practices to be economical practices as they can save some money by using less paper and electricity. According to P4, they embrace practices that are profitable. Therefore, for manufacturing SME employers any practice that is economical encourages them to adopt the practices.

Thus, the factor of "**Financial benefits**" highlighted by all SME employers interviewed (100%) is confirmed to be the most motivational factor for adopting green practices. This finding is in alignment with the existing GSCM studies. A study by Ariffin et al. (2015) reports a similar finding while investigating the drivers of green manufacturing practices in Malaysian SMEs. Subsequently, Mathiyazhagan and Haq (2013) also affirmed "increasing anticipated business benefits" as an important motivational factor for Indian industries to adopt GSCM. In contrast, the studies on Indian manufacturing SMEs did not identify this as a key factor. Thus, financial benefit is assumed to be an important finding disclosed by this study in the context of Indian manufacturing SMEs.

(2) **Cost saving**

All the manufacturing SMEs' employers interviewed highlighted that they were motivated to adopt GSCM practices due to cost savings. Since these GSCM practices help reduce cost through various kinds of activities, SMEs are much motivated to adopt such practices within their organisations. This was highlighted by one of the participants as below:

> I try to purchase from the suppliers closer to me due to which lot of transportation cost is saved and time is saved. I even save warehousing cost by making the product only when there is order. This also saves my office space (P2)

P2 performs a lot of green activities that help his firm to save money. According to him, he saved transportation costs and warehousing costs by performing green activities. Similarly, other participants also highlighted that they saved their company cost by adopting GSCM practices. For instance, some conformed that they reduced electricity cost by keeping their machinery switched off when not in use, and some stressed on using fewer papers for printing by switching to the latest technologies for communication.

"**Cost saving**" forms the second most important motivational factor that is encouraging SME employers to adopt GSCM practices based on the responses from the employers. This factor was also discussed by Drohomeretski et al. (2014) in their research where they recognised reducing cost as the major motivational factor that encouraged the adoption of GSCM practices in Brazilian automobile industries. Similar results were obtained by Dhull and Narwal (2016) in their review-based paper which recongnised "desire to reduce cost" as one of the important internal drivers of GSCM in India. Furthermore, this factor was also recognised by Gandhi et al. (2018) who were researching on Indian manufacturing SMEs. Hence, finding of this study matches what is published in existing literature.

(3) **Self-realisation**

Self-realisation is another factor that motivates the Indian manufacturing SME employers. Most of the employers (60%) highlighted that they have self-realised the environmental issues and taken initiatives to go green. This is evident from the comment made by one of the participants:

> Truly speaking these green practices are not so common in India and so whatever we do is out of our own understanding and willingness not due to any other thing. (P5)

P5 stated that he practicsed green based on his own understanding and willingness. He understood the importance of these practices but he could only perform those activities within his ability as there is no proper support from the government. Hence, according to P5 self-realisation helped him to practice green.

The factor "**Self-realisation**" was discussed in the literature by Ariffin et al. (2015) in the context of Malaysian SMEs and in the work by Dhull and Narwal (2016) in Indian context. However, this motivational factor was not considered by researchers researching on the Indian manufacturing SME context. Since self-realisation is found to be one of the important reasons for SME employers to adopt GSCM practices in their organisations, this finding can be considered unique to the body of literature pertaining to manufacturing SMEs.

(4) **Collective peer effort**

Four of the SME employers interviewed stressed on getting encouragement from the people around them. According to them, when you are a part of the society, success can be achieved with collective efforts. Hence, peer efforts encourage the participants to practise green around them. This was apparent from the statement by P4:

> …..like I said earlier, we don't get much Government support in our organisation. The sewage condition in our lane is not good so it was our association and the people around us who gather together to clean the lane and make it a better place to work. So we joined hands to work on this, I mean the sewage work.. (P4)

According to P4, the surroundings of their factory were not good. The sewage conditions were very poor so all the employers in the market joined hands to make their factory surrounding a better place as they were suffering from huge loss of business since their customers were not happy to enter the market place. Thus, he stated that collective efforts help to encourage them to practise green and to make their surrounding a better place to work.

Thus, the results reveal that "**collective peer effort**" is one of the main factors that have motivated some employers to adopt GSCM practices. A study by Hu and Hsu (2010a, b) considered "manpower involvement" as one of the 20 critical factors for implementing GSCM practices in Taiwanese electrical and electronic industries. Similarly, a review-based research by Jain and Sharma (2014) listed "accumulated experience" as one of the influential factors. Thus, collective peer effort is a finding of this study that is in alignment with the existing studies in the wider context of the globe. However, none of the existing studies focusing on Indian manufacturing SME sector have considered this factor. Hence, this factor can be considered as a unique contribution of this study.

The next section discusses the external factors identified by this study.

7.2 External Factors

(1) Government Pressure

All the employers interviewed said that they were driven by government rules and regulations, or the different programmes initiated by the government. The government in India is making huge effort to increase the awareness among the people of India. They are enforcing strict company laws and regulations to reduce pollution and make the environment of India better to live in. Hence, some employers considered that government rules were forcibly imposed on them while some considered such rules as motivation for them to practise green. This was evident from the statement provided by the participant below:

> There is pressure from the Government where we are working. We try to use fewer polys and less polythene which are harmful for the environment. (P2)

According to P2, due to government rules, they have reduced the use of polythene bags in order to save the environment. Hence, government pressures are considered useful in motivating Indian SMEs to practise green. Filtering GSCM literature on Indian manufacturing SMEs revealed that "**government pressure**" was also discussed by Kumar et al. (2015) as "government rules and regulations" in their research on identifying drivers of micro, small and medium enterprises. Hence, this confirms that there is existing literature on Indian manufacturing SMEs context.

(2) **Government Influence**

The government of India is the most influential factor in motivating business organisations to be environment friendly. They have started creating a lot of awareness among the people by starting different programmes and policies. One such programme, named Swatch Bharat Abhiyan (Keep India Clean Mission), focused on making the country clean and a better place to live. The government started this programme by involving all the popular celebrities who were looked upon and followed by the public. Due to their involvement, general public started following their footsteps and made this programme a big success from the very beginning. Most of the employers (80%) interviewed discussed that they get motivated by this factor. This is apparent from the comment made by one of the participant as follows:

> Our current Prime Minister is creating awareness about cleanliness, about keeping our country clean by advertising and all other means. (P4)

P4 said that the awareness has been created by the government through advertising and publicising about the adoption of green issues. He appreciated the work done by the government in creating awareness. Hence, the work undertaken by the government is noticeable as most of the participants interviewed confirmed that they have adopted GSCM practices due to the awareness created by the government programmes.

Government influence was discussed in the literature by many scholars (Dhull and Narwal 2016; Walker et al. 2008; Zhu et al. 2007). However, this factor, in the studies particularly focusing on Indian manufacturing SMEs, was not identified. Thus, this factor is considered an important finding by this study.

(3) **Customer Demands**

Customer satisfaction is always a priority for all businesses. Three employers interviewed cited of getting motivated by the demands created by their customers. This was put in words by one of the participants as:

> For me the major driving factor is the customers as I do as per the demand of my customers, we have pressure from our foreign clients for eco-product. (P5)

For P5 satisfying customer demands is one of the most important driving factors. According to him, he is doing business only to serve his customers and he will do whatever is required to fulfil his customers' demands. To be noted here P5 was talking about his foreign clients, not Indian customers. Hence, pressure from customers to produce green products is an important motivational factor for the SMEs. Although, "**Customer demand**" was also identified in the study by Kathiresan and Ragunathan (2017) that confirms the literature on Indian SMEs, it is essential to note that the customer demand discussed here was created by foreign customers rather than Indian customers. This indicates that the demand from foreign customers for eco-products may also create pressure or motivation for Indian SMEs, which is a new finding from this study.

(4) **Pressure from Other Sources**

Two employers conveyed of being motivated to adopt GSCM practices by the awareness created by other sources in India, for instance, NGOs, actors, sportspersons, colleagues and friends. This factor was also recognised by Mathiyazhagan et al. (2015) in their study on Indian manufacturing SMEs as "pressure from non-governmental organisations for eco-friendly products". Hence, this factor is in alignment with the existing studies on Indian SMEs.

However, in this study it was evident from the statement given by one of the participants:

> …we see actors and actresses advertising on televisions and media everyday. So these things are also motivating…. (P4)

P4 felt that he was motivated to practise green in his factory by the different advertisements on the television. Various actors were involved in creating awareness on green issues in India. Hence, these advertisements help in motivating many people in India.Similarly, Indian manufacturing SME employers are also motivated by the awareness created by such sources.

(5) **Pressure to Sustain in the Market**

The pressure of sustaining in the market is an important motivation for the Indian manufacturing SME employers (100%). Since these SMEs work on a small scale, survival becomes an important factor and they do whatever is required to sustain in the market. This was placed in words by one of the participants as:

> since I run my business on my own I have the pressure of sustaining in the market so I do what market says, I produce accordingly. (P3)

P3 stated that the pressure of sustaining in the market keeps him motivated to perform well. For surviving, he also does what the market demands, including the demand for greener environment, and hence produces accordingly. Thus, the pressure of sustaining is an important motivational factor for SME employers.

"**Pressure to sustain in the market**" was discussed in previous literature by Ariffin et al. (2015) and Dhull and Narwal (2016); whereas, this factor was not highlighted in the studies focusing particularly on Indian manufacturing SMEs. Hence, this factor can be considered as a unique factor identified by this study in relation to Indian manufacturing SMEs.

8 Results and Contributions

This study identified that employers or owners of SME businesses are the main decision-makers for adopting GSCM practices in their organisations. Based on the information gained from them. This study explored a number of new internal factors—***"self-realisation" and "collective peer effort";*** and some new external factors—***"pressure to sustain in the market" and "government influence"*** that were not considered in exisiting literature in GSCM among the Indian manufacturing SMEs, and these should be unique contributions to the body of literature. Thus, the researcher believes that the main objective of this research, i.e. exploring the key factors motivating Indian manufacturing SME employers to adopt GSCM practices within their organisations, is achieved.

Whilst some the findings of this study are in alignment with the existing literature it was found that a few important drivers discussed in existing literature were not identified by this study. For instance, drivers related to suppliers, eco-designing related issues, etc. Kumar et al. (2015) and Kathiresan and Ragunathan (2017) also examined these motivational factors and concluded that these factors are not considered by SMEs yet. Hence, this confirms that the results of this study are in alignment with the studies conducted using quantitative methods. Thus, it can be concluded that the Indian manufacturing SME employers have not been motivated by factors such as suppliers, eco-design, etc., and these factors may not be the most important for SMEs in India.

Referring to the research gaps, the studies conducted by Malviya and Kant (2015) and Kumar and Garg (2017) called for in-depth research in similar areas that would help explore more robust results. Since there is only limited studies in GSCM using qualitative methods this study should have contributed to the research domain.

Vijayvargy et al. (2017) also recommended to have a detailed study on a particular industry in Indian context to have an elaborated understanding of GSCM adoption behaviours of different-sized organisations. They focused on general industries of emerging economy and discussed different aspects (adoption, practices and role) of GSCM. However, this research has focused on a detailed review of one aspect, i.e. motivational factors that encourage Indian manufacturing SMEs to adopt and implement GSCM practices.

What is reasonably unique and interesting about this study is the way that has captured the interpretive accounts of SME employers and presented their experiences and expressions that provided a detailed understanding of the topic under the research.

Given the fact that no existing studies on GSCM have evaluated this aspect of the manufacturing sector in such detail, the results from this study are both novel and significant. The researcher to the best of her knowledge can confirm that the findings provided in the chapter represent the research participants' viewpoints.

This study reviewed the existing literature on drivers/motivations of GSCM among Indian SMEs which formed the basis of the current study. Based on the literature

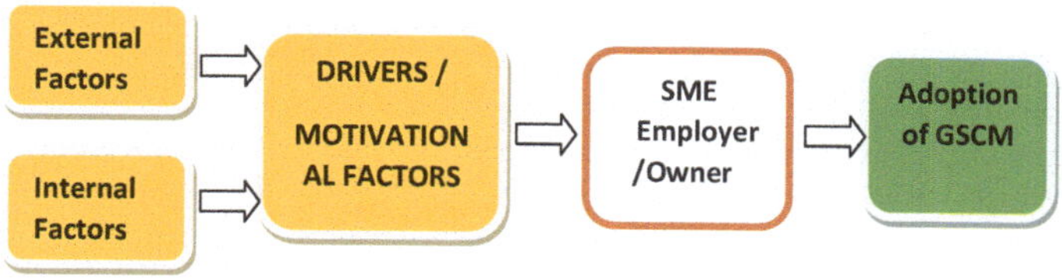

Fig. 3 Final framework for adopting GSCM practices in SMEs

review and the emprical study of this research, the researcher has developed a framework for GSCM as illustrated in Fig. 3. The framework should be able to help Indian SME practitioners and decision-makers for mapping the processes related to GSCM and understanding the motivatioonal factors that can initiate the adoption of GSCM practices within their organisations.

9 Conclusions

It is encouraging to find that Indian SMEs have started thinking about the environment and that they feel content when they perform something to protect the environment. Hence, some SMEs have started giving priority to environmental practices within their businesses. It can be expected that this trend will soon get popular among all the SMEs in near future.

In conclusion, although this study has identified some new factors motivating SME employers to adopt GSCM practices, the main factors that can motivate them better in implementing such practices are financial benefits and cost saving which are also discussed by Mathiyazhagan et al. (2018).

The main limitation of this study is that it has only focused on Indian manufacturing SME sector and has used small samples. However, this study has come up with a number of motivational factors that should be able to encourage Indian SMEs to adopt GSCM. Further studies can use similar methods to find new factors in different industrial contexts with bigger simples when possible. Nevertheless, this study will should help the policymakers and the decision-makers to better understand the GSCM implementation process and help them to increase the adoption of GSCM in the firms. Furthermore, the GSCM framework developed in this study is aimed should be able to guide them to take appropriate measures to help organisations to successfully implement GSCM practices.

References

Abdullah, M. I., Sarfraz, M., Qun, W., & Javaid, N. (2018). Drivers of green supply chain management. *Scientific Journal of Logistics, 14*(4), 437–447.

Agi, M. A. N., & Nishant, R. (2017). Understanding influential factors on implementing green supply chain management practices: An interpretive structural modelling analysis. *Journal of Environmental Management, 188,* 351–363. (Elsevier Ltd.).

Ahi, P., & Searcy, C. (2013). A comparative literature analysis of definitions for green and sustainable supply chain management. *Journal of Cleaner Production, 52,* 329–341.

Ali, A., Bentley, Y., Cao, G., & Habib, F. (2017). Green supply chain management—Food for thought? *International Journal of Logistics: Research and Applications, 0*(0), 1–17.

Ariffin, R., Ghazilla, R., Sakundarini, N., Hanim, S., Rashid, A., Ayub, S., Olugu, E. U., et al. (2015). Drivers and barriers analysis for green manufacturing practices in Malaysian SMEs: A preliminary findings. In *12th Global Conference on Sustainable Manufacturing* (Vol. 26, pp. 658–663).

Balasubramanian, S., & Shukla, V. (2017). Green supply chain management: An empirical investigation on the construction sector. *Supply Chain Management: An International Journal, 22*(1), 58–81.

Batra, S., & Chanana, B. (2015). Greening the supply chain: A way towards a new way. *International Journal of Bussiness Quantitative Economics and Applied Management Research, 2*(3), 23–31.

Brassaw, B. (2017). *The state of recycling in India: Slow improvements.* https://earth911.com/business-policy/india-recycling/.

Bryman, A., & Bell, E. (2011). *Business research methods* (Vol. 3), Oxford University Press. https://doi.org/10.4135/9780857028044.

Côté, R. P., Lopez, J., Marche, S., Perron, G. M., & Wright, R. (2008). Influences, practices and opportunities for environmental supply chain management in Nova Scotia SMEs. *Journal of Cleaner Production, 16*(15), 1561–1570.

Dashore, K., & Sohani, N. (2013). Green supply chain management—Barriers & drivers: A review. *International Journal of Engineering Research & Technology, 2*(4), 2021–2030.

de Oliveira, U. R., Espindola, L. S., da Silva, I. R., da Silva, I. N., & Rocha, H. M. (2018). A systematic literature review on green supply chain management: Research implications and future perspectives. *Journal of Cleaner Production, 187,* 537–561.

Dhull, S., & Narwal, M. S. (2016). Drivers and barriers in green supply chain management adaptation: A state-of-art review. *Uncertain Supply Chain Management, 4,* 61–76.

Diabat, A., & Govindan, K. (2011). An analysis of the drivers affecting the implementation of green supply chain management. *Resources, Conservation and Recycling, Elsevier, 55*(6), 659–667.

Drohomeretski, E., Costa, S. G., & Lima, E. P. (2014). Green supply chain management drivers, barriers and practices within the Brazilian automotive industry. *Journal of Manufacturing Technology Management, 25*(8), 1105–1134.

Environmental Performance Index, 2018. (2018). Environmental Performance Index.

Epoh, L. R., & Mafini, C. (2018). Green supply chain management in small and medium enterprises: Further empirical thoughts from South Africa. *Journal of Transport and Supply Chain Management, 12,* 1–12.

Fang, C., & Zhang, J. (2018). Performance of green supply chain management: A systematic review and meta analysis. *Journal of Cleaner Production, 183,* 1064–1081. (Elsevier Ltd.).

Gandhi, S., Mangla, S. K., Kumar, P., & Kumar, D. (2015). Evaluating factors in implementation of successful green supply chain management using DEMATEL: A case study. *International Strategic Management Review, 3,* https://doi.org/10.1016/j.ism.2015.05.001.

Gandhi, N. S., Thanki, S. J., & Thakkar, J. J. (2018). Ranking of drivers for integrated lean-green manufacturing for indian manufacturing SMEs", *Journal of Cleaner Production, 171,* 675–689 (Elsevier Ltd.).

Gardas, B., & Narkhede, B. (2013). Exploring the green supply chain management: A technical review. *International Journal of Application or Innovation in Engineering & Management, 2*(5), 441–450.

Geng, R., Mansouri, S. A., Aktas, E., & Yen, D. A. (2017). The role of Guanxi in green supply chain management in Asia's emerging economies: A conceptual framework. *Industrial Marketing Management, 63,* 1–17. (The Authors).

Govindan, K., Diabat, A., & Madan Shankar, K. (2015). Analyzing the drivers of green manufacturing with fuzzy approach. *Journal of Cleaner Production, 96,* 182–193.

Govindan, K., Muduli, K., Devika, K., & Barve, A. (2016). Investigation of the influential strength of factors on adoption of green supply chain management practices: An Indian mining scenario. *Resources, Conservation and Recycling, 107*, 185–194 (Elsevier B.V.,).

Gunasekaran, A., Subramanian, N., & Rahman, S. (2015). Green supply chain collaboration and incentives: Current trends and future directions. *Transportation Research Part E: Logistics and Transportation Review, 74,* 1–10.

Hijaz, S., Al-Hujran, O., Al-Debei, M. M., & Abu-Khajil, N. (2015). Green supply chain management and SMEs: A qualitative study. *International Journal of Business Information Systems, 18*(2), 198–220.

Homburg, C., Klarmann, M., Reimann, M., & Schilke, O. (2012). What drives key informant accuracy? *Journal of Marketing Research, 49,* 594–608.

Hu, A. H., & Hsu, C.-W. (2010a). Critical factors for implementing green supply chain management practice an empirical study of electrical and electronics industries in Taiwan. *Management Research Review, 33*(6), 586–608.

Hu, A. H., & Hsu, C.-W. (2010b). Critical factors for implementing green supply chain management practice. *Management of Environmental Quality: An International Journal, 33*(6), 586–608.

Jaggernath, R. (2015). Green supply chain management. *World Journal of Entrepreneurship, Management and Sustainable Development, 11*(1), 37–47.

Jain, V. K., & Sharma, S. (2014). Drivers affecting the green supply chain management adaptation: A review. *IUP Journal of Operations Management, 13*(1), 54–63.

Jayaram, J., & Avittathur, B. (2015). Green supply chains: A perspective from an emerging economy. *International Journal of Production Economics, 164,* 234–244. (Elsevier).

Jia, F., Zuluaga-cardona, L., Bailey, A., & Rueda, X. (2018). Sustainable supply chain management in developing countries: An analysis of the literature. *Journal of Cleaner Production, 189,* 263–278. (Elsevier Ltd.).

Joshi, R., & Ahmed, S. (2016). Status and challenges of municipal solid waste management in India: A review. *Cogent Environmental Science, Cogent, 2*(1), 1–18.

Kathiresan, G., & Ragunathan, S. (2017). Impact of drivers for the implementation of green concept in small and medium sized (Smes) leather industries of northern Tamilnadu. *Rasayan Journal of Chemistry, 10*(3), 723–728.

Khushbu, V., & Shah, H. G. (2014). Barriers of green supply chain management: A literature review. *International Journal of Engineering Research & Technology, 3*(5), 1657–1665.

Kumar, D., & Garg, C. P. (2017). Evaluating sustainable supply chain indicators using fuzzy AHP: Case of Indian automotive industry. *Benchmarking: An International Journal, 24*(6), 1742–1766.

Kumar, N., Stern, L. W., & Anderson, J. C. (1993). Conducting interorganizational research using key informants. *Academy of Management Journal, 36,* 1633–1651.

Kumar, M., Kumar, V., & Ram, P. (2015). Factors affecting Green SCM implementation in MSME's. *International Journal of Innovative Science, Engineering & Technology, 2*(12), 631–645.

Laosirihongthong, T., Adebanjo, D., & Tan, K. C. (2013a). Green supply chain management practices and performance. *Industrial Management & Data Systems, 113*(8), 1088–1109.

Laosirihongthong, T., Prajogo, D. I., & Adebanjo, D. (2013b). The relationships between firm's strategy, resources and innovation performance: Resources-based view perspective. *Production Planning & Control*, 1–16.

Lee, S. M., Rha, J. S., Choi, D., & Noh, Y. (2013). Pressures affecting green supply chain performance. *Management Decision, 51*(8), 1753–1768.

Luthra, S., Garg, D., & Haleem, A. (2015). An analysis of interactions among critical success factors to implement green supply chain management towards sustainability: An Indian perspective. *Resources Policy, Elsevier, 46,* 37–50.

Luthra, S. B, Garg, D., & Haleem, A. (2016a). The impacts of critical success factors for implementing green supply chain management towards sustainability: An empirical investigation of Indian automobile industry. *Journal of Cleaner Production, 121*, 142–158.

Luthra, S., Garg, D., & Haleem, A. (2016b). The impacts of critical success factors for implementing green supply chain management towards sustainability: An empirical investigation of Indian automobile industry. *Journal of Cleaner Production, 121,* 142–158. (Elsevier Ltd.).

Malviya, R. K., & Kant, R. (2015). Green supply chain management (GSCM): A structured literature review and research implications. *Benchmarking: An International Journal, 22*(7), 1360–1394.

Mangla, S. K., Govindan, K., & Luthra, S. (2017). Prioritizing the barriers to achieve sustainable consumption and production trends in supply chains using fuzzy analytical hierarchy process. *Journal of Cleaner Production, 151,* 509–525. (Elsevier Ltd.).

Mathiyazhagan, K., & Haq, A. N. (2013). Analysis of the influential pressures for green supply chain management adoption-an Indian perspective using interpretive structural modeling. *International Journal of Advanced Manufacturing Technology, 68*(1–4), 817–833.

Mathiyazhagan, K., Govindan, K., NoorulHaq, A., & Geng, Y. (2013). An ISM approach for the barrier analysis in implementing green supply chain management. *Journal of Cleaner Production, 47,* 283–297. (Elsevier Ltd.).

Mathiyazhagan, K., Diabat, A., Al-Refaie, A., & Xu, L. (2015). Application of analytical hierarchy process to evaluate pressures to implement green supply chain management. *Journal of Cleaner Production, Elsevier, 107,* 229–236.

Mathiyazhagan, K., Datta, U., Bhadauria, R., Singla, A., & Krishnamoorthi, S. (2018). Identification and prioritization of motivational factors for the green supply chain management adoption: Case from Indian construction industries. *Opsearch, Springer India, 55,* 202–219.

Matthew, J. (2017). Supply chain management (SCM) utilization based on SCM drivers in manufacturing industries. *Jurnal Pengurusan, 50,* 1–16.

Patton, M. (2002). *Qualitative research and evaluative methods* (3rd ed.). London: Sage Pub. Ltd.

Rajeev, A., Pati, R. K., Padhi, S. S., & Govindan, K. (2017). Evolution of sustainability in supply chain management: A literature review. *Journal of Cleaner Production, 162,* 299–314. (Elsevier Ltd.).

Rengarajan, R., Gnanaprakasam, C., & Selvan, K. G. (2015). A study on the adoption of green supply chain initiatives among small and medium enterprises and its benefits for the environment sustainability. *International Journal of Transformation in Business Management, 5*(II), 1–11.

Rozar, N. M., Mahmood, W. H. W., Ibrahim, A., & Razik, M. A. (2015). A study of success factors in green supply chain management in manufacturing industries in Malaysia. *Journal of Economics, Business and Management, 3*(2), 287–291.

Sambrani, V. N., & Pol, N. (2016). Green supply chain management: A literature review. *Journal of Supply Chain Management, 13*(4), 7–16.

Sarkis, J. (2003). A strategic decision framework for green supply chain management. *Journal of Cleaner Production, 11*(4), 397–409.

Sarkis, J. (2012). A boundaries and flows perspective of green supply chain management. *Supply Chain Management: An International Journal, 17*(2), 202–216.

Sarkis, J., Zhu, Q., & Lai, K. H. (2011). An organizational theoretic review of green supply chain management literature. *International Journal of Production Economics, 130*(1), 1–15.

Seman, N. A. A., Zakuan, N., Jusoh, M. A., & Arif, M. S. M. (2012). Green supply chain management: A review and research direction. *International Journal of Managing Value and Supply Chains, 3*(1), 1–18.

Seth, D., Shrivastava, R. L., & Shrivastava, Sanjeev. (2016). An empirical investigation of critical success factors and performance measures for green manufacturing in cement industry. *Journal of Manufacturing Technology Management, 27*(8), 1076–1101.

Seuring, S., & Muller, M. (2008). From a literature review to a conceptual framework for sustainable supply chain management. *Journal of Cleaner Production, 16*(15), 1699–1710.

Sharma, V. K., Chandana, P., & Bhardwaj, A. (2015). Critical factors analysis and its ranking for implementation of GSCM in Indian dairy industry. *Journal of Manufacturing Technology Management, 26*(6), 911–922.

Singh, P., & Kumar, V. (2015). Quantitative analysis of drivers affecting green supply chain management in Rajasthan SMEs. *International Journal of Science and Research (IJSR), 4*(4), 2384–2390.

Singh, M., Brueckner, M., & Padhy, P. K. (2015a). Environmental management system ISO 14001: Effective waste minimisation in small and medium enterprises in India. *Journal of Cleaner Production, 102,* 285–301.

Singh, N., Jain, S., & Sharma, P. (2015b). Motivations for implementing environmental management practices in Indian industries. *Ecological Economics, 109,* 1–8 (Elsevier B.V.,).

Small and Medium Business Development Chambers of India, 2015. (2018). About MSMEs in India. Retrieved July 24, 2016, from http://www.smechamberofindia.com/About_MSMEs.aspx.

Soda, S., Sachdeva, A., & Garg, R. K. (2015). GSCM: Practices, trends and prospects in Indian context. *Journal of Manufacturing Technology Management, 26*(6), 889–910.

Soda, S., Sachdeva, A., & Garg, R. K. (2016). Implementation of green supply chain management in India: Bottlenecks and remedies. *The Electricity Journal, 29*(4), 43–50 (Elsevier Inc.,).

Srivastava, S. K. (2007). Green supply-chain management: A state-of-the-art literature review. *International Journal of Management Reviews, 9*(1), 53–80.

Tachizawa, E. M., Gimenez, C., & Sierra, V. (2015). Green supply chain management approaches: Drivers and performance implications. *International Journal of Operations & Production Management, 35*(11), 1546–1566.

TERI. (2013). *TERI Environmental Survey, 1.* https://doi.org/10.1017/CBO9781107415324.004.

The Economic Times of India. (2015). Indian SMEs getting green guidance from Europe. *The Economic Times.* Retrieved July 21, 2016, from http://economictimes.indiatimes.com/small-biz/sme-sector/indian-smes-getting-green-guidance-from-europe/articleshow/46581204.cms.

Uma, D. P. (2013). Role of Smes in economic development of India. *Asia Pacific Journal of Marketing & Management Review, ISSN, 2*(6), 2319–2836.

Verma, D., Dixit, R. V & Singh, K. (2018). Green supply chain management : A necessity for sustainable development. *The IUP Journal of Supply Chain Management, XV*(1), 40–58.

Vijayvargy, L., Thakkar, J., & Agarwal, G. (2017). Green supply chain management practices and performance: The role of firm-size for emerging economies. *Journal of Manufacturing Technology Management, 28*(3), 299–323.

Walker, H., Di Sisto, L., & McBain, D. (2008). Drivers and barriers to environmental supply chain management practices: Lessons from the public and private sectors. *Journal of Purchasing and Supply Management, 14*(1), 69–85.

Worku, H. A., & Virdi, S. S. (2019). Green supply chain management practices in Ethiopia' s leather and leather product industry. *IUP Journal of Supply Chain Management, XVI*(2), 25–43.

Zailani, S., Jeyaraman, K., Vengadasan, G., & Premkumar, R. (2012). Sustainable supply chain management (SSCM) in Malaysia: A survey. *International Journal of Production Economics.*

Zhu, Q., Sarkis, J., & Lai, K. (2007). Green supply chain management: Pressures, practices and performance within the Chinese automobile industry. *Journal of Cleaner Production, 15*(11), 1041–1052.

Chapter 9
The Influence of Ethical Practice on Sustainable Supplier Selection in the Furniture Industry

Arvind Upadhyay, Waleed Alhuzaimi, Vinaya Shukla, and Shaheda Nur

Abstract This study was carried out with an objective to investigate the sustainable supplier selection and ethics influence within the furniture industry. Literature has shown that corporates have increasingly adopted the ethical standards provided they also achieve economic sustainability. The current study carried out in the furniture industry aimed to achieve four objectives: (1) to appreciate the role of ethics in selection of a sustainable supplier; (2) to evaluate and assess different methods used in the selection of suppliers; (3) to appreciate the role of ethical practices in supply chain management; and (4) to explore the important ethical practices within the furniture industry. In the furniture industry, organizations are keen to involve top management in pushing for ethical practices that enhance sustainability within supply chain management. This includes the presence of environmental policies for sustainability and carrying out corporate social responsibility activities in order to boost it. Despite the cost still being a major factor for corporates, understanding the importance of ethical practice in such an industry is becoming appreciated as more rules and standards become standard in these corporate sectors. Thirty-one participants were interviewed in four groups. The major conclusions supported the appreciation of the role of ethics in influencing sustainable supplier selection and cite the significance of adopting ethical practices in the furniture industry. The study found out that the process of selecting suppliers is flexible, and that supplier selection is integrated with ethical practices. It boosts the organization's image, reputation and competitiveness.

A. Upadhyay (✉)
Brighton Business School, University of Brighton, M134, Mithras House, Brighton, UK
e-mail: a.upadhyay@brighton.ac.uk

W. Alhuzaimi
Naval Commander, Kingdom of Saudi Arabia

V. Shukla
Business School, Middlesex University London, London NW4 4BT, UK

S. Nur
University of the West of Scotland London Campus, London SE1 6NP, UK

U. Ramanathan and R. Ramanathan (eds.), *Sustainable Supply Chains: Strategies, Issues, and Models*, https://doi.org/10.1007/978-3-030-48876-5_9

Keywords Supplier selection · Sustainability · Sustainable supplier · Ethics influence in furniture industry

1 Introduction

Corporates have progressively allied their functions with the ethical principles, and the purpose of this intensification is to attain economic sustainability (Goebel et al. 2012; Kleindorfer et al. 2005). For most of the organizations, it becomes an obligation to maintain their existence because the traditional concept of such business has now changed. The organizations have comprehended a modern concept for their business, which considers the overall development of the business and the society. In traditional business, there was no consideration of the ethical issues for business, employees, environment and society. In modern business, such considerations become their prime concern. Thus, sustainability and ethical practices have become increasingly significant for many organizations (Marc and Hossam 2012), and organizations are now trying to practice codes of conduct. These include safety and health of employees, legal and regulatory policies, restricting under-age workers, environmental protection, ethical standards, right and respect for employees. These are practised not only within the organization but also within partner organizations and stakeholders (Koplin et al. 2007). In selecting a sustainable supplier, a company can evaluate the suppliers through their environment costs, green design, pollution control, recycling, resource consumption, etc. The furniture industry is always more concerned with ethical and environmental issues. They practice sustainable manufacturing, carbon management, eco-design, ethical timber, etc. as their ethical responsibility. Though the furniture industry is now trying to minimize the environmental costs, it should be more concerned about the environment and sustainability as a whole.

1.1 Background of the Sustainability in the Furniture Industry

The furniture industry encompasses all the activities and companies involved in manufacturing, designing, distributing and selling furniture and household equipment. The furniture industry has rapidly grown in a few decades all over the world. Ethical consideration and sustainability play crucial roles in the furniture industry. They restrict unethical practices in procurement, manufacturing and distributing by the suppliers. Sustainability ensures a suitable and friendly work environment for the workers and improvement in the business process. It restricts the suppliers from any activities that may have an adverse effect on business reputation. Thus, the business can get standard raw materials for manufacturing. Sustainability encourages the furniture industry to recycle and incorporate green design, which minimizes wastage and secures proper utilization of resources. Now furniture businesses design the furniture in such a way that it is alternative to the traditional business and process. The modern furniture business reduces the environmental costs, increases employment ethics, secures social development and increases the profit for the business.

Suppliers are an inseparable part of any industry. Without them, an industry cannot run their businesses. To maintain the sustainability policy in any business, there is a need to build a supplier group. If the suppliers are unethical and ignore the environmental issues, it becomes a curse for the industry. Ethical and legal restrictions over the suppliers help in building a sustainable industry. The furniture industry has rapidly grown in a few decades all over the world. If the industry ignores the ethical and environmental issues, there is an incremental negative environmental impact on society. In almost every country, this industry has followed the legal rules and regulations regarding its business operations. Procurement processes have also been monitored and controlled by the business organizations. Such steps taken by the furniture businesses are blessings for the industry, economy, society and environment.

Though ethical consideration and sustainability have a great influence on the environment and society as well, there is still a lack of insight on the significance and role of ethics in the selection of sustainable suppliers. In addition, no previous studies have clearly discussed the appraisal methods used in the selection of suppliers and the importance of ethical practices in the furniture industry. In this research, the researcher has tried to examine the impact of different ethical practices on sustainable supplier selection in the furniture industry. This research will also help to address many important issues that influence the selection of sustainable suppliers, identification of the methods used to select sustainable suppliers, assessment of the roles of sustainability and ethical practices in supplier selection, and finding some of the ethics used in practice within the furniture industry. In order to do this, the remainder of the paper has been divided as follows: Sect. 2 presents the research objectives; Sect. 3 describes the related literature regarding the assessment of the roles of sustainability and ethical practices in supplier selection. Section 4 covers the research methodology, Sect. 5 discusses the research findings and analysis and research limitations and Sect. 6 concludes and highlights our research findings and makes some recommendations for future research.

2 Research Objectives

The key aim of this study is to define how sustainable suppliers are chosen in the furniture industry and the role performed by ethical exercises in this procedure. However, the other objectives of the research are:

1. To evaluate the impacts of ethics in the selection of sustainable suppliers.
2. To assess the different methods used in the selection of the suppliers.
3. To examine the function of sustainability and ethical practices in supply chain management.

4. To study the significance of ethical practices in the furniture industry.

3 Literature Review

3.1 Sustainable Supplier Selection

To select the right supplier is very crucial for any company and organizations need to follow some specific standard procedures or protocols whilst selecting the suppliers (Goebel et al. 2012). Moreover, it is very important for any organizations to affiliate their jobs with business ethics and sustainability standards (Goebel et al. 2012; Kovacs 2009). Without practising the particular standard method, the company would not be able to select the right suppliers for them (Meera 2014; Kovacs 2009). Previously, the purchasing managers used certain criteria such as quality, cost and on-time delivery in appraising the appropriate supplier (Schiele 2007). However, these conventional features will not explicate the ethical exercises of such suppliers, and this needs to be examined further by a well-defined code of conduct and policies, like—did they break any set principles within the corporation? (Blowfield 2000).

3.2 Sustainable Supplier Selection Methods

A direct correlation has been found between the selection of sustainable suppliers and the success of the supply chain in several past papers. Three main decision-making methods for sustainable supplier selection have been recommended since the 1980 s to solve the issues regarding the selection of a supplier. These methods are Analytical Network Process (ANP), Analytical Hierarchy Process (AHP) and Technique for Order Preference by Similarity to Ideal Solutions (TOPSIS). Saaty (1980) pointed out that AHP is considered a useful method used to solve the issues regarding the selection of suppliers. The AHP method helps to identify the weights of every criterion and their sub-criterion through which more precision can be ensured in ranking the suppliers of a company.

ANP is the next supplier selection method that is used as an additional approach to the AHP method. Multi-Criteria Decision-Making (MCDM) is affected a lot by the ANP method. This method helps to categorize the record of probable risks in accordance with the relative significance of a company. Saaty (2005) maintained that the ANP network framework was constructed on the judgment of some experts that demands an intrinsic perception of the decision problem. Based on this, specific knowledge and skills were required to establish the scope of the problems; a fundamental building block to effect expected outcomes. Hwang and Yoon constructed the TOPSIS method in 1981 that is considered as the third method. Sen and Yang (1998) pointed out that this method is used to gain scores of ranking and position for the different options in order of their appropriateness. According to, it is considered

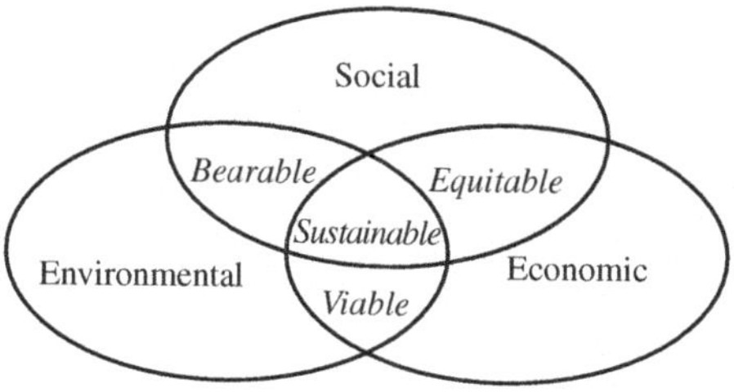

Fig. 1 Elements of sustainability *Source* Molamohamadi (2013)

a standard approach to examine different options at a time based on cost and benefit criteria. The practitioners bringing the best outcomes prefer this method.

These approaches are very important in the supplier selection procedures as explained above. The consideration of sustainability must be incorporated into every method in order to select suitable suppliers with environmental and social sustainability aspects in mind.

3.3 Sustainability and Ethical Practices

The compulsion to integrate social and environmental aspects in supply chain management has increased. These two issues have been considered as principles in choosing a supplier. The components believed to best describe the ethical practice in corporate businesses have changed significantly over time. Previously, compliance with the conventional guidelines was perceived as an ethical practice but this concept has altered due to diverse ecological changes and apprehensions in business. For this research paper, the idea of Svensson (2009) has been implemented which described that sustainability and ethical practices are closely related (Svensson 2009) to each other, and sustainability not only maximizes the profit for business but also ensures a green environment by managing financial, environmental and social risk. This notion has been accredited to the principle of the triple bottom line, which incorporates the three elements of sustainability. These include social, economy and environment (Molamohamadi 2013) (Fig. 1).

It is the duty of the procurement department of an organization to make effective decisions about the process of choosing a sustainable and reliable supplier. This process allows an organization to reduce manufacturing costs, enhance the quality of production and increase the level of customer satisfaction.

3.4 Ethics in the Furniture Industry

The furniture industry is one of the most profitable, and a broad reformed code of ethics for this sector is badly needed in the twenty-first century. Currently, green supply chain management has emerged and subsidized to environmental sustainability. Green Supply Chain Management is outlined as an exercise of managing raw material elements from the supplier and manufacturer to the consumers by safeguarding that this practice does not deplete the atmosphere (Hu 2010). This involves devotion, training and responsiveness from both the top management and suppliers to grasp, realize and approve a firm code of ethics to certify this can be attained.

Governments executed rules and regulation through diverse associations maintaining the ecosystem by supporting the latest law act that tightens the contamination and demolition of natural resources. Suppliers understand the impact of not conforming to such laws, and therefore they have devised a certain code of ethics that ought to be respected by all shareholders in the business. In their study, Baird and Rowen (2010) described that application of ethics has made the furniture industry very reliable for the usage of raw materials and coupled with directorial prevention of the dumping of by-products has nurtured the sustainability effort.

3.5 Research Hypothesis

The researcher has identified the following hypotheses in this research:

> ***H1a***. *Sustainable supplier selection is mostly related to the compliance level along with the ethical codes.*
> ***H1b***. *The sustainable supplier selection is persistent with the requirements of the sustainable environment authority to which the top management level comply.*
> ***H2a***. *The sustainable supplier selection practices are fundamentally connected to the level of incentive and their adherent to principles and practices.*
> ***H2b***. *Incentives escalate the level of compliance with business codes and sustainable environment exercises.*

4 Research Methodology

4.1 Research Design

A qualitative research approach has been chosen for the study. A qualitative method concentrates on socially constructed reality where it believes that the experience, knowledge and expertise of the researcher are crucial to generate more plausible and generalizable research outcome (Amaratunga and Baldry 2001). Further, a qualitative study is more imperative and effective in providing more in-depth results than the quantitative method of the study (Adamides et al. 2012). A randomization technique has been applied in the interview segment. In total, 31 mid-level managerial staff were selected from each involved section. The inclusion principles were: (1) the

A) Supplier Selection

1. Do you have a standard procedure you follow while selecting a supplier?
   YES ☐ NO ☐
   If YES, please give more details about the procedure.
2. What are the criteria used in choosing a supplier? (kindly asnswer in ascending order)
3. Does your company ethics code compliance have a role in selecting a sustainable supplier? If yes, explain.

B) Sustainability

1. How often do you face issues with supplier sustainability? kindly list them.
2. How are sustainability issues addressed within your organization?
3. Social corporate responsibility is one pillar of sustainability. What are the activities does your company promotes to ensure social sustainability?
4. How do you ensure sustainable environmental policies are applied in your supplier's organization?

C) Ethical Practices

1. Does the top level management show concern about ethics?
   If YES, How?
2. How can failure to adhere to code of ethics affect the supply chain management?
3. Do you have discipline measures in place to address the breach of code of ethics? If yes, list them?
4. What is the motive to apply the code of ethics in your organisation?

Fig. 2 The interview form questions

participant must have the experience of working with the firm for over 2 years, and (2) participants must have knowledge of the supplier selection procedures.

4.2 Data Collection

In this research, the data collection has been conducted by using face to face and telecommunication interviews based on the semi-structured model. The interview issues were concentrated on: (a) supplier selection, (b) sustainability and (c) ethical practices. The research is conducted with a furniture firm in the United Kingdom. Interviews were carried out across all management levels (Fig. 2).

4.3 Data Analysis

The descriptive analysis technique has been adopted for this research. All of the interview questions have been examined thoroughly in order to extract the impact of principles in the furniture industry and in the selection of sustainable suppliers. A descriptive analysis tool SPSS has been used to observe the subsequent questions A2, A3 and C4. Irreversibly, the study concentrates on the impact of ethics on supplier selection in the furniture industry how, where and when ethical exercise is being functional during these complicated courses.

4.4 Ethics

Whilst carrying out this analysis, the examiner has confirmed that all information will be executed with dignity and confidentiality. However, each of the participants has signed a consent form, prior to conducting the data collection process making sure they have their conscious participation in the interview and no concern of confidentiality breach.

4.5 Background of the Main Case Study

A medium-sized furniture company founded in March 2004 with 94 employees has been chosen for this study. The company has planned to expand its operation worldwide and is investing outside the UK, China, Poland and India.

5 Research Findings and Analysis

From different management levels, around 31 employees were interviewed. Employees of the company were divided into four groups, which were dispersed into diverse areas, globally where the business works. The following table shows the details of the group of employees, the number of interviewees and their responses (Table 1).

Table 1 Group and the number of group subjects

Group number	Number of interviewees	Responses gathered
Group 1	6	6/6
Group 2	7	7/7
Group 3	8	8/8
Group 4	10	10/10

5.1 *Research Findings*

Interviews were grouped into three themes as shown (Fig. 3).

5.1.1 Supplier Selection

Do You Follow Any Standard Procedure While Selecting a Supplier? if the Answer Is Yes, Kindly Briefly Explain the Procedure

Each interviewee was asked whether there is any standard procedure present and followed whilst selecting the supplier. After collecting all the responses, it has found that almost all of the 31 respondents confirmed that there is no standard way to select the suppliers.

Should There Be Any Criteria While Selecting a Supplier?

The interviews responded in a different way to answer this particular question. Based on the responses of the interviewees, these include the following variables such as reputation, product characteristics, deliverability, cost, product sustainability, reliability and quality. The criterion of reputation and sustainability incorporate ethics and raw materials, correspondingly. The variables to select the supplier are ranked as follows (Fig. 4).

The most important variables for selecting the suppliers in furniture industries have cost, deliverability and reliability responded by 24% of the interviewees. 19% of respondents said that sustainability of raw materials is very important to them, 16% of the respondent answered that they will not negotiate in terms of the quality, 13% of the respondent said that product characteristics are important to them and the remaining 4% of respondent said that suppliers reputation is very important to them to select the supplier they want.

Fig. 3 Interview categories

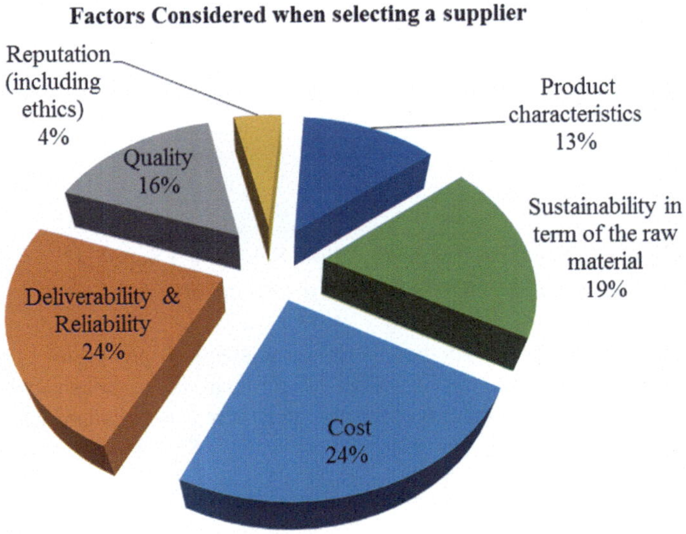

Fig. 4 Criteria considered when selecting a supplier

Fig. 5 Ethics code compliance role in supplier selection

Does the Company's Code of Ethics Comply in Selecting a Sustainable Supplier? if the Answer Is Yes, Explain

To facilitate the selection of a sustainable supplier, a company's code of ethics has a very vital role, which was agreed by 77.42% of respondents. The remaining 22.58% of respondent said that ethics has no role in selecting the supplier. The furniture companies evaluated the scenario by conducting a questionnaire named, "Professional Quality Questionnaire" (PQQ) to cover the ethical issues such as employees' rights, salary, health and safety (Fig. 5).

The ethical standards described in the code of ethics covers the issues such as health and safety, the security of employment rights, equality, corruption and fair trade in the developing countries.

5.1.2 Sustainability

How Often Issues like Supplier Sustainability are faced by the Companies? Kindly Explain

Group members of the groups 1, 3 and 4 have reported issues that they have faced because of the sustainability. But the members of group 2 argued that the issues related to supplier sustainability are country-specific.

Do the Issues like Sustainability Ever Addressed Within the Organizations? if the Answer Is Yes, Explain?

The group members from groups 1, 2 and 3 said that they have not faced any issues regarding sustainability in their organizations. Adding to that, they also said that the key method to handle sustainability is by retaining a decent correlation with the providers. Group members from group 4 believe that the issues related to sustainability can be addressed if the organization maintains the same values, beliefs, fairness and equality within the ethical environment with the supplier.

What are the activities your company takes to promote and ensure social sustainability?

By offering placement programmes, a job opportunity to local peoples, subsidizing donations, volunteering programme supporting the young individuals, coaching the young generation and resourcing the merchandises from a sustainable provider the company safeguards sustainability.

Adding to that, appreciably the company keeps interacting with the local agencies and offers them internships across the trade functions. It is important for the company to keep a very good tie with the local schools for establishing the "inspirational mentoring" programmes" which will contribute to the employment of new generations.

For instances, as per the group 4 respondents, the firm managing 3.5-tonne vehicle that creates the least possible Green House Gas (GHG). One of the most important policies of the company is to ensure that their suppliers maintain the highest exercises such as reprocessing the leftovers and controlling landfill waste. Group 2 interviewees explained that the company demands sustainability certification from their suppliers. Group 3 and group 1 interviewee also described that the company instructs the new staffs on the subjects of bio-health and safety to endorse enhanced sustainability in their operation.

5.1.3 Ethical Practice

Does the Top-Level Management Show Interest in Terms of Ethics? if the Answer Is Yes, Explain?

The research also revealed that top-level managements such as the board of directors and the senior managers are cautious about ethical issues. All the interviewees responded positively to the senior-level management commitment to ethical practice in the company.

From the analysis, it has found that the ethical practice can be used to guide organizations' relations with customers, dealers and personnel.

How Does a Failure to Adhere to the Code of Ethics Affect Supply Chain Management?

As said by group 1 interviewee, the failure of the suppliers to follow the code of ethics leads to adversarial consequences of the firm status and the product in turn. Group 2, 3, and 4 interviewees agreed with this result adding that the organization can lose confidence in its provider and in turn this might affect the trustworthiness and interactions with providers.

Do the Companies Have Disciplinary Measures to Address the Breaching of Ethical Terms? if Yes, List Them?

Members of all groups confirmed that there are disciplinary assessments to handle any breach of ethical terms. They also added that the organization has a written warning approach to address issues of breaching the code of ethics.

What Are the Motives of an Organization to Apply the Code of Ethics?

The rationale behind the code of ethics is image enhancement and humanitarian perspective (Fig. 6).

This finding represents the grounds on which the corporate image stands and its ethical presence in the behaviour of the employees in any organization. About 80.65% of the respondents agreed that the purpose of the organization in applying the code of ethics is to gain a humanitarian perspective while the remaining 19.35% supported the image enhancement.

Fig. 6 Motive to apply a code of ethics

5.2 *Research Analysis (RA)*

Research analysis has been divided into three parts such as supplier selection, sustainability and ethical practices. These are described as follows.

5.2.1 Supplier Selection

The finding of the research revealed that there was no specific standard procedure or/protocol confirmed when choosing the providers in the firm which varies notably from the outcome stated by Goebel et al. (2012). They claimed that companies are progressively developing benchmarks to affiliate their tasks to principles and sustainability. Conversely, the current study supports the outcomes identified by Meera (2014) that supplier assortment is a chronological method of assessment and identification through which an association can choose the best supplier for their raw ingredients. Consequently, there is no need for a specific standard or process to implement, but the use of varying techniques. Lack of a particular standard procedure in choosing suppliers does not prohibit the company from executing an ethical standard to assess the suppliers (Kovacs 2009).

The outcome of the criteria to select suppliers comprises cost, deliverability and dependability, excellence, reputation, product characteristics and sustainability (Schiele 2007). This report confirmed that the cost was considered the substantial feature and diminishing the cost in suppliers leads to the competitive advantage (Hu 2010). Reliability was also found to be a significant factor, Svensson (2009) argues, and it leads to the expansion of sustainable exercises.

5.2.2 Sustainability

The research shows that a company with a code of ethics hardly encounters any issues regarding sustainability (Hu 2010). The research finds that the suppliers who maintain some level of relationship with the company incorporate sustainability issues. Adding to that if the company can guarantee that the suppliers follow the identical ethical practices the company follows, it leads to sustainability. Code of ethics attended by a firm ensures supplier relationships and compliance with the set of ethical standards (Kovacs 2009).

The study recommends that the company considers that social corporate responsibility is a pillar to secure sustainability. The activities such as internship programmes, job opportunities to local people, sponsorship and apprenticeships are the examples of social sustainability. The findings of this research are similar to the research by Manufacturing Skills of Australia (2012) on social responsibility.

All the companies that participated in the study assured that they agreed with all the policies in terms of environmental safety. The environmental issues that are noticed throughout the research are recycling aspects, reduction of GHG emissions, ensuring employees' safety and health as well as reduction of the waste. The findings support the report by Darnall (2008) that environmental issues affect the sustainability of furniture firms. The study also supports Jennings (2005) findings. Jennings asserted that the company must ensure that the suppliers are environmentally sustainable.

5.2.3 Ethical Practice

The study has shown that top-level management is involved in confirming the ethical practice in a company (Weaver et al. 2005). Senior-level management is also responsible for regulating the ethical behaviour that needs to be used in the organization (Weaver et al. 2005; Goebel et al. 2012). The senior-level management and the procurement manager are able to select a supplier who conforms to the code of business conduct in the industry (Moore 2004). The study also revealed that organizational culture is needed in order to encourage ethical behaviour and guides the supplier selection and sustainability.

As stated by Darnall (2008), the point that explains the sustainability concerns in a company has shifted from brand image to a competitive advantage. This research has found out that failure to maintain the code of ethics affects the supply chain management in terms of organization image, reputation and brand. Unethical practices in the business can harm the good reputation and value of the firm (Carter and Jennings 2004; Handfield et al. 2002). Daniels (2006) highlighted that the purpose of having the ethics and sustainability focus has shifted from corporate image to achieving the competitive edge. This research, however, maintains that the company's biggest aim to apply the code of ethics is to gain a humanitarian perspective followed by image enhancements.

When there is any contravention of the code of ethics, there are certain measures and practices that will be taken in agreement with the findings, which are very similar to the findings, by Green (1996). Existing written notices and removal are good practices to establish the code of ethics even further.

5.3 Summary of Findings

In light of the above-mentioned outcomes and evaluation, the study has found that the sustainability in the supplier selection has an undeviating link to the subsistence and enactment of ethics code and has proved the research hypothesis. Additionally, the research has shown that most of the top-level management has a high level of responsiveness and training about code ethics implementation, and they believe that it is important to gain a competitive advantage in the market. New paradigm shifts in companies' cultures happen though designing more eco-friendly strategies, offering internship programmes and job opportunities to local people, sponsoring charities, introducing volunteering programmes helping young people, training the young generation and resourcing the products from a sustainable supplier, ensuring the organizations' sustainability.

This finding supports the second part of the hypothesis H1b. The level of top management compliance with the sustainable environmentalism expert requirements regulates the choice of the sustainable supplier.

The principles for choosing the providers consider the following features such as expenditure, trustworthiness, deliverability and excellence of the providers. It indicates that the choice to attain sustainable providers is associated with some objects, and it supports the hypotheses H2a. The method of choosing sustainable providers has openly connected to the degree of enticements and their acknowledgement to moral exercises.

Moreover, the existence of motives in subsequent ethical exercises and environmental exertions upturns compliance and sustainable plan adherence. This is demonstrated in the H2b hypotheses.

5.4 Research Limitations

The limitations of this study are as follows:

- Ethical practice and sustainable supplier selection in the furniture industry is a quite new addition which will take some time for the company to embrace, put in action and launch in their policies.
- Due to not having any dedicated department to handle the sustainability issues in the company, the researcher finds some difficulties in collecting data from a consolidated platform.
- Further multi-layered research with companies from small-to-large size, locally and internationally is required to develop further insight on gaps in the code of ethics identifying where development should focus in the future.

6 Conclusions and Recommendations

To conclude, the company needs to carry out ethical exercises in their organization, as these activities assist the organization to select environmentally sustainable suppliers, shrink their manufacturing costs, enhance the quality of production and increase the level of customer satisfaction. The research has found that there is no standard way to select the suppliers but rather it is dependent on the supplier requirements (Meera 2014). The research has also shown that different variables such as reputation, product characteristics, deliverability, cost, product sustainability, reliability and quality are important criteria while selecting a supplier. There were mixed views with regard to the level of importance given to a company's code of ethics in the selection of a sustainable supplier. 77.42% of the respondents agreed with this issue, and the remaining 22.58% of respondent said that ethics has no role in selecting the supplier. There were also mixed outcomes on the matter of suppliers' sustainability where the members of groups 1, 3 and 4 have reported that they have faced issues because of the sustainability. Nevertheless, group 4 argued that the issues related to supplier sustainability are country-specific. All the interviewees responded positively to the issue that top-level management such as the board of directors and the senior managers are concerned about ethics. After analyzing all the outcomes of this research, the researcher identified some room for taking actions on some specific issues. The researchers make the following recommendations:

6.1 Recommendations

- In the furniture businesses, the supplier's evaluation process should be flexible enough to make it simple and straightforward to accept the ethical considerations in different scenarios while trading with the suppliers. Moreover, an organization, operating business in multiple geographical places, should develop efficient measures to achieve the existing conditions without compromising the supplier selection and sustainability.
- The failure of the suppliers to comply with the code of ethics could lead to an adverse effect on the company reputation and the brand, in turn, may affect the reliability and relationships with suppliers. Hence, there should be disciplinary measures like written warning and dismissal to handle any breach of ethical terms.
- The business organizations should assimilate the CSR (Corporate Social Responsibility) actions with the issues and problems regarding sustainability. Arranging social companions to sustainability issues helps the various parties to build collectively in an accord to gain sustainability and maintain the organization.
- Finally, ongoing and thorough training and development programmes are needed for executives, personnel and stakeholders as part of regularization and setting behaviour to achieve sustainability in the organization.

References

Adamides, E. D., Papachristos, G., & Pomonis, N. (2012). Critical realism in supply chain research: Understanding the dynamics of a seasonal goods supply chain. *International Journal of Physical Distribution and Logistics Management, 42*(10), 906–930.

Amaratunga, D., & Baldry, D. (2001). Case study methodology as a means of theory building: Performance measurement in facilities management organisations. *Work Study, 50*(3), 95–104.

Baird, N. & Rowen, S. (2010). Lean and green: How sustainable practices are changing retail. *Retail Systems Research.*

Blowfield, M. (2000). Ethical sourcing: A contribution to sustainability or a diversion? *Sustainable Development, 8*(1), 191–200.

Carter, C., & Jennings, M. (2004). The role of purchasing in corporate social responsibility: A structural equation analysis. *Journal of Business Logistics, 25,* 145–186.

Daniels, C. (2006, January 16). *Companies Coy on Eco-Management.* The New Zealand Herald.

Darnall, N. (2008). Environmental management systems and green supply chain management: Complements for sustainability? *Business Strategy and the Environment, 5*(1), 30–45.

Goebel, P., Carsten, R., Richard, P., & Christina, S. (2012). The influence of ethical culture on supplier selection in the context of sustainable sourcing. *International Journal of Production Economics, 6*(140), 7–17.

Green, K. (1996). Purchasing and environmental management: Interactions, policies and opportunities. *Business Strategy and the Environment, 5*(3), 188–197.

Handfield, R., Walton, S., Sroufe, R., & Melnyk, S. (2002). Applying environmental criteria to supplier assessment: A study in the application of the analytical hierarchy process. *European Journal of Operational Research, 141,* 70–87.

Hu, A. H. (2010). Critical factors for implementing green supply chain management practice. *Management Research Review, 10*(1), 1–10.

Jennings, C. (2005). The role of purchasing incorporate social responsibility. *Journal of Business Logistics, 2*(1), 145–156.

Kleindorfer, P., Singhal, K., & van Wassenhove, L. (2005). Sustainable operations management. *Production and Operations Management, 14,* 482–492.

Koplin, J., Seuring, S., & Mesterharm, M. (2007). Incorporating sustainability into supply management in the automotive industry—The case of the Volkswagen AG. *Journal of Cleaner Production, 15,* 1053–1062.

Kovacs, G. (2009). Corporate environmental responsibility in the supply chain management. *Journal of Cleaner Production, 4*(3), 571–1578.

Manufacturing Skills Australia. (2012). Sustainabilities Issues in Furniture. Retrieved September 26, 2016, from http://sustainabilityskills.net.au/wp-content/uploads/2012/07/Sustainability-issues-in-furniture_June12.pdf.

Marc, A. R., & Hossam, A. K. (2012). Sustainable manufacturing and design: Concepts, practices and needs. *Sustainability, 4,* 154–174.

Meera, B. (2014). Environmental sustainability through green supply chain management practices among Indian manufacturing firms. *International Journal of Scientific and Research Publications, 5*(1), 1–8.

Molamohamadi, Z. (2013). Supplier selection in a sustainable supply chain. *Journal of Advanced Management Science, 6*(1), 271–281.

Moore, R. (2004). The methods used to implement an ethical code of conduct and employee attitudes. *Journal of Business Ethics, 4*(3), 244–255.

Saaty, T. L. (1980). *The analytic hierarchy process.* New York: McGraw-Hill.

Saaty, T. L. (2005). The analytic hierarchy and analytic network processes for the measurement of intangible criteria and for decision-making.

Schiele, H. (2007). Supply-management maturity, cost savings and purchasing absorptive capacity. *Journal of Purchasing and Supply Chain Management, 5*(3), 274–293.

Svensson, G. (2009). A corporate model of sustainable business practices: An ethical perspective. *Journal of World Business, 6*(2), 1–10.

Weaver, G., Trevinio, L., & Agle, B. (2005). Ethical role models in organisations. *Organisational Dynamics, 34,* 313–330.

Chapter 10
Reducing Edible Food Waste in the UK Food Manufacturing Supply Chain Through Collaboration

Guangming Cao, Pramitkumar Shah, and Usha Ramanathan

Abstract While a third of food produced is wasted at the pre-consumer stage in the UK food manufacturing supply chain (FMSC) and has had significant negative economic and environmental impacts, many challenges remain in how to reduce edible food waste. This chapter addresses the problem of whether and to what extent FMSC collaboration could lead to the reduction of edible food waste. Evidence in the literature suggests that despite an increasing attention having been paid to reduce edible food waste, there is a scarcity of studies that focus on the relationship between FMSC collaboration and the reduction of edible food waste. Consequently, the aim of this chapter is to develop a research model that explains the relationships among FMSC collaboration, collaborative effectiveness and the reduction of edible food waste. The model is underpinned by the relation view and has been empirically tested with 122 survey responses from food manufacturing firms, using structural equation modelling. The findings indicated that FMSC collaboration has a positive effect on collaborative effectiveness, which in turn results in the reduction of edible food waste during production, processing and storage. Thus, an important implication of this chapter is that the UK FMSC members would benefit from closely collaborating with their supply chain partners to achieve greater collaborative effectiveness and thereby reducing edible food waste.

Keywords FMSC collaboration · Collaborative effectiveness · Edible food waste reduction · Relational view

---

G. Cao
Ajman University, Ajman, UAE

P. Shah
University of Bedfordshire, Luton, UK

U. Ramanathan (✉)
Nottingham Trent University, Nottingham, UK
e-mail: usha.ramanathan@ntu.ac.uk

U. Ramanathan and R. Ramanathan (eds.), *Sustainable Supply Chains: Strategies, Issues, and Models*, https://doi.org/10.1007/978-3-030-48876-5_10

1 Introduction

According to the United Nations' Food and Agriculture Organisation (FAO), edible food (EF) waste refers to wholesome edible material intended for human consumption, arising at any point in the food supply chain that is instead discarded, lost, degraded or consumed by pests; and the FAO claims that around one-third of EF is wasted or lost at the pre-consumer stage (FAO 2013). The UK generates about 10 million tonnes of EF waste and 18% (WRAP 2018) to 30% (Parfitt et al. 2010) of the waste is associated with food processing, distribution and retail in the UK food manufacturing supply chain (FMSC). Thus, it is not surprising that EF waste generated at the stage of manufacturing has drawn significant attention because of its negative impact on the UK environment, society and economics (Papargyropoulou et al. 2014; WRAP 2018).

One key reason for EF being wasted at the pre-consumer stage in the UK FMSC is the lack of coordination of FMSC operational- and logistic-related activities (e.g. Eksoz et al. 2014; Gokarn and Kuthambalayan 2017) as FMSC is more complicated and difficult to manage than other supply chains (Gadde and Amani 2016) as food is a perishable commodity (Mithun Ali et al. 2019). Although different practical approaches have been taken to reduce EF waste in the UK FMSC, significant reduction in EF waste is yet to be seen. It is estimated that EF waste in the UK FMSC will increase by 26% by 2020 (Gadde and Amani 2016).

Despite the importance of reducing EF waste in the FMSC, there is a paucity of studies on this topic (Mena et al. 2014; Redlingshöfer et al. 2017). Nevertheless, there is evidence in the literature to suggest that collaboration among the FMSC members might help build an esprit de corps to unite efforts thereby reducing EF waste (Kaipia et al. 2013; Göbel et al. 2015). However, such an approach is still a subject of debate (Kouwenhoven et al. 2012; Papargyropoulou et al. 2014) as previous studies have shown inconclusive results (e.g. Göbel et al. 2015; Ali et al. 2016). More relevant research is needed (Piboonrungroj 2012; Quinn 2012; Liljestrand 2017).

Thus, this chapter seeks to develop an understanding of the impact of FMSC collaboration on EF waste reduction. The key research questions are

- What are the key dimensions of FMSC collaboration and its effectiveness?
- To what extent is FMSC collaboration influencing collaborative effectiveness and finally EF waste reduction?

Answering these questions could add to the existing literature regarding the impact of FMSC collaboration on EF waste reduction, which remains to be under-researched (Mena et al. 2011; WRAP 2018). In particular, this chapter seeks to draw on the relational view to represent a multi-dimensional construct models to provide insights into the antecedents to and the consequences of FMSC collaboration and the key dimensions of FMSC collaboration and collaborative effectiveness, which is missing from prior studies (Piboonrungroj 2012). Additionally, answering the above research questions could help FMSC practitioners to better manage their collaborative activities thereby reducing EF waste.

The remainder of this chapter is structured as follows. Section 2 provides an overview of the FMSC and its key characteristics and EF waste along the FMSC in the UK. Section 3.2 reviews the literature and discusses the development of the research hypotheses. Section 4 explains the research method including research constructs, the associated measurements and the data collection process. Section 5 presents data analysis and results. It is followed by discussion in Sect. 5.1 and conclusion in Sect. 6.

2 FMSC and Its Key Characteristics and EF Waste in the UK FMSC

2.1 FMSC and Its Key Characteristics

FMSC, being a part of the overall food supply chain that consists of a network of organisations from farm suppliers, farmer, marketers/storage, processors, wholesalers/distributors, retailers, to final consumers, "transforms the raw food products supplied by the primary producers into the product that meet consumer requirement" (Dani 2015). The UK FMSC is the second largest and fastest growing manufacturing supply chain industry in the UK, with a net worth of £96 billion and accounting for 19% of total UK manufacturing supply chain by turnover (Rhodes 2015). It plays a vital role in the food supply chain.

UK FMSC comprises a variety of sectors and processes and can be classified by the different types of manufacturing activities, where the activities deal with different types of food products: meat, fish, fruit and vegetables, fats and oils, milk products, grain mill products, animal feeds and other food products. It is becoming an interconnected complex system with a large variety of relationships among alliances, horizontal and vertical cooperation, and forward and backward integration. FMSC is frequently characterised by demand uncertainty, supply uncertainty, price uncertainty, and process uncertainty; while its products are characterised by, for example, perishability, quality variation, seasonality and bulkiness (Gokarn and Kuthambalayan 2017). Thus, FMSC has specific inherent characteristics which differ from other supply chains and is more complex and difficult to manage (Gadde and Amani 2016).

2.2 FMSC Collaboration and Its Effectiveness

Today, FMSC faces vexing challenges such as the major issues of food recalls, food safety and traceability (Dani 2015), poor stock management, product damage, demand uncertainty, shelf life management and packaging design (Mena et al. 2011). FMSC members working and operating alone are no longer sufficient to resolve these common issues and to achieve the desired goals (Matopoulos et al. 2007). Thus, understanding collaboration as "two or more independent companies work jointly to plan and execute operation towards common goals and mutual benefits" (Cao and Zhang 2011), FMSC collaboration has become a necessity rather than an option (Matopoulos et al. 2007). That is, chain members become involved and actively work together in coordinating activities which span the boundaries of their organisation (Matopoulos et al. 2007).

By working with their upstream or downstream partners, FMCS members may improve their collaborative effectiveness, "a supernormal profit jointly generated in an exchange relationship that cannot be obtain by either firm in isolation and can only obtain through the joint contribution of the collaborative partners" (Cao and Zhang 2012). As a result, FMCS members could more effectively address the changing needs such as improving visibility, safety, accurate forecasting and monitoring (Banchuen et al. 2015) and/or enhancing demand replenishment, collaborative forecasting and shared distribution (Barratt 2004).

2.3 EF Waste in UK FMSC

The UK FMSC contributed the second largest proportion of EF waste in their supply chain and logistics operation, accounting for 18% (WRAP 2018) to 31% of the total UK EF waste generated (Parfitt et al. 2010). This has a high negative impact on the UK economy because the reduction of EF waste could save the UK FMSC £300 million a year (Whitehead et al. 2013; DEFRA 2015). Besides, EF waste has a damaging effect on the UK environment such as climate change (Papargyropoulou et al. 2014).

While the FMSC members had increased attention on environmental preservation and focused on reducing various types of food waste such as packaging waste, energy waste, processing water waste, landfill waste and pollution during transportation and cooking processing, there is not any notable achievement in the EF waste reduction in the UK FMSC (Mena et al. 2011, 2014).

Evidence in the literature suggests that a wide range of factors exist for EF waste in the manufacturing environment (Verghese et al. 2018), such as organisational inefficiencies of supply chain operators, forecasting, ordering, promotional planning and packaging (Mena et al. 2011; Canali et al. 2014), management concerns (Papargyropoulou et al. 2014), lack of information sharing (Kaipia et al. 2013), poor logistics management (Liljestrand 2017), uncertainty in demand, competition and product seasonality (Mena et al. 2011; Gokarn and Kuthambalayan 2017), and/or shelf life constraints, weather fluctuation, and longer lead times for imported food products (Mena et al. 2011). However, there is no evidence to show which factors are the most significant reasons for EF waste (Canali et al. 2014).

3 Theoretical Background and the Research Hypotheses

3.1 Theoretical Background

In order to answer the research questions regarding the link from FMSC collaboration to EF waste reduction, this section discusses the theoretical background and develops the research model.

Prior studies suggested that in order to understand FMSC collaboration, the relational view helps explain the nature that FMSC members need to come together in coordinating their activities to solve "messy problems" which cannot be resolved in any other way (e.g. Soosay and Hyland 2015). According to Dyer and Singh (1998), the relational view emphasises that a firm's competitive advantage can be gained from the relationship between firms or critical resources that may span firm boundaries and be embedded in inter-firm resources. In the context of FMSC, the relational view suggests that when FMSC members allocate more value to each other in terms of, for example, providing undistorted information for demand forecast, integrating knowledge, providing skills through joint training programmes and nuanced information about utility functions, they can deal with unexpected changes in the environment more effectively, thereby gaining "supernormal profit jointly generated in an exchange of relationship that cannot be generated by either firm in isolation and can only be created through the joint contribution of the collaborative partners" (Cao and Zhang 2012). Thus, the relational view emphasises collaborative advantage which can be extracted only from shared resources among the collaborative partners (Turkmen 2013). Subsequently, the relational view will be used to underpin this research.

3.2 Research Hypotheses

Based on the relational view, this section develops hypotheses to explain the relationships among FMSC collaboration, collaboration effectiveness and EF waste reduction.

While the literature on supply chain collaboration puts more emphasis on buyer–supplier relationship and less on the coordinating supply chain and the logistical processes, Cao and Zhang (2011) suggested that prior studies tend to focus on examining the collaboration benefits that directly improve a firm's performance than the collaborative effectiveness such as joint value creation process or relational benefits or common benefits. However, Piboonrungroj (2012) confirmed that without explicitly considering collaborative effectiveness, supply chain collaboration could not directly improve a firm's performance. Lavie (2006) mentioned that while each member provides a subset of resources such as forecast information, knowledge, and training, the collective resources of all members form the basis of collaboration; which leads to collaborative effectiveness such as accurate forecasting, reduced uncertainty and improved product safety; which in turn helps improve individual member's performance in terms of, for example, reduced spillage, improved forecast planning and improving the employee's skills. This is because collaboration is a function of the extracted value of all combined and shared resources among the collaborative partners (Cao and Zhang 2011). Similarly, Ataseven and Nair (2017) suggested that collaboration enables supply chain firms to attain relational benefits or collaborative effectiveness through streamlining supply chain processes, and also by coordinating logistical activities with business partners which may indirectly lead to improving organisations' operations efficiencies.

However, insufficient attention has been paid to FMSC collaboration, though FMSC faces vexing challenges of poor stock management, poor temperature control management, product damage, lack of traceability, reduced agility, demand uncertainty, shelf life management and packaging design (Mena et al. 2011; Dani 2015; WRAP 2018). There is evidence in the literature to suggest that FMSC structure, demand characteristics and the nature of the product all impinge greatly on the operational process and logistical activities (Matopoulos et al. 2007; Morita et al. 2015). Thus, it is likely that the success of FMSC collaboration relies on the coordination of the FMSC operational and logistical activities through exchange of high-quality information and RFID and knowledge integration (Wiengarten et al. 2010).

By synthesising prior literature on supply chain collaboration, this research proposes that FMSC collaboration can be understood in terms of five main components: joint demand forecasting, joint training programme, smart packaging, knowledge integration and using RFID. First, it is proposed that joint demand forecasting plays a key role in FMSC collaboration while Mena et al. (2014) identified that poor forecasting is one of the most common causes for generating the high level of EF waste in the UK FMSC. From the relational view, it can be argued that when FMSC partners are combining their forecast-related information, they will be able to improve the chain's ability to generate consensus and accurate forecasts for the seasonal, promotional and short-shelf life of perishable food products, which would help to efficiently manage current inventory status, reduce lead time and cost (Eksoz et al. 2014), better match demand and supply of product correctly, thereby reducing over-production of EF (Mena et al. 2014).

Second, Badea et al. (2015) suggested that joint staff training in collaborative supply chain is crucial for the supply chain members to achieve collaborative goals and improve the supply chain efficiency as joint staff training improves employees' proficiency and the supply chain's competency to meet the different risk factors that can harm the entire supply chain processes. In the context of FMSC, Shinbaum et al. (2016) stated that joint training programmes give experience and enhance the existing skills of employees to do something in the correct procedure; otherwise, due to ineffective employee training, more than one-third of all food could be wasted in the process. Additionally, Park et al. (2010) suggested that continuous and repetitive joint training programmes would be required to enable FMSC collaboration to improve the logistical operations and procedures.

Third, smart packaging design, aiming at improving, combining or extending the traditional function of storing, protecting and information about the food product

(Vanderroost et al. 2014), can be used to balance between the environmental impact of the packaging itself, and the packaging's ability to reduce waste of the packaged product (Conte et al. 2015). As a result, smart packaging design can significantly reduce the amount of EF waste because it can prolong the shelf life of food; provide heavy interaction between the user and product; low resources needed for transportation, handling and storage; and generate a low amount of litter (Conte et al. 2015). Furthermore, smart packaging design may depend on FMSC collaboration, which involves monitoring food product as it moves through the supply chain and also the environment in which it is kept (Watson et al. 2015). For example, when food product data is recorded and sent back to manufacturers on quality, safety, shelf life and logistic efficiency-related information (Verghese et al. 2013; Watson et al. 2015); a food manufacturing company could then match the remaining shelf life of each food product with the remaining transport duration options during stock rotation, which leads to reducing EF waste related to poor shelf life management (Jedermann et al. 2014).

Fourth, knowledge integration refers to the process of transferring knowledge, both tacit and explicit, across organisational boundaries and applying the resultant knowledge to solve problems, which would be a source of supply chain collaborative effectiveness (Crook et al. 2008) and/or collaborative advantage (Koufteros et al. 2005). Due to the complexities of FMSC structure and potential mismatch of management styles, it is important for FMSC members to coordinate knowledge transfer, to overcome mismatched management styles and to build knowledge architectures (Mishra and Shah 2009). Collectively, FMSC members could generate new and relevant knowledge (knowledge exploration) through assimilation of existing knowledge (knowledge exploitation) to find the best way to reduce EF waste in their FMSC operation (Cao and Zhang 2012). As a result, collective knowledge could enable FMSC members to better understand the reality in which specific stages and locations, the EF waste occurs and find out why it occurs (Kumar et al. 2017).

Fifth and finally, using radio-frequency identification (RFID), "a technology that enables large amounts of information to be stored on chips (tags/transponders) that can be read at a distance by readers, without requiring line of sight scanning" (Nash 2010), allows supply chain members to have timely information across the various stages of the supply chain to rapidly intervene in targeted situations (Piramuthu 2005), thereby improving the efficiency and security of the entire supply chain (Lee et al. 2011). Instead of a labour-intensive barcode system, RFID employs an automated scan of the food products to manage out-of-stock, restocking and replenishment tasks, and most importantly, improve the ability of FMSC members to track and trace the source of contamination in their FMSC network, which ultimately improves food product safety (Unnevehr 2000) and leads to reduced EF waste (Mena et al. 2014). For example, Sahin et al. (2002) indicated that using RFID to continuously monitor perishable items throughout the FMSC process gives complete visibility of the remaining shelf life of food products and the location of the oldest stock, which could help to reduce expiry date related EF waste in the FMSC (Grunow and Piramuthu 2013).

As a result, FMSC collaboration could lead to collaborative effectiveness in terms of accurate forecasting to generate the correct demand for the product through undistorted and agile information sharing (Eksoz et al. 2014), better promotional planning by FMSC members' sharing all related information (Mena et al. 2011), improving product safety, improving temperature monitoring regularly (Mena et al. 2011), better inventory control to improve stock rotation according to demand variability through, and better shelf life management.

Consequently, FMSC collaboration and its effectiveness will lead to the reduction of over-production, processing, and storage-related EF waste. Over-production EF waste constitutes higher cost to food manufacturing companies as raw materials, ingredients, water, labour and energy are wasted given that the prepared food no longer has an end customer and is scrapped as commercial waste and diverted to landfill (Darlington et al. 2009). Processing EF waste or "process control waste" (Mena et al. 2014) is generated in the different operational processes of the FMSC from, for example, poor housekeeping procedures, process inherent losses, poor conformity, and cross-contamination of food product due to operator neglect, and inappropriate handling of product during distribution processes (Darlington et al. 2009; Mena et al. 2014). Lastly, storage EF waste means the amount of EF product wasted in the multiple storage point at the food manufacturer, such as EF product is waiting for order to arrive (Chabada et al. 2012; Mena et al. 2014).

Drawing on the relational view and the literature, the following hypotheses could be developed:

H1: FMSC collaboration is positively associated with collaborative effectiveness.
H2a: FMSC collaboration is positively associated with over-production EF waste reduction.
H2b: FMSC collaboration is positively associated with processing EF waste reduction.
H2c: FMSC collaboration is positively associated with storage EF waste reduction.
H3a: collaborative effectiveness is positively associated with over-production EF waste reduction.
H3b: collaborative effectiveness is positively associated with processing EF waste reduction.
H3c: collaborative effectiveness is positively associated with storage EF waste reduction.

4 Research Methods

The hypotheses were tested using PLS-SEM, which is suggested to be well suited for research situations where theory is less developed and formative constructs are part of the structural model (Hair et al. 2013). Since there is little research on FMSC and the present study handles both reflective and formative constructs such as FMSC collaboration and its effectiveness, PLS-SEM is a suitable method to empirically test the hypotheses.

4.1 Measures of Constructs

The constructs listed in Table 1 were measured using scales adopted or further adapted from relevant items that were validated across a variety of studies. Both reflective and formative measurement models were used. While a clear structure which represents FMSC collaboration is lacking (Kumar and Nath Banerjee 2012), this study draws on prior studies to measure this construct using five dimensions: joint demand forecasting, knowledge integration, using RFID, smart packaging design and joint training programmes. Collaborative effectiveness is measured using six items: accurate forecasting, better promotional planning, improving product safety, improving temperature monitoring, better inventory control and better shelf life management. EF waste is measured in terms of processing, over-production and storage. Additionally, firm size and job titles are used as a control variable.

4.2 Sample and Data Collection

In this study, primary data was collected from a sample of UK FMSC firms identified from the FAME database. 1253 target respondents were identified based on both their job titles (such as CEO, directors, logistic manager, operational manager and general manager) and the availability of e-mail addresses. Thus, this study used a non-probability sampling, which is consistent with research on supply chain operational and logistics. As observed by Wagner and Kemmerling (2010), leading journals in this area and supply chain operational and logistic researchers "hardly ever deploy their survey to the entire population of interest".

A total of 1253 e-mail surveys were sent through Qualtrics software with personalised cover letters and the aim of the study. To improve response rate, three rounds of e-mails were sent over four weeks. A total of 122 usable responses were received, which represents a 9.7% response rate, which is comparable with previous supply chain collaboration studies, such as 6% in Cao and Zhang (2011) and 9.7% in Kumar and Banerjee (2014). Additionally, according to Hair et al. (2016), a sample of 122 respondents is satisfactory for our research model: as the maximum number of arrows pointing at a construct is five in this research; the minimum sample requirement is 70 in order to detect a minimum R-squared value of 0.25 in any of the constructs for a significant level of 5%.

Table 1 Measurement scales

Construct	Measurements items	References
Joint demand forecasting (JDF)	• Our FMSC partners can forecast and plan collaboratively with us through the integrated information system • We can depend on our supply chain partners to provide us with a good market forecast and planning information • We plan volume demands for the coming seasons together with our FMSC partners	Rajaguru and Matanda (2013), Singhry et al. (2015)
Knowledge integration (KI)	• We and our partners provide resources to each other to explore new ideas and innovations • When we and our partners get new ideas, we communicate with each other straight away • We and our partners have regular meetings to encourage knowledge dissemination • We and our partners combine our expertise to jointly solve task-related challenges	Crook et al. (2008), Cao and Zhang (2011), Hudnurkar et al. (2014)
Smart packaging design (SPD)	• Use a range of packaging indicators, such as thermal sensor, intelligent (smart) tag, and microchip; to provide information about the condition of packed food • We and our partners serve smaller packing of food products • Use the Active Packaging system, such as modified atmosphere packaging, oxygen scavengers, moisture absorbers, aseptic packaging, and carbon dioxide production; to slow down the oxidation of certain food components • Well-designed packaging provides better protection to the food product as it moves through the supply chain, such as during distribution or transit	Mahalik and Nambiar (2010), Verghese et al. (2013)
Using RFID technology (RFID)	• We and our partners currently use RFID technology for on-time replenishment • We and our partners are currently using an RFID system for tracking the food product throughout the FMSC • We and our partners are currently using RFID technology for improving cost efficiency, e.g. through improved asset visibility which reduces stock loss • We and our partners are using RFID technology for supply chain operations integration	Lee et al. (2011), Hudnurkar et al. (2014), Chen (2015)
Joint training programme (JTP)	• We and our partners jointly organise food-related courses for employees, such as food management certification • We and our partners jointly organise food-related training sessions for employees to learn the correct procedures and their importance • We and our partners organise joint training programmes to enhance existing skills among all levels of employees • We and our partners see training as an important way of helping the company to achieve its goals • We and our partners frequently update our joint training programme	Kim et al. (2013), Kumar and Rahman (2015), Shinbaum et al. (2016)

(continued)

Table 1 (continued)

Construct	Measurements items	References
Collaboration effectiveness	• Accurate forecasting (AF)—We and our partners potentially increase our profitability through accurate forecasting • Better promotion planning (BPP)—We and our partners potentially increase our profitability through accurate forecasting • Improving product safety (IPS)—We and our partners jointly analyse previous promotions • Improving temperature monitoring (ITM)—We and our partners introduced a logging system to record temperatures at both ends of the food chain • Better inventory control (BIC)—We and our partners are better at integrating warehouse management • Better shelf life management (BSLM)—We and our partners are reducing processing time in the supply chain to maximise the available life of food products (minimum life on receipt)	Mena et al. (2011, 2014), Verghese et al. (2013), Betz et al. (2015)
Processing EF waste (PEFW)	My company has achieved a significant reduction of EF waste that is generated • Due to spillage in our processing stage • Due to the poor conformity of the food product, such as quality, appearance, flavours • From wrong labelling, such as wrong date code, wrong ingredient and nutritional data information, wrong price and promotional stickers • Due to frequent changes in the production schedules in our processing stages	Darlington et al. (2009), Beretta et al. (2013), Mena et al. (2014)
Over-production EF waste (OPEFW)	My company has achieved a significant reduction of EF waste that is generated • By planning errors, such as forecast error, promotion error, and poor stock management • During seasonality and special days, such as Christmas, Easter etc • Due to weather uncertainty or variability • During promotional events	Darlington et al. (2009), Beretta et al. (2013), Mena et al. (2014)
Storage EF waste (SEFW)	My company has achieved a significant reduction of EF waste that is generated • Due to cannibalisation (new product 'eats' up the sales of and demand of an existing product) of the food product • From the expiry dates of food products • Due to the recall of food products from markets	Parfitt et al. (2010), Mena et al. (2011, 2014)

4.3 Respondents

The respondents' characteristics can be understood in terms of their organisational positions and years of industry experience. The respondent's jobs included 45% were in CEO level positions, 20% of operational directors, 17% of supply chain operation and logistic managers, 13% of other general managers and 5% other manager categories. Regarding respondents' work experience in the FMSC, 32% had more than 20 years; 18% had 15–19 years; 21% had 10–14 years; 17% had 5–9 years; and 12% had less than 5 years.

Of all respondents, 47% worked in medium-sized food manufacturing companies with 50 to 250 employees; 33% belonged to small companies with fewer than 50 employees, and 20% were from large food manufacturing companies with more than 251 employees.

4.4 Common Method and Non-response Bias

Harman's single-factor test was conducted in this study to determine if common method bias was a main concern. The analysis result revealed that the first factor explained 30.95% of the variance, thus common method bias was not a serious problem. Non-response bias was assessed through comparing the early respondents and late respondents because late respondents are most similar to non-respondents (Wagner and Kemmerling 2010). The results show that there is no statistical significant difference ($p > 0.005$) between the two respondent groups.

4.5 Evaluation of the Research Model and Hypotheses Testing

The reflective measurement model was evaluated by considering the internal consistency, indictor reliability, convergent validity and discriminant validity, as indicated by Table 2. The formative measurement model was evaluated in terms of multicollinearity, the indicator weights, significance of weights and the indictor loadings (Hair et al. 2014). All the tests were satisfactory.

The significance and relevance of the structural model relationships are shown in Fig. 1 and Table 3 which represent the hypothesized relationships among the constructs. The results show that FMSC collaboration is strongly related to collaborative effectiveness. Surprisingly, FMSC collaboration is not directly and significantly related to over-production EF waste, processing EF waste and storage EF waste; while collaborative effectiveness is significantly related to over-production EF waste, processing EF waste and storage EF waste. In order to further understand the relationship between FMSC collaboration and over-production EF waste, processing EF waste or storage EF waste, the mediation analysis was conducted, which showed that the association between FMSC collaboration and EF waste in relation to over-production, processing and storage is fully mediated through collaborative effectiveness.

Table 2 Convergent validity and internal consistency reliability

Construct	Indicators	Loading	Indicator reliability	Cronbach's alpha	Composite reliability	AVE
Joint demand forecast (JDF)	JDF_1	0.846	0.715	0.827	0.897	0.743
	JDF_2	0.908	0.824			
	JDF_3	0.83	0.689			
Knowledge integration (KI)	KI_1	0.822	0.676	0.873	0.913	0.724
	KI_2	0.884	0.781			
	KI_3	0.862	0.743			
	KI_4	0.833	0.693			
Smart packaging design (SPD)	SPD_1	0.886	0.785	0.878	0.917	0.736
	SPD_2	0.927	0.86			
	SPD_3	0.872	0.76			
	SPD_4	0.734	0.54			
Using RFID (RFID)	RFID_1	0.926	0.857	0.966	0.975	0.909
	RFID_2	0.974	0.948			
	RFID_3	0.957	0.915			
	RFID_4	0.954	0.91			
Joint training programme (JTP)	JTP_1	0.93	0.864	0.917	0.942	0.805
	JTP_2	0.931	0.866			
	JTP_3	0.785	0.616			
	JTP_4	0.932	0.868			
Processing EF waste (PEFW)	PEFW_1	0.793	0.628	0.819	0.878	0.644
	PEFW_2	0.746	0.565			
	PEFW_3	0.824	0.678			
	PEFW_4	0.845	0.714			
Over-production EF waste (OPEFW)	OPEFW_1	0.880	0.774	0.890	0.924	0.752
	OPEFW_2	0.885	0.783			
	OPEFW_3	0.874	0.764			
	OPEFW_4	0.828	0.685			
Storage EF waste (SEFW)	SEFW_1	0.926	0.857	0.915	0.946	0.855
	SEFW_2	0.913	0.833			
	SEFW_3	0.934	0.872			

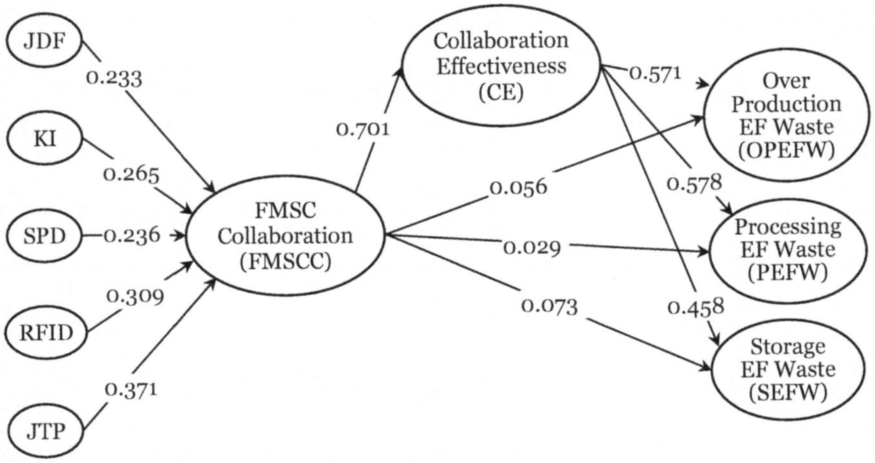

Fig. 1 Structural model

Table 3 Hypotheses testing results

	Hypothesized path	Standard path coef.	P value	Empirical evidence
Hypothesis 1	FMSCC → CE	0.701	0.000	Significant
Hypothesis 2a	FMSCC → OPEFW	−0.056	0.766	Not significant
Hypothesis 2b	FMSCC → PEFW	−0.029	0.847	Not significant
Hypothesis 2c	FMSCC → SEFW	−0.073	0.542	Not significant
Hypothesis 3a	CE → OPEFW	0.571	0.000	Significant
Hypothesis 3b	CE → PEFW	0.578	0.000	Significant
Hypothesis 3c	CE → SEFW	0.458	0.000	Significant

5 Discussion and Conclusion

5.1 Discussion

Research suggests that about 10 million tonnes of EF waste, ranging from 18% (WRAP 2018) to 30% (Parfitt et al. 2010), are generated in the UK FMSC, which has significant negative impact on the UK environment, society and economics (Papargyropoulou et al. 2014; WRAP 2018). Despite the importance of reducing EF waste in the FMSC, there is a paucity of studies on this topic (Mena et al. 2014; Redlingshöfer et al. 2017), although there is some evidence in literature suggests that lacking coordination of FMSC operational- and logistic-related activities is one key reason for the EF waste in the UK FMSC (e.g. Eksoz et al. 2014; Gokarn and Kuthambalayan

2017). Thus, this study drew on the relational view and examined the key dimensions of FMSC collaboration and its effectiveness, and the extent to which FMSC collaboration influences collaborative effectiveness and finally EF waste reduction.

With respect to the dimensions of FMSC collaboration, this synthesised prior studies on supply chain collaboration and proposed a five-dimensional construct model of FMSC collaboration, including joint demand forecasting, using RFID technology, smart packaging design, joint training programme and knowledge integration. In addition, this study has emphasized collaborative effectiveness which resides across the firm's boundaries through collaboration between the FMSC members, not within an individual firm (Cao and Zhang 2012). Although previous studies attempted to conceptualise collaborative effectiveness, its operationalisation in the context of FMSC collaboration has not been undertaken so far. In this study, six different dimensions were identified for collaborative effectiveness: accurate forecasting, better promotional planning, improving product safety, improving temperature monitoring, better inventory controls and better shelf life management.

Regarding the relationship between FMSC collaboration and collaborative effectiveness, this study's finding shows that collaborative effectiveness is significantly enhanced through the collaboration between the FMSC members. While this finding is consistent with prior research on supply chain collaboration in general (e.g. Cao and Zhang 2011), it demonstrates that in the specific context of FMSC and from the relational view, FMSC members involved in intense collaborative efforts through relation-specific assets, share knowledge and combine complementary resources that can forge a better idiosyncratic inter-firm relationship, which leads to collaborative effectiveness (Turkmen 2013). This is believable as, for example, knowledge integration between the supply chain partners facilitates the problem solving approach efficiently in the operational process to improve product quality and shorten the lead time (Mishra and Shah 2009). Through joint planning, efficient promotion planning and on-time replenishment through demand information sharing can be better executed (Sridharan and Simatupang 2009). Joint expertise sharing and the quality, timeliness and usefulness of demand information shared with each other generate visibility that leads to meaningful operational benefits and collaborative effectiveness (Mishra and Shah 2009). Using RFID, a food manufacturing company can effectively reduce out-of-stock items and lead times, thereby improving transportation efficiency, saving labour and tracing items in a more practical environment (Cui et al. 2017). Through smart packaging design, FMSC members monitor the food product and storage conditions of the perishable food products, which helps improve product safety. Joint training programmes enhance the employee's food handing procedure and improve knowledge about cross-contamination, which leads to increased food safety.

With regard to the relation between FMSC collaboration and EF waste reduction, the findings from this study show that there is a statistically insignificant direct relationship between FMSC collaboration and all three types of EF waste reduction. This finding could be understood in the wider context of supply chain collaboration and its impacts, which remains to be fully explained (Mishra and Shah 2009) as prior studies have shown mixed findings. While several previous studies (Piboonrungroj 2012) showed that supply chain collaboration had no significant direct effect on firm performance, others obtained a positive relationship (e.g. Cao and Zhang 2011). However, Piboonrungroj (2012) argued that Cao and Zhang (2011)'s study focused on competitive advantage for an individual firm, not a collaborative advantage. Furthermore, this study differs from prior studies in that this study investigated the relationship between FMSC collaboration and EF waste reduction, rather than the impact of supply chain collaboration on overall firm performance or competitive advantage. Nevertheless, this study reveals a strong positive relationship between FMSC collaboration and EF waste reduction through the mediation of collaborative effectiveness. This means that FMSC members cannot generate internal rent (private benefits) without the relational rent.

With respect to the relationship between collaborative effectiveness and EF waste reduction, the finding from this study reveals that there is a statistically significant relationship between collaborative effectiveness and all three types of EF waste reduction. This suggests that through, for example, improving product safety and temperature monitoring, FMSC members can reduce EF waste due to lack of freshness, change in aroma and flavour, microbial spoilage and temperature fluctuations during the preparation and distribution of perishable food (Lorite et al. 2017); by decreasing inventory inaccuracies and maintaining proper stock rotation, inventory waste and out-of-date-related EF waste can be reduced (Cui et al. 2017).

5.2 *Implications*

This study has several important implications. Conceptually, this study used the relational view to understand the role of FMSC collaboration and collaborative effectiveness on EF waste reduction. By adopting the relational view in the specific context of FMSC, this study focused on FMSC collaboration than individual organisations. This helps to understand that relationship-specific assets can improve coordination of tasks or activities between FMSC members in order to accomplish mutual goals, and also examined how FMSC members enable themselves to gain collaborative effectiveness, leading to EF waste reduction. As a result, this study introduced the conceptual framework that shows the relationship between FMSC collaboration, collaborative effectiveness and EF waste. More specifically, this study has developed the instruments of FMSC collaboration, collaborative effectiveness and EF waste; and has validated them statistically. Those instruments can be used for collaboration formation and evaluating causes and effects of collaboration in FMSC. The constructs of the conceptual framework and their related measurement scales specifically offer a more nuanced view of FMSC collaboration, and also provide a rich and structured understanding of FMSC collaboration, which has not been adequately addressed in the extant literature (Matopoulos et al. 2007; Cao and Zhang 2012). Additionally, this study indicates with empirical evidence that FMSC collaboration has a positive direct effect on collaborative effectiveness and a positive and indirect effect on EF waste reduction through the mediation of collaboration collaborative effectiveness. This study has thus provided a novel conceptual framework and empirical evidence to explain the under-researched relationship among FMSC collaboration, its effectiveness and EF waste reduction.

The study's practical implications are that without generating collaborative effectiveness (relational rent), FMSC members cannot reduce EF waste. Therefore, FMSC practitioners should be willing to engage in collaborative knowledge transfer, joint learning processes, demand information sharing, regularly updating inventory status and training employees so that they could reduce EF waste by enhancing accurate forecasting, better promotional planning, improving product safety, improving temperature monitoring regularly, better inventory control, and better shelf life management.

Additionally, reducing EF waster has important environmental, social and economic implications. The environmental impact of EF waste reduction means helping to preserve world's natural resources for the generations to come, protect soil from degradation, decrease the pressure for land conversion into agriculture, reduce the emission of methane and carbon dioxide which make a significant contribution to tackling climate change (Papargyropoulou et al. 2014). Socially, EF waste reduction could improve human health by bringing down the cost of food to communities so more people can afford more nutritious food products (Papargyropoulou et al. 2014). Moreover, reduction in EF waste means reduced greenhouse gas emissions, which leads to reduced air pollution, and that leads to reducing the incidence of health problems, such as cancer, asthma and other cosmetic problems (Chen 2014). Finally, economically, EF waste reduction means less energy, raw material and human capital that are used in FMSC and businesses which can cut EF waste disposal costs and also compliance with environmental legislation becomes cheaper and more straightforward.

5.3 Limitation and Future Research

Any conclusions drawn from this study should be considered in light of several limitations, some of which provide avenues for future research. This study uses a non-probability sampling, though widely accepted by research on supply chain collaboration; future research could adopt probability sampling in order to improve generalizability. This study did not consider horizontal collaboration in the FMSC; so future research could consider the combination of vertical and horizontal collaboration, which would have more value for FMSC's managers. Thirdly, the nature of FMSC collaboration and its impact on collaborative effectiveness and EF waste reduction may take a long time; thus, future research could adopt a longitudinal study, which may yield supplementary insights. Finally, this study was conducted in the context of UK FMSC; future research should use the same hypothesised structural relationship in different countries to determine any country-specific facilitating and inhibiting factors; and also compare the level of collaboration across countries.

6 Conclusion

The aim and objective of this research were to examine the relationship among FMSC collaboration, collaborative effectiveness and EF waste reduction underpinned by the relational view. The study's results revealed that FMSC collaboration is positively contributing to EF waste reduction only through the mediation of collaborative effectiveness. The implication is that without achieving collaboration effectiveness, FMSC members are unlikely to reduce the EF waste in their supply chain operation and logistic operations. Therefore, this study strongly encourages FMSC members to coordinate with their partners, which will help in coordinating their supply chain's operational and logistical activities to reduce EF waste, which would have positive environmental, economic and social impacts.

References

Ali, M. H., Tan, K. H., Makhbul, Z. M., & Ngah, A. H. (2016). Augmenting Halal food integrity through supply chain integration. *Journal Pengurusan (UKM Journal of Management), 48.*

Ataseven, C., & Nair, A. (2017). Assessment of supply chain integration and performance relationships: A meta-analytic investigation of the literature. *International Journal of Production Economics, 185,* 252–265.

Badea, A., Prostean, G., Hutanu, A., & Popa, S. (2015). Competency training in collaborative supply chain using KSA model. *Procedia-Social and Behavioral Sciences, 191,* 500–505.

Banchuen, P., Sadler, I., & Shee, H. (2015). Supply chain collaboration aligns order-winning strategy with business outcomes. *IIMB Management Review.*

Barratt, M. (2004). Understanding the meaning of collaboration in the supply chain. *Supply Chain Management: An International Journal, 9*(1), 30–42.

Beretta, C., Stoessel, F., Baier, U., & Hellweg, S. (2013). Quantifying food losses and the potential for reduction in Switzerland. *Waste Management, 33*(3), 764–773.

Betz, A., Buchli, J., Göbel, C., & Müller, C. (2015). Food waste in the Swiss food service industry—Magnitude and potential for reduction. *Waste Management, 35,* 218–226.

Canali, M., Östergren, K., Amani, P., Aramyan, L., Sijtsema, S., Korhonen, O., & OConnor, C., et al. (2014). Drivers of current food waste generation, threats of future increase and opportunities for reduction.

Cao, M., & Zhang, Q. (2011). Supply chain collaboration: Impact on collaborative advantage and firm performance. *Journal of Operations Management, 29*(3), 163–180.

Cao, M., & Zhang, Q. (2012). *Supply chain collaboration: Roles of interorganizational systems, trust, and collaborative culture*, Springer Science & Business Media.

Chabada, L., Dreyer, H. C., Romsdal, A., & Powell, D. J. (2012). Sustainable food supply chains: Towards a framework for waste identification. In *Paper Presented at the IFIP International Conference on Advances in Production Management Systems.*

Chen, A. (2014). New research quantifies health benefits of reducing greenhouse gas emissions. http://newscenter.lbl.gov/2014/11/18/new-research-quantifies-health-benefits-of-reducing-greenhouse-gas-emissions/.

Chen, R.-Y. (2015). Autonomous tracing system for backward design in food supply chain. *Food Control, 51,* 70–84.

Conte, A., Cappelletti, G., Nicoletti, G., Russo, C., & Del Nobile, M. (2015). Environmental implications of food loss probability in packaging design. *Food Research International, 78,* 11–17.

Crook, T. R., Giunipero, L., Reus, T. H., Handfield, R., & Williams, S. K. (2008). Antecedents and outcomes of supply chain effectiveness: An exploratory investigation. *Journal of Managerial Issues, 20*(2), 161–177.

Cui, L., Deng, J., Liu, F., Zhang, Y., & Xu, M. (2017). Investigation of RFID investment in a single retailer two-supplier supply chain with random demand to decrease inventory inaccuracy. *Journal of Cleaner Production, 142*, 2028–2044.

Dani, S. (2015). *Food supply chain management and logistics: From farm to fork*, Kogan Page Publishers.

Darlington, R., Staikos, T., & Rahimifard, S. (2009). Analytical methods for waste minimisation in the convenience food industry. *Waste Management, 29*(4), 1274–1281.

DEFRA. (2015). Food statistics pocketbook 2015, Department for Environment, Food and Rural Affairs (DEFRA). https://www.gov.uk/government/uploads/system/uploads/attachment_data/file/526395/foodpocketbook-2015update-26may16.pdf.

Dyer, J. H., & Singh, H. (1998). The relational view: Cooperative strategy and sources of interorganizational competitive advantage. *Academy of Management Review, 23*(4), 660–679.

Eksoz, C., Mansouri, S. A., & Bourlakis, M. (2014). Collaborative forecasting in the food supply chain: A conceptual framework. *International Journal of Production Economics, 158*, 120–135.

FAO. (2013). *FAO, 2013. Food wastage footprint: Impacts on natural resources.* Rome: Food and Agriculture Organization of the United Nations.

Gadde, L.-E., & Amani, P. (2016). Food supply in a network context: An alternative framing and managerial consequences in efforts to prevent food waste. *British Food Journal, 118*(6), 1407–1421.

Göbel, C., Langen, N., Blumenthal, A., Teitscheid, P., & Ritter, G. (2015). Cutting food waste through cooperation along the food supply chain. *Sustainability, 7*(2), 1429.

Gokarn, S., & Kuthambalayan, T. S. (2017). Analysis of challenges inhibiting the reduction of waste in food supply chain. *Journal of Cleaner Production, 168*, 595–604.

Grunow, M., & Piramuthu, S. (2013). RFID in highly perishable food supply chains—Remaining shelf life to supplant expiry date? *International Journal of Production Economics, 146*(2), 717–727.

Hair, J. F., Hult, G., Ringle, C., & Sarstedt, M. (2014). *A primer on partial least squares structural equation modeling (PLS-SEM)*, Sage.

Hair, J. F., Ringle, C. M., & Sarstedt, M. (2013). Partial least squares structural equation modeling: rigorous applications, better results and higher acceptance. *Long Range Planning, 46*, 1–12.

Hair Jr, J. F., Hult, G. T. M., Ringle, C., & Sarstedt, M. (2016). *A primer on partial least squares structural equation modeling (PLS-SEM)*, Sage Publications.

Hudnurkar, M., Jakhar, S., & Rathod, U. (2014). Factors affecting collaboration in supply chain: A literature review. *Procedia—Social and Behavioral Sciences, 133*, 189–202.

Jedermann, R., Nicometo, M., Uysal, I., & Lang, W. (2014). Reducing food losses by intelligent food logistics, The Royal Society.

Kaipia, R., Dukovska-Popovska, I., & Loikkanen, L. (2013). Creating sustainable fresh food supply chains through waste reduction. *International Journal of Physical Distribution & Logistics Management, 43*(3), 262–276.

Kim, K. K., O'Bryan, C. A., Crandall, P. G., Ricke, S. C., & Neal, J. A. (2013). Identifying baseline food safety training practices for retail delis using the Delphi expert consensus method. *Food Control, 32*(1), 55–62.

Koufteros, X., Vonderembse, M., & Jayaram, J. (2005). Internal and external integration for product development: The contingency effects of uncertainty, equivocality, and platform strategy. *Decision Sciences, 36*(1), 97–133.

Kouwenhoven, G., Nalla, V. R., & von Losoncz, T. L. (2012). Creating sustainable businesses by reducing food waste: A value chain framework for eliminating inefficiencies. *International Food and Agribusiness Management Review, 15*(3).

Kumar, D., & Rahman, Z. (2015). Sustainability adoption through buyer supplier relationship across supply chain: A literature review and conceptual framework. *International Strategic Management Review, 3*(1), 110–127.

Kumar, G., Banerjee, R., Meena, P., & Ganguly, K. K. (2017). Joint planning and problem solving roles in supply chain collaboration. *IIMB Management Review*.

Kumar, G., & Banerjee, R. N. (2014). Supply chain collaboration index: An instrument to measure the depth of collaboration. *Benchmarking: An International Journal, 21*(2), 184–204.

Kumar, G., & Nath Banerjee, R. (2012). Collaboration in supply chain: An assessment of hierarchical model using partial least squares (PLS). *International Journal of Productivity and Performance Management, 61*(8), 897–918.

Lavie, D. (2006). The competitive advantage of interconnected firms: An extension of the resource-based view. *Academy of Management Review, 31*(3), 638–658.

Lee, J., Palekar, U. S., & Qualls, W. (2011). Supply chain efficiency and security: Coordination for collaborative investment in technology. *European Journal of Operational Research, 210*(3), 568–578.

Liljestrand, K. (2017). Logistics solutions for reducing food waste. *International Journal of Physical Distribution & Logistics Management, 47*(4), 318–339.

Lorite, G. S., Selkälä, T., Sipola, T., Palenzuela, J., Jubete, E., Viñuales, A., et al. (2017). Novel, smart and RFID assisted critical temperature indicator for supply chain monitoring. *Journal of Food Engineering, 193*, 20–28.

Mahalik, N. P., & Nambiar, A. N. (2010). Trends in food packaging and manufacturing systems and technology. *Trends in Food Science & Technology, 21*(3), 117–128.

Matopoulos, A., Vlachopoulou, M., Manthou, V., & Manos, B. (2007). A conceptual framework for supply chain collaboration: Empirical evidence from the agri-food industry. *Supply Chain Management: An International Journal, 12*(3), 177–186.

Mena, C., Adenso-Diaz, B., & Yurt, O. (2011). The causes of food waste in the supplier–retailer interface: Evidences from the UK and Spain. *Resources, Conservation and Recycling, 55*(6), 648–658.

Mena, C., Terry, L. A., Williams, A., & Ellram, L. (2014). Causes of waste across multi-tier supply networks: Cases in the UK food sector. *International Journal of Production Economics, 152*, 144–158.

Mishra, A. A., & Shah, R. (2009). In union lies strength: Collaborative competence in new product development and its performance effects. *Journal of Operations Management, 27*(4), 324–338.

Mithun Ali, S., Moktadir, M. A., Kabir, G., Chakma, J., Rumi, M. J. U., & Islam, M. T. (2019). Framework for evaluating risks in food supply chain: Implications in food wastage reduction. *Journal of Cleaner Production, 228*, 786–800.

Morita, M., Machuca, J. A. D., Flynn, E. J., & Pérez de los Ríos, J. L. (2015). Aligning product characteristics and the supply chain process—A normative perspective. *International Journal of Production Economics, 161*, 228–241.

Nash, T. A. (2010). RFID technology and its impact on the supply chain.

Papargyropoulou, E., Lozano, R., K. Steinberger, J., Wright, N., & Ujang, Z. B. (2014). The food waste hierarchy as a framework for the management of food surplus and food waste. *Journal of Cleaner Production, 76*(0), 106–115.

Parfitt, J., Barthel, M., & Macnaughton, S. (2010). Food waste within food supply chains: Quantification and potential for change to 2050. *Philosophical Transactions of the Royal Society B: Biological Sciences, 365*(1554), 3065–3081.

Park, S.-H., Kwak, T.-K., & Chang, H.-J. (2010). Evaluation of the food safety training for food handlers in restaurant operations. *Nutrition research and practice, 4*(1), 58–68.

Piboonrungroj, P. (2012). *Supply chain collaboration: Impacts and mediation on firm performance*, Cardiff University.

Piramuthu, S. (2005). Knowledge-based framework for automated dynamic supply chain configuration. *European Journal of Operational Research, 165*(1), 219–230.

Quinn, F. J. (2012). Collaboration: Still a hot topic. *Supply Chain Management Review, 16*(5), *3*(16), 5.

Rajaguru, R., & Matanda, M. J. (2013). Effects of inter-organizational compatibility on supply chain capabilities: Exploring the mediating role of inter-organizational information systems (IOIS) integration. *Industrial Marketing Management, 42*(4), 620–632.

Redlingshöfer, B., Coudurier, B., & Georget, M. (2017). Quantifying food loss during primary production and processing in France. *Journal of Cleaner Production, 164,* 703–714.

Rhodes, C. (2015). Manufacturing: Statistics and policy—Parliament UK. http://researchbriefings.parliament.uk/ResearchBriefing/Summary/SN01942.

Sahin, E., Dallery, Y., & Gershwin, S. (2002). Performance evaluation of a traceability system. An application to the radio frequency identification technology. In *2002 IEEE International Conference on Paper presented at the Systems, Man and Cybernetics.*

Shinbaum, S., Crandall, P. G., & O'Bryan, C. A. (2016). Evaluating your obligations for employee training according to the food safety modernization act. *Food Control, 60,* 12–17.

Singhry, H. B., Rahman, A. A., & Ng, S. I. (2015). Measurement for Supply chain collaboration and supply chain performance of manufacturing companies. *International Journal of Economics & Management, 9.*

Soosay, C. A., & Hyland, P. (2015). A decade of supply chain collaboration and directions for future research. *Supply Chain Management: An International Journal, 20*(6), 613–630.

Sridharan, R., & Simatupang, T. M. (2009). Managerial views of supply chain collaboration.

Turkmen, H. (2013). *Scientific review of the relational view theory and its contribution to critical sourcing decisionmaking*, University of Twente.

Unnevehr, L. J. (2000). Food safety issues and fresh food product exports from LDCs. *Agricultural Economics, 23*(3), 231–240.

Vanderroost, M., Ragaert, P., Devlieghere, F., & De Meulenaer, B. (2014). Intelligent food packaging: The next generation. *Trends in Food Science & Technology, 39*(1), 47–62.

Verghese, K., Lewis, H., Lockrey, S., & Williams, H. (2013). The role of packaging in minimising food waste in the supply chain of the future. *CHEP Australia, 3,* 50.

Verghese, K., Lockrey, S., Rio, M., & Dwyer, M. (2018). DIRECT, a tool for change: Co-designing resource efficiency in the food supply chain. *Journal of Cleaner Production, 172,* 3299–3310.

Wagner, S. M., & Kemmerling, R. (2010). Handling nonresponse in logistics research. *Journal of Business Logistics, 31*(2), 357–381.

Watson, J. A., Wysocki, A. F., & Bucklin, R. A. (2015). *The role and impact of technology on supply-chain management in the food industry.* UF/IFAS Extension: Agricultural and Biological Engineering Department.

Whitehead, D. P., Parfitt, D. J., Bojczuk, M. K., & James, K. (2013). Estimates of waste in the food and drink supply chain. http://www.wrap.org.uk/sites/files/wrap/Estimates%20of%20waste%20in%20the%20food%20and%20drink%20supply%20chain_0.pdf.

Wiengarten, F., Humphreys, P., Cao, G., Fynes, B., & McKittrick, A. (2010). Collaborative supply chain practices and performance: Exploring the key role of information quality. *Supply Chain Management: An International Journal, 15*(6), 463–473.

WRAP. (2018). Food surplus and waste in the UK—Key facts. Retrieved 16 June, 2019, from http://www.wrap.org.uk/sites/files/wrap/Food%20Surplus%20and%20Waste%20in%20the%20UK%20Key%20Facts%2014%205%2019.pdf.

The manufacturer's authorised representative in the EU is Springer Nature Customer Service Centre GmbH, Europaplatz 3, 69115 Heidelberg, Germany. If you have any concerns regarding our products, please contact ProductSafety@springernature.com

Printed and bound by CPI Group (UK) Ltd, Croydon, CR0 4YY
30/09/2026
02225648-0011